Fragmentation in Semi-Arid and Arid Landscapes

Fragmentation in Semi-Arid and Arid Landscapes

Consequences for Human and Natural Systems

edited by

Kathleen A. Galvin
Colorado State University,
Fort Collins, CO, U.S.A.

Robin S. Reid
International Livestock Research Institute,
Nairobi, Kenya

Roy H. Behnke Jr.
Macaulay Land Use Research Institute,
Aberdeen, Scotland, U.K.

and

N. Thompson Hobbs
Colorado State University,
Fort Collins, CO, U.S.A.

Springer

A C.I.P. Catalogue record for this book is available from the Library of Congress.

ISBN 978-1-4020-4905-7 (HB)
ISBN 978-1-4020-4906-4 (e-book)

Published by Springer,
P.O. Box 17, 3300 AA Dordrecht, The Netherlands.

www.springer.com

Photo Credits:

East Africa (cattle, wildebeest, factory)
Photo by Cathleen J. Wilson

Central Asia (mountains with yurt and livestock) - Kyrgyz nomads' summer camp,
Pamir mountains, Tajikistan
Photo by Carol Kerven, June 2005

Printed on acid-free paper

DEDICATION

'the marble index of a mind for ever
Voyaging through strange seas of Thought, alone.'

That was William Wordsworth's response to a statue of Sir Isaac Newton, and I suspect that Wordsworth got it right about Newton. This book is dedicated to Jim Ellis, who also took some voyages of scientific discovery. In comparison to a colossal mind like Newton's, these were modest trips, but real ones nonetheless, and Jim remains the best creative scientist I have known or am ever likely to know well.

The remarkable thing about Jim was that he didn't travel alone; he took a lot of us with him a lot of the way. He had the ability to see the unexpected and the simple, and the character to lead people in investigating these possibilities. The authors of the chapters in this book are just a few of the people who worked with Jim, were personally influenced by him, and regarded him with affection and admiration. That we reside on five continents tells you something about the breadth of the man's intellectual ambitions, and his ability to assemble diverse people to match those ambitions.

As always in Jim's work, many have willingly contributed, especially Kathy Galvin who did the practical things that got this book into print. But this is Jim's book. He identified the research issues, got the funding, assembled the collaborators and, in the weeks before his death, brought us together to agree to produce this book. I hope that what follows does justice to his scientific legacy. Despite our efforts, this work necessarily lacks the final leavening, that unexpected insight into the significant and obvious, which was Jim's special kind of contribution. I know that there are thoughts I will not think without him.

A provincial Ethiopian hotel is a good place from which to write this dedication. It's the kind of place where you pay when you check in, but it has pretensions: The sheets are clean and every room has a telephone that never works. At dusk on a Sunday evening the whole parade is here – the kids on the play equipment in front of the bar, the lovers at the bar, the business man and the plump girl in the rooms behind. Jim travelled well. We travelled together in Asia, but I think he would have smiled on this African scene. Alert, judging but not judgmental, amused – a good field man and companion.

So forget Wordsworth's strange and lonely seas. There is, I believe, a children's book with the title 'Science Can Be Fun.' Jim proved it could be so.

Good scientist, good friend, good friend in science, missed.

Roy Behnke
Awassa, Ethiopia

TABLE OF CONTENTS

FOREWORD

Casual readers of the title of this book might be forgiven for thinking that it is a little esoteric, far-removed from the pressing day-to-day concerns of humans and wildlife in the drylands of the world. But they could not be more wrong. It addresses an issue of the utmost practical importance in the world today, yet does so on the basis of exciting new theory about how the world operates.

Of the billion or so human beings who now live in the world's arid and semi-arid lands, a majority depend on natural resources for their livelihoods. These natural resources include livestock and their forage, as well as the wild biota that creates opportunities for tourism or subsistence harvesting. Arid and semi-arid lands are spread over a third of the world's land surface, from Colorado to the Kalahari, the Sahel to the Simpson, the Altai Steppe to Amboseli. Notwithstanding their diversity, these lands are broadly characterised by low productivity, management at large scales, and great climate variability – in short, by high spatial and temporal heterogeneity.

This book is about the implications of that high spatial and temporal heterogeneity for life, management and policy in arid and semi-arid lands. Over centuries, these lands have been subjected to colonisation and development modelled on experiences from centres of human power in less heterogeneous regions, particularly Europe. The result has been institutions and infrastructure imposed without appreciating that heterogeneity might cause arid and semi-arid lands to function in a fundamentally different way. In particular, the widespread paradigm of subdivision and intensification has fragmented the landscapes, and, with them, the capacity of both humans and wildlife to take advantage of variation in space to help cope with unavoidable variations in time. The fragmentation caused by intensified land use is presumed to benefit net regional productivity. The time has come to understand where this is a legitimate trade-off, and where it is an imposed prejudice which is undermining the future of a vast chunk of the world.

So this is a very practical book for the drylands. But the analysis of fragmentation also demands a fascinating and cutting-edge collusion between the ecological and social sciences. It forces us to think beyond the straitjacket of averages to concentrate on the impacts of variability about the mean; and not just in one dimension while the rest "are held constant", but in space and time simultaneously.

Taken together, the papers that follow face up to this challenge in a wide variety of different systems across the world, with a variety of disciplinary foci, and linking field data with modelling and theory. They summarise results from an international effort to learn from case studies across the world in regions with different levels of spatial and temporal heterogeneity, combined with different social, environmental and historical contexts. This, the biocomplexity 'SCALE' project, was the original brainchild of Jim Ellis, sadly no longer with us, ably carried through by the project team represented in the following pages. Jim's work sought to determine whether (and which) arid and semi-arid lands really function in a different way to other regions of the world; he carried the banner for the idea of non-equilibrial ecosystem dynamics. Today, in this book and elsewhere, the question of whether there is a truly differentiated 'science of desert living' to be uncovered is being taken up more generally, to speak to the future of these regions.

Faced with global change – in climate, in economic systems, in governance – at an ever-increasing pace, at least 200 million inhabitants of drylands are believed to be vulnerable to losing their livelihoods and even their lives over the coming decades. With confidence and coherence, this book contributes theory, understanding and implications to help reduce this vulnerability in relation to the poorly-grasped threats arising from fragmentation.

It is a pleasure to recommend this excellent and thought-provoking read.

Mark Stafford Smith,
Australia, July 2006

LIST OF CONTRIBUTORS

Adanje, Samuel A. Kenya Wildlife Service, P.O. Box 40241, Nairobi, Kenya

Agatsiva, Jasphat L. Department of Resource Surveys and Remote Sensing, P.O. Box 47146, Nairobi, Kenya

Alimaev, Iliya I. Department of Pasture and Fodder, Scientific Centre for Animal Production and Veterinary Research, 51 Jandosov St., Almaty, Kazakhstan

Ash, Andrew J. CSIRO Sustainable Ecosystems, 306 Carmody Rd, St Lucia, Q 4067, Australia

Bagine, Richard Kenya Wildlife Service, P.O. Box 40241, Nairobi, Kenya

Behnke, Roy H. Macaulay Institute, Craigiebuckler, Aberdeen AB15 8QH U.K.

Boone, Randall B. Natural Resource Ecology Laboratory, Colorado State University, Fort Collins, CO 80523, USA

BurnSilver, Shauna B. Natural Resource Ecology Laboratory and Department of Anthropology, Colorado State University, Fort Collins, CO 80523, USA

Chuluun, Togtohyn Environmental Remote Sensing and Geographic Information System Laboratory, National University of Mongolia, Ulaanbaatar 210646, Mongolia

Coughenour, Michael B. Natural Resource Ecology Laboratory, Colorado State University, Fort Collins, CO 80523, USA

Ellis, James E. deceased

Galvin, Kathleen A. Department of Anthropology and Natural Resource Ecology Laboratory, Colorado State University, Fort Collins, CO 80523, USA

Gichohi, Helen African Wildlife Foundation, P.O. Box 48177, Nairobi, Kenya

Gross, John E. National Park Service, 1201 Oakridge Drive, Suite 150, Fort Collins, CO 80525-5589, USA

Hobbs, N. Thompson Department of Forest, Rangeland, and Watershed Stewardship and Natural Resource Ecology Laboratory, Colorado State University, Fort Collins, CO 80523, USA

Kifugo, Shem C. International Livestock Research Institute, P.O. Box 30709, Nairobi, Kenya

Knapp, Linda M. Department of Anthropology and Natural Resource Ecology Laboratory, Colorado State University, Fort Collins, CO 80523, USA

Kristjanson, Patti International Livestock Research Institute, P.O. Box 30709, Nairobi, Kenya

Kruska, Russell S. International Livestock Research Institute, P.O. Box 30709, Nairobi, Kenya

Kshatriya, Mrigesh International Livestock Research Institute, P.O. Box 30709, Nairobi, Kenya

Lackett, Jill M. Natural Resource Ecology Laboratory, Colorado State University, Fort Collins, CO 80523, USA

McAllister, Ryan R. J. CSIRO Sustainable Ecosystems, Davies Lab, PMB PO Aitkenvale, Q 4814, Australia

Nkedianye, David International Livestock Research Institute, P.O. Box 30709, Nairobi, Kenya

Ogutu, Joseph O. International Livestock Research Institute, P.O. Box 30709, Nairobi, Kenya

Ojima, Dennis Natural Resource Ecology Laboratory, Colorado State University, Fort Collins, CO 80523, USA

Reid, Robin S. International Livestock Research Institute, P.O. Box 30709, Nairobi, Kenya

Said, Mohammed Y. International Livestock Research Institute, P.O. Box 30709, Nairobi, Kenya

Stokes, Chris J. CSIRO Sustainable Ecosystems, Davies Lab, PMB PO Aitkenvale, Q 4814, Australia

Thornton, Philip K. International Livestock Research Institute, Nairobi, Kenya and Institute of Atmospheric and Environmental Sciences, School of Geosciences, University of Edinburgh, Edinburgh, UK

Worden, Jeffrey International Livestock Research Institute, P.O. Box 30709, Nairobi, Kenya

ACKNOWLEDGEMENTS

We would like to thank the US National Science Foundation for funding the project, Biocomplexity, Spatial Scale, and Fragmentation: Implications for Arid and Semi-arid Ecosystems (SCALE) - Grant No. DEB-0119618, which supported much of the research that forms the basis for the chapters in this volume.

PART I. INTRODUCTION AND FRAMEWORK

Chapter 1

GLOBAL SIGNIFICANCE OF EXTENSIVE GRAZING LANDS AND PASTORAL SOCIETIES: AN INTRODUCTION

Robin S. Reid[1], Kathleen A. Galvin[2], and Russell S. Kruska[1]
[1]International Livestock Research Institute, P.O. Box 30709, Nairobi, Kenya; [2]Department of Anthropology and Natural Resource Ecology Laboratory, Colorado State University, Fort Collins, CO 80523, USA

1. EXTENT, STRUCTURE, AND FUNCTION OF GLOBAL GRAZING LANDS[1] AND PASTORAL SOCIETY

1.1 Extent and use of grazing lands by people, livestock and wildlife

More of the land surface of the earth is used for grazing than for any other purpose (FAO 1999, WRI 2000, Asner et al. 2004, Ojima and Chuluun, Chapter 8). Although livestock and wildlife graze in forests and woodlands, we focus here on the lands where most herding peoples and their livestock graze: in 'open' grazing lands, which include savannas, grasslands, prairies, steppe, and shrublands (Asner et al. 2004). These grazing lands cover 61.2 million km^2 or 45% of the earth's surface (excluding Antarctica), 1.5 times more of the globe than forest, 2.8 times more than cropland and 17 times more than urban settlements (see Figure 1-1)[2]. These lands range from extremely dry (hyper-arid) to very wet (humid) and represent 78% of the land area grazed by livestock (Asner et al. 2004). Extending from the

K. A. Galvin et al. (eds.), Fragmentation in Semi-Arid and Arid Landscapes: Consequences for Human and Natural Systems, 1–24.

equator to near the poles, grazing lands cover most of some continents (77% of Australia, 61% of Africa, 49% of Asia, but only 18% of Europe), and particularly dominate lands outside the tropics. About two-thirds of all grazing lands are in developing countries. Pastoral people and their livestock, with herds composed of diverse species, ranging from reindeer in the north to alpaca in the south, share the more thinly populated grazing lands with a much wider variety of wild grazers and browsers, from kangaroos to elephants to bison. In conservation areas around the globe and in much of the cold grazing land around the north pole (with some exceptions), wildlife often graze these lands alone, without livestock.

We focus in this book on 'extensive' grazing lands, where human populations are low enough (< 20 people/km^2) that people, livestock, and wildlife can co-exist on the land (Figure 1-1). Although these extensive lands make up 91% of all grazing lands, they support only 24% of all people in grazing lands. People also use these lands for tourism, recreation, hunting, and foraging (Reid et al. 2004b), even though herding and wildlife grazing are the most common livelihoods. 'Intensive' grazing lands (≥ 20 people/km^2, Figure 1-1) are those dominated by livestock and sometimes scattered crops, often with very few wildlife. They cover only 9% of global grazing lands but support 76% of all people who inhabit these lands.

In dry and cold grazing lands (see hyper-arid, arid, semi-arid, and cold grazing lands, Figure 1-1), herding is the major human land use, rather than

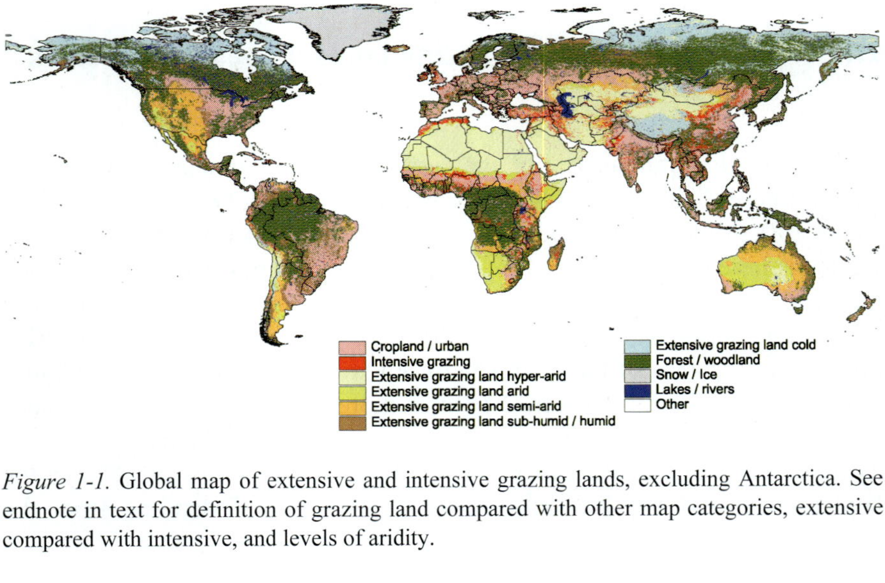

Figure 1-1. Global map of extensive and intensive grazing lands, excluding Antarctica. See endnote in text for definition of grazing land compared with other map categories, extensive compared with intensive, and levels of aridity.

farming or hunting, because grazing is an efficient and reliable way to turn sunlight into human food in extreme and variable land environments. More than half (60%) of the world's grazing land is dry and relatively warm, where rainfall cannot keep pace with evaporation back into the atmosphere, and thus are constantly or seasonally water scarce (see hyper-arid to semi-arid grazing lands, Figure 1-1). Africa, the Arabian peninsula, Peru, and south and central Asia are home to the driest (hyper-arid) grazing lands, while the large grazing lands in western North America and Australia are arid and semi-arid. The Americas and Africa are home to most of the wettest (sub-humid/humid) grazing lands (Figure 1-1). But grazing lands also include places where consistent cold temperature limits pasture growth more than rainfall: in high mountain pastures like the Andes or Tibetan Plateau or in the circumpolar north (see cold grazing lands, 16% of the total, Figure 1-1). In Figure 1-1, cold grazing lands do not include those that have very cold winters but warm summers with good pasture production (southern Mongolia, western US, for examples). Most of the case studies in this book focus on extensive grazing lands that are dry (Chapters 4, 7, 9-12), are partly cold (Mongolia, Chapter 8), or have cold winters and productive summers (Chapters 5-8), although some include wetter or intensive grazing lands as a minor component (Chapters 6, 9, 11).

Although extensive grazing lands are thinly populated, herding societies are ethnically diverse and herders produce a significant proportion of the world's livestock from diverse herds. Ethnically, pastoral people are very diverse, with over 70 different linguistic/cultural groups of pastoralists in eastern Africa alone (Murdock 1959). More than half of the world's pastoralists are in Africa (55%), with 20% in Asia, 15% in the Americas and 10% in Australia (Child et al. 1984). Even though extensive grazing lands support only 3% of the world's people, they keep 35% of the world's sheep, 23% of the goats, and 16% of the cattle and water buffalo. In Africa, pastoral herders, compared with settled farmers, produce from 50-75% of all the milk, beef, and mutton produced on the continent (de Haan et al. 1997) much of which is consumed in higher potential areas (Gill 1999). Similarly in Iran, although only 1.5% of the people are nomadic herders, they hold a full 25% of the national herd (Koocheki and Gliessman 2005). Globally, the main livestock species that herders keep are taurine (humpless) and zebu (humped) cattle, donkeys, goats, and sheep. Regionally, pastoral people herd alpaca and llama in the Andes, the Bactrian camel and horse in east-central Asia, the dromedary (one-humped camel) in Africa and West Asia, reindeer in northern Eurasia, water buffalo in south Asia, and yak in Asia (Blench 2000). Pig herding used to be common in the Middle East and Europe and ducks and geese are still herded in India (Blench 2000).

But not all herders keep livestock for the same reasons. One major distinction is that some pastoralists are more oriented toward livestock production for home consumption, while others are more oriented toward sales in the market. This distinction affects the composition of their herds, how well they track changes in pastures, and how many livestock they sell. In Asia and Africa, most herders keep animals for food first and the market second; the opposite is true for ranchers in Australia, Argentina, and North America. Subsistence herders, who feed their families from livestock, need to produce edible animal products (milk, meat, ghee) as many days of the year as possible, but must balance these needs with desires for additional animals or animal products such as live animals, wool, or hides to sell for cash (Behnke 1983, 1985). To achieve this balance, they focus on a milking herd of females and often a herd with two or more species (e.g., sheep, yaks) to avoid the risk of herding just one. More people can be supported on milk than on meat alone because milk is more efficient than meat in turning grass into human food (Spedding 1971, Galvin et al. 2004). This kind of production depends on human energy in the form of labor substituting for the need for outside energy and capital. With large families and thus more labor, subsistence herders can 'track' changes in pastures more closely than a few ranchers, and thus can support more animals on the same pasture than a rancher can (Western 1982, Niamir-Fuller 1999). Also, many of these herders live with wild predators and need the labor to protect their livestock. Because of the need for labor, some of these subsistence herders are in more heavily populated, intensive grazing lands (see Figure 1-1), which are predominantly (93%) in developing countries (our calculations).

The opposite tends to be true for ranchers in places like Australia, Argentina, or North America. These largely commercial herders need to produce the most valuable and marketable product: meat, with some livestock products being reserved for subsistence or home consumption. Livestock graze on huge tracts of land, require significant investments of capital, and are tended by very few people. Note that these regions appear as almost purely 'extensive' in Figure 1-1. Ranchers tend to specialize on one species of animal, and have male-dominated herds. In these grazing lands, ranchers usually attempt to reduce or eliminate predators. We present examples from herders in sites with weakly developed markets (Chapters 7-12) and those with well developed markets (Chapters 4-6) in this book.

1.2 Productivity and ecosystem services in grazing lands

Commercial and subsistence herders live in vast grazing lands that, despite their aridity, can be deceptively productive. Charles Darwin puzzled over the abundance of animals supported in extensive savannas (grassland with sparse tree cover) when recounting his voyage on the HMS Beagle: "…there can be no doubt of its being sterile country…covered by a poor and scanty vegetation. Now if we look to the animals inhabiting these wide plains, we shall find their number extraordinary great and their bulk immense….I confess it is truly surprisingly how such a number of animals can find support in a country producing so little food" (Darwin 1909-1914:43). What Darwin did not know is that some savannas produce remarkable amounts of forage, much of it readily digestible for a grazer (Bourlière and Hadley 1983). For example, even though a forest obviously supports much more plant mass than a savanna does (up to 10 times more), some wet savannas produce as much plant matter in a year as a forest does (and thus are 10 times more productive). Some forests add only about 10% to their total weight each year, while savannas can reproduce 150% of their weight each year. And much more of that plant mass is edible in savannas than in forests. Leaves, which are often the best quality food for grazers, make up about 2% of the forest plants in Ivory Coast of west Africa, but they make up 15-60% of nearby savannas.

Many grazing lands are not only productive, but also rich in diversity, a storehouse of carbon, and provide a wide range of ecosystem services in addition to food. Some of these ecosystem services benefit local pastoralists, but many benefit humankind generally (like carbon sequestration and biodiversity). Vegetation in grazing lands supports many of the earth's remaining large wild herbivores, including the remnants of the great Pleistocene herds in eastern and southern Africa (Owen-Smith 1987). This vegetation also prevents erosion and reduces the amount of dust that forms in these lands, which can travel from one continent to another (Shinn et al. 2000). Although tropical forests obviously store large amounts of carbon aboveground (212 Gt C), tropical savannas have a greater potential to store carbon belowground than any other ecosystem (264 Gt C, IPCC 2000). Although biodiversity in grazing lands is usually lower than it is in forests, at least for African plants, rainforests are only 14% richer in species (about 2020 plant species/10,000 km^2) than savannas (about 1750 plant species/10,000 km^2) (Menaut 1983). In South America, the cerrado of Brazil and temperate grasslands of Argentina and Uruguay are particularly rich in plant species (Groombridge 1992, Solbrig 1996, Blench and Sommer 1999). Grazing lands in uplands and on mountains are often the major source of water for lowlands. And grazing lands provide people with an array of less recognized ecosystem services: plant production (e.g., forage, woodfuels, medicines), climate regulation, water regulation and provision, biochemical provision (e.g., nutrients), soil conservation, nutrient cycling, and pollination (MEA 2005). The peoples who live in these drylands are also storehouses of knowledge about local conditions and also how to live and thrive in the face of climatic variability.

1.3 Strategies to manage climatic variability and access resource heterogeneity

How have animals and herders managed to persist in harsh and varying environments in these vast grazing lands? The core reason is that wildlife and herders have developed a range of ways, over many millennia, to adapt to or avoid the risks that threaten their survival. These risks include unpredictable climate, disease, competition for water and forage, and predation. The biggest risk that herders and wildlife face, climate, includes low and varying rainfall, and often deep and long-lasting snow cover, which affects their access to forage and water. For example, lush green pastures can give way to barren sand in dry regions or deep snow in cold regions in a matter of weeks. Rain commonly falls in patches, creating a mosaic of green and brown patches across the landscape. There are also wide swings in forage availability from year to year depending on rain and snow. In dry regions, the variability of rainfall increases as rainfall decreases (Conrad 1941), which increases risk for grazers and herders. Further, in parts of Africa and South America, rain falls in two seasons rather than one, which serves to spread rainfall throughout the year (Farmer 1986, Ellis and Galvin 1994), favoring production of livestock over crops (Marshall and Hildebrand 2002). And in temperate grazing lands, rain often comes either in the summer or the winter, which present different challenges to herders, livestock, and wildlife.

In these variable environments, mobility is critical. Movement allows herders and wildlife to access forage and water that is unevenly distributed in space and varies over time (Coughenour, Chapter 3; Scoones 1995, Niamir-Fuller 1999)[3]. Access to this 'landscape heterogeneity' by grazing animals will be a major theme of this book (see Hobbs et al., Chapter 2). Some of this heterogeneity in forage and water is created by inherent (and often slowly changing) features of particular landscapes. For example, soils, elevation, and topography create heterogeneous patches of different types of vegetation that livestock and wildlife can exploit (Hobbs et al., Chapter 2). The distribution of this heterogeneity affects how much grazing animals must move to satisfy their feeding requirements and how many grazing animals a landscape can support (Coughenour, Chapter 3). Heterogeneity is also created by faster changes over time, particularly by local patches or large scale gradients in rainfall that shift the location of abundant forage and water over time. One particularly important aspect of this heterogeneity is the existence of rare, 'key resources', like wetlands or wetter hill-slopes, which create an ecological 'safety net' for herders and animals in the dry season or drought (Scoones 1991, Illius and O'Connor 1999). Access to key resources 'often determines whether or not herders survive harsh years without massive livestock losses' (Little 2003:22); this applies also to wild grazers. Presence of these key resources clearly allows grazing landscapes to support more wildlife and livestock than if the key resource were absent. Moreover, Wang et al. (2006) demonstrated that increasing heterogeneity in temperate grazing lands attenuated feedbacks of population density on population growth rate - which, in theory, will allow larger populations to be supported in heterogeneous landscapes than those where resources are more homogeneous.

Herders and their livestock access the complex resources of landscapes, including those that are rare or ephemeral, by moving. These movements may require dispersing entire households on daily, seasonal, and annual time scales (McCabe et al. 1999, Niamir-Fuller 1999, Ritchie and Olff 1999, Boone et al. 2005). Wildlife also move on similar time scales to access these resources (Senft et al. 1987, Coughenour 1991, Bailey et al. 1996, Fryxell et al. 2004). These movements take the form of the few, large and long seasonal migrations of wildlife, people and livestock, sometimes over hundreds of kilometers, that still exist in some places of the world (caribou, wildebeest, Fulani herders, reindeer). Much more common are smaller and shorter seasonal movements of people, livestock, and wildlife over tens of kilometers, as herders and wildlife retreat to warmer and wetter pastures as the dry or cold season advances. The most common are daily movements to reach abundant forage and water, located in different parts of grazing landscapes. The consequence of restricting mobility and access to landscape heterogeneity, by cutting up landscapes into smaller pieces, physically or socially, will be a major theme of this book.

Mobility offers a fundamentally important way to reduce risks of food shortages for people, wildlife, and livestock using grazing lands where resources vary in time and space. To further reduce exposure to climatic (and other) risks, herders create extensive 'social safety nets'. In Australia 'agistment' arrangements allow private ranches to be managed cooperatively among owners (Stokes et al., Chapter 4; McAllister et al. 2006). Agistment compensates for fragmentation of land by restoring elements of spatial connectivity through formal and informal agreements. Similarly, collectives of herders in China and Mongolia persist in harsh regions even though the grazing land is being privatized (Neupert 1999; Ojima and Chuluun, Chapter 8). Collective arrangements help to solve grazing rights conflicts, determine how long herders can graze in certain pastures and can even help in fencing the outermost boundaries (Banks et al. 2003). In Africa different types of customary rules control access to resources, labor and pasture rental. For instance, pastoral safety nets have been the primary means of dealing with drought in the Sahel of west Africa (Niamir-Fuller 1999). Maasai pastoralists in east Africa maintain kin-based networks for mutual assistance to ensure the survival of households during crisis, which includes access to pasture and water (Potkanski 1997).

Movement is also the key to ecological impact: extensive herding and wildlife migrations, with regular seasonal movements, disperse the impacts of grazing impact in space and allow recovery of seasonally used pastures in time (Coughenour, Chapter 3). Concentration of livestock around water points, for example, creates denuded areas with thin plant cover and diminished plant and animal diversity (Tolsma et al. 1987, Andrew 1988, James et al. 1999, de Leeuw et al. 2001). When the activities of people outside reserves confine elephants within reserves, tree cover and biodiversity can be lost (Cumming et al. 1997, Western and Maitumo 2004). As pastoralists settle, the distance livestock travel to reach forage often contracts, concentrating impact on plants around settlements (Verlinden et al. 1998, Turner 1999b). Movement, on the other hand, creates complexity, heterogeneity and diversity that may be important to the resilience of grazing ecosystems (Coughenour, Chapter 3).

Although movement mitigates the risks created by climate variability, it can create significant problems for pastoralists. Some problems arise because of what has been called the 'paradox of pastoral land tenure'. In this contradiction, herders need both secure access to pasture (or land) and water, but also the flexibility to move in response to unpredictable events caused by climate, politics, or other social factors (Fernández-Giménez 2002). The classic way to address insecurity of land ownership is to establish fixed boundaries on land and set up rules that grant access to the land within these boundaries to individuals (private property) or particular social groups (common property) (Ostrom 1990, Bromley 1992). But because of the paradox described above, other ways of managing land with more flexible boundaries may be more appropriate (Turner 1999a), particularly through the use of social capital (Galvin, Chapter 15).

Movement also can have high political and social costs in modern nation states. Movement weakens the ability of herders (and wildlife) to secure access to particular pieces of land and particular water points because they use them temporarily and move on. Governments and the settled public often think pastoralists are backwards, primitive and undeveloped (Blench 2000). This is partially because herders have less access to education and communication, so they have less opportunity to influence policy than their settled and farming neighbors. Pastoralism is often viewed by national governments as a livelihood of episodic crisis because of the tolls taken on people and animals by intermittent drought (Blench 2000), despite the fact that drought is a normal part of the dynamics of these dryland systems. Part of the reason for this poor understanding of pastoralism is because most people alive today grew up in farming or urban landscapes (Leneman and Reid 2001). Agricultural policy usually supports farming over herding for the same reason (Horowitz and Little 1987).

2. CURRENT CHANGES IN PASTORAL SOCIETY AND GRAZING LANDS

Grazing lands around the world evolved from open lands supporting wildlife and hunter-gatherers 10,000 years ago to the often fragmented and more contracted grazing landscapes we see today. One of the major changes in grazing lands is the expansion of farming and settlement into drier areas over the last few millennia, which has pushed herders, hunters, and wildlife out of much of the wettest, most productive grazing land over time. About 35-50% of the wetter (semi-arid and dry sub-humid) portions of former grazing land are now plowed for irrigated and rainfed crops and about 2-4% settled for towns and cities (MEA 2005). Croplands in these drylands are particularly evident in eastern Brazil, the Guinean savanna zone of west Africa, Spain, central North America, and India (MEA 2005). Less than 10% of the driest (hyper-arid and arid) grazing lands are cropped or settled globally. In east Africa, farmers now plant crops in about 60-70% of former forest, about a third of all woodlands and bushlands, 23% of grasslands and only 1-3% of deserts and semi-deserts (Reid et al. 2005). This trend of crop expansion into former grazing land continues today (Neupert 1999, Little 2003, Geist and Lambin 2004). In addition, creation of conservation areas has reserved portions of grazing lands for exclusive use by native biota (Fratkin and Mearns 2003, Brockington et al. 2006). We calculate these protected areas cover 10.9% of extensive grazing lands, 4% of intensive grazing lands, 10.7% of forests/woodlands, and 3% of croplands/urban areas globally.

In subsistence herding societies of the developing world, the number of livestock per capita is falling. This is caused by two situations. In some cases, human populations are growing, while livestock populations vary, but do not grow. In others, human populations vary, but livestock populations fall. In either case, the result is that people have fewer numbers of livestock from which to make their living. This, in turn, sometimes results in increased poverty and sometimes in increased diversification and intensification of livelihood strategies, or both. Pastoral population growth is not universal, but it is common in Africa and parts of Asia, which results in fewer livestock per person (Neupert 1999, Fratkin and Roth 2005). This has serious implications for food security, especially in those very dry areas where there are few options for other livelihood strategies and where people do not have the education, skills, and wealth to move to other economic endeavors. Following the demise of communism in the former Soviet countries of Central Asia in the early 1990s, a similar decline in the number of livestock per person emerged. While human populations remained fairly steady or grew relatively slowly, livestock populations plummeted (Behnke 2003, Lunch 2003). During the last decade, the numbers of livestock began to grow. In any case, in regions of both Central Asia and in Africa the consequences of these dynamics have included human food insecurity, hunger, and malnutrition (Kerven 2003, Fujita et al. 2004). The declining number of livestock per person has, for the most part, not occurred in the industrialized livestock sector like those in North America and Australia and human food security is not a resultant phenomenon, but issues of economic viability are. Globalization of markets has made it an imperative for these groups to consolidate and increase their operations to deal with economies of scale.

Transformation of land ownership (tenure) from common to private property has had significant impacts on pastoral society and grazing lands (Galaty 1994). Although this is an historical process in much of North America and Australia, much of the grazing land in the rest of the world is starting to move into private holdings, or has done so recently. In Africa, privatization is pervasive and is occurring for many reasons including the perception, and sometimes realization, of more equitable access to land, of more access to services such as markets and education, and greater participation in local economies (Rutten 1992). In Central Asia and China, on the other hand, livestock have just recently become privately-owned (livestock were owned by the state in Soviet times), but land is slowly moving towards privatization, especially in the grazing lands of China (cf Banks 2003). It is often the key resource areas (dry or cold season refuges) that first get privatized or expropriated for other uses, particularly agriculture. Areas that are less productive are often privatized later. One effect of losing a key resource area for people and animals is that they lose the buffer from extreme events that the key resource provided. This process has conferred neither prosperity nor resilience on most pastoralists. For example, pastoralists historically coped with drought by accessing grazing refuges or key resource areas. Privatization of these areas prevents such coping, allowing drought to have devastating effects because people have lost access to those resources, causing livestock and wildlife losses and human hunger (Illius and O'Connor 2000). Privatization usually takes place in grazing lands that are wet, are closer to urban centres and/or contain significant key resources that are essential for the success of crop cultivation (Galaty 1994). Being close to urban centers or having infrastructure that can support agricultural markets, both for livestock and crops, is important and is increasingly becoming essential for pastoral livelihoods.

The marketplace is changing dramatically for pastoralists. In some places, livestock are becoming more of a commodity and in others the private market is a new phenomenon, such as in the former Soviet block. In industrialized markets, ranchers are now directly dealing with a global market. In Central Asia, Australia, and North America, livestock have been part of the market economy for the last 100 years or so but in very different ways. After the Soviets invaded the Central Asian states, livestock and livestock products were owned by the state. Individuals did not participate directly in the market. Since independence, however, collapse of the collectives that supported pastoral production has resulted in individual ownership of livestock and direct market commercialization of livestock, a state that many people are finding difficult to navigate (Williams 2002, Behnke 2003). For industrial ranching, there has been a trend towards declining terms of trade for the beef industry with input costs rising and output prices unchanged or falling (CIE 1997, Beutler 2003). The result has been either a consolidation of properties or intensification of production systems or both.

Although livestock keepers in North America and Australia have sold livestock for decades, the marketing of livestock as the primary household livelihood strategy (rather than keeping animals and selling their products) is a fairly recent phenomenon in Africa and parts of Asia. This does not mean that herders did not sell livestock; on the contrary, pastoralists have long traded or sold livestock to purchase grain, other foods and commodities, especially during droughts and in dry seasons (Sato 1997, Swift and Hamilton 2003). However, pastoralists are increasingly entering the market (Fratkin et al. 1999, Fratkin 2004). There are two reasons for this trend. First, pastoralists need to sell more animals to buy enough food as a result of the fewer numbers of livestock per person. Some pastoralists are also selling more livestock for cash and reinvesting back into their herds in terms of veterinarian services and different types of livestock. Second, more livestock in the markets coincide with a global trend towards increased livestock consumption (Misselhorn 2005, ILRI 2006); consumption of livestock products may double by the year 2020 (Delgado et al. 1999). This global change contributes to the increased marketing of livestock (Fratkin 2004).

Where herders have sufficient capital, they have intensified the production of livestock by developing water, re-seeding grazing lands with more productive grass species, fencing lands to control grazing patterns, and investing in greater veterinary care (see Stokes et al., Chapter 4). This 'enclosure' of the grazing lands is still uncommon on a global scale, but very common in particular places like Australia, North America, and Argentina (Blench 2000). Here, pastoralists substitute capital (in the form of boreholes, fencing) for the labor needed in subsistence pastoralism, and change the composition of their herds to meet the demands of the market (as described above).

Pastoral families often choose to settle and move their households and livestock less often as they become more connected to markets, social services, and particularly if they start to own specific parcels of land (Salzman 1980, Fratkin and Roth 2005). Moreover, marketing of livestock, privatization of land, and sedentarization of people are almost always interrelated. Behnke (2003), for example, found that the shift in Kazakhstan in the 1990s to a settled form of livestock keeping was due to the depressed economic conditions that accompanied the adoption of commercial objectives. Similar conditions prevailed in Mongolia following independence (Janzen 2005). Sedentarization has had profound effects on people including changing lifestyle expectations, such as, education for children and diversity of livelyhoods. Land privatization, fence construction, and protected lands policy are some of the factors that accompany sedentarization, particularly in Africa, such as in Kenya and Tanzania.

These changes in pastoral society alter the vegetation in grazing lands in some places and not in others. Where climate is highly variable and human populations are low, livestock populations rarely impact vegetation because recurrent drought forces herders to move frequently, or causes major die-offs of livestock (Ellis and Swift 1988, Behnke et al. 1993, Vetter 2005). Movements or die-offs of livestock allow vegetation in grazing lands to recover. However, in wetter grazing lands or where herders are sedentary, heavy livestock grazing can change the composition of the vegetation, particularly discouraging herbaceous plants like grasses and leafy herbs in favor of woody plants like shrubs and trees (Asner et al. 2004, Vetter 2005, Rohde et al. 2006). Even in very dry lands, heavy livestock grazing can cause a shift in the composition of forage from those palatable for livestock to those that are less palatable (Hiernaux 1998). What is clear is that the argument supporting degradation of grazing lands by pastoralists (Mabbutt 1984, GLASOD 1990) has been overstated (Ellis and Swift 1988, Nicholson et al. 1998, Geist and Lambin 2004), but is not resolved yet (Illius and O'Connor 1999, Sullivan and Rohde 2002, Hein and De Ridder 2006). But equally clear is that pastoralists do see undesirable change in their grazing lands when grazing is too heavy (e.g., Desta and Coppock 2004). The remaining issues include how extensive undesirable change is, who judges that change (Warren 2002), and how policy can support (or stop weakening) efforts by local communities to manage change.

Over the last few decades, the expansion of shrublands and woodlands into grasslands and savannas has occurred across the globe (Skarpe 1991, Archer 1994, Scholes and Archer 1997, Van Auken 2000), although not everywhere (Witt et al. 2006). This trend is caused by humans for four reasons: carbon dioxide emissions favor C_3 plants which are usually woody, light to heavy grazing by livestock on grasses encourages woody plants to expand, heavy grazing reduces grass fuel thus reducing fires that would kill woody plants, and nitrogen pollution favors woody plants (Archer et al. 1995, Van Auken 2000, Bond et al. 2003, Asner et al. 2004). After grazing stops, woody plants can remain in grazing lands over the long term (Asner et al. 2004). The origin of the causes of this shift are both local and global; while grazing can be managed locally, greenhouse gas and nitrogen pollutants usually originate far away in more industrialized systems. Whatever the cause, more woody plants makes grazing less suitable for wild and domestic animals that subsist on grass (wildebeest, bison, cattle, sheep) and more suitable for animals that browse woody plants (giraffe, impala, moose, goats, and camels). This shift can also have major consequences for biodiversity, cycling of nutrients, vegetation productivity, carbon sequestration, and other ecosystem attributes (Gill and Burke 1999, Asner et al. 2004).

3. THE FOCUS OF THIS BOOK

Many of the human-driven changes described above cause the process that is the subject of this book: fragmentation of grazing lands. Although we focus here on the ways that land tenure and intensification drive fragmentation, we also describe the role of sedentarization, marketing, technology, human population growth, policy and other factors (Behnke, Chapter 13 for overview; Chapters 4-12 for case studies). We suggest that the application of ideas of private land ownership and agricultural intensification, so clearly advantageous in mesic environments, has restricted the movements of people, livestock and wildlife across drier landscapes, thereby limiting their access to resources that fluctuate over time. Moreover, the application of policy that locates markets, schools and other social services at central places, like towns, forces pastoral families to settle, becoming less mobile. Thus, people and animals have diminished options to reach forage and water whose locations change in space and time, seasonally, and throughout drought cycles. This has had negative effects on ecological processes that sustain these natural systems and human economies. Declining human welfare has often resulted in the need for inputs to offset the effects of fragmentation. We explore these issues in this book.

We organize this book as follows. The idea of fragmentation is explored in Chapter 2 as it applies to dry, pastoral systems throughout the world. Hobbs and colleagues develop a case that fragmentation arises from different natural, social, and economic conditions worldwide but create similar outcomes for human and natural systems. Fragmentation is the isolation of land or habitat (Villard 2002). Related concepts such as habitat loss, spatial scale, and heterogeneity are defined, and general sources of fragmentation are explained. The case is made that fragmentation of grazing lands is a virtually universal outcome of modern systems of land tenure with important consesquences for both ecosystems and people. It sets the stage for the rest of the book. This chapter is followed by a discussion of the importance of spatial scale, movement and heterogeneity in grazing ecosystems (Coughenour, Chapter 3).

We then explore fragmentation of grazing lands at nine sites around the world, including Dalrymple Shire, Australia (Stokes et al., Chapter 4), the Northern Great Plains, USA (Lackett and Galvin, Chapter 5), Jackson Valley, Wyoming, USA (Lackett and Hobbs, Chapter 6), Kazakhstan (Alimaev and Behnke, Chapter 7), Mongolia (Ojima and Chuluun, Chapter 8), the Athi-Kaputiei Plains, Kenya (Reid et al., Chapter 9), southern Kajiado, Kenya (BurnSilver et al., Chapter 10), Ngorongoro Conservation Area, Tanzania (Galvin et al., Chapter 11), and the North-West Province, South Africa (Galvin et al., Chapter 12; see all sites in Figure 1-2). These sites are predominantly in arid and semi-arid grazing lands and parts of some sites are sub-humid (e.g., Ngorongoro highlands). These nine case studies examine how fragmentation occurs, the patterns that result, and the consequences of fragmentation for ecosystems and the people who depend on them for their livelihoods. This is followed by a set of synthesis chapters including one on the drivers of fragmentation (Behnke, Chapter 13), a synthetic comparison of heterogeneity among ecosystems (Boone et al., Chapter 14), and the responses of pastoralists to fragmentation (Galvin, Chapter 15). These themes are briefly discussed below.

Landscapes are divided into parts through physical barriers, such as fences, and administrative barriers, such as political boundaries and social norms (Boone and Hobbs 2004, Reid et al. 2004a). Fragmentation also occurs as land is converted from one land cover type to another, particularly cropping (Hobbs et al., Chapter 2; Coughenour, Chapter 3). Thus, there are numerous causes for fragmentation of grazing land and the case studies in the book point this out. The manner in which these variables interact

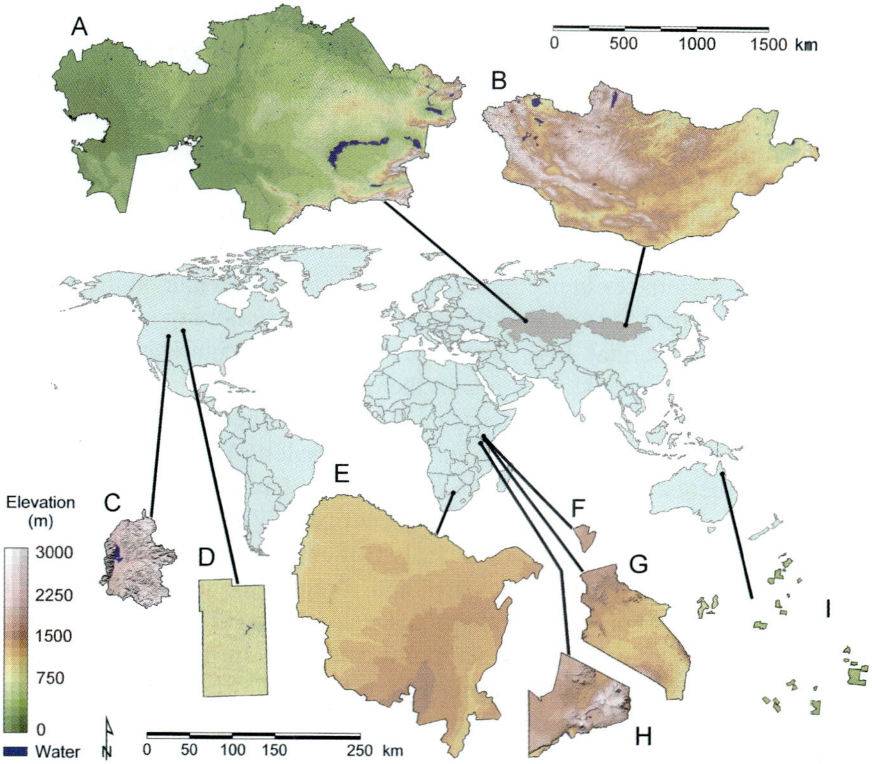

Figure 1-2. Areas for which case studies (chapters 4-12) were prepared. They include: A) the Republic of Kazakhstan, B) Mongolia, C) the US National Elk Refuge plus surrounding areas, D) selected counties in the US Northern Great Plains, E) the North-West Province of South Africa, F) Athi-Kaputiei Plains, Kenya, G) southern Kajiado District, Kenya, H) Ngorongoro Conservation Area, Tanzania, and I) Dalrymple Shire, Queensland, Australia. Sites are mapped using two scales: one for the sites above the world map (A, B) and one for the sites below the world map (C through I). Elevation (color) and topography (lightness) of sites are depicted.

to drive fragmentation is complex because the variables do not influence fragmentation in the same way and their importance shifts over time and from case to case. Behnke (Chapter 13) makes sense of these divergent causes of fragmentation by viewing fragmentation as driven by divergent interests and interest groups. He argues for a better understanding of the institutional and economic factors that drive fragmentation and the impact of fragmentation on economic inequality in pastoral societies. Fragmentation has consequences for ecosystems and human well-being.

As described above, herders and wildlife move in response to variability in the quantity and quality of forage and water at different scales in time and space. In Chapter 3, Coughenour addresses issues of movement disruption for animals. Habitat loss, habitat fragmentation, and sedentarization all change movement of domestic and wild herbivores. Livestock and wildlife may be affected when fragmentation decreases their foraging efficiency, by weakening their ability to adapt to spatial and temporal variations in forage availability and by increasing risks of local extinction. Movement arguably increases ecosystem resilience and decreases ecosystem vulnerability. Thus, herbivore movement is central to the structure and functioning of ecosystems and this is why habitat loss, fragmentation and sedentarization threaten ecosystems and the people who inhabit them.

The degree of landscape heterogeneity, as described above, determines the array of both forage and water choices available to herders and wildlife. As land fragments, those choices are compressed. This is particularly true in dry ecosystems where inherent landscape heterogeneity is overlain with highly variable rainfall that adds considerable, but often transient, heterogeneity in the amount and quality of both water and forage. Boone, BurnSilver and Kruska (Chapter 14) provide measures of heterogeneity for a variety of sites by quantifying the spatial and temporal variation in forage. They also develop a metric of social heterogeneity by assessing infrastructural variation in pastoral regions. These analyses provide clear warnings of where the loss of system heterogeneity through fragmentation would require the most external inputs to maintain pastoral livelihoods in the system.

But fragmentation differs in direction and persistence of fragmentation from one place to another. There are several systems where fragmentation is increasing but still reversible, such as in Kajiado, Kenya (BurnSilver et al., Chapter 10), while there are others where pastoralists are choosing to re-aggregate previously fragmented landscapes, such as in Dalrymple Shire, Australia (Stokes et al., Chapter 4). Globally, these might include most of the 'extensive' grazing lands in Figure 1-1. We also find examples where fragmentation is advanced, chronic, and unlikely to be reversed such as in Kitengela, Kenya (Reid et al., Chapter 9). Globally, these might include most of the 'intensive' grazing lands in Figure 1-1. There are several compelling reasons to fragment grazing lands which include securing ownership of land and water, ability to cultivate, taking advantage of services such as education, and natural causes such as disease incidence and climate change among others (Behnke, Chapter 13). Irreversible fragmentation tends to occur in wetter grazing lands or where outside inputs, like markets and social services, are available and affordable. However, for most extensive grazing lands there are few viable economic strategies other than extensive livestock production. So where fragmentation is chronic in these drylands, how can fragmentation be slowed or reversed? How can people continue to access the forage and water they need, once the landscape becomes fragmented? Galvin (Chapter 15) explores the factors that support pastoralists to slow or reverse fragmentation through social capital. Social capital is seen here as the glue which embodies the myriad ties and institutions, both formal and informal, through which households gain access to resources. They include cooperative networking and use of social organizations (common in east Africa) as well as institutions beyond the community such as local and state government (common in Central Asia and in commercial livestock production systems such as in Australia). This chapter also covers possible policy responses that will support pastoral adaptations to fragmentation.

In the chapters that follow, it will become clear that fragmentation, first described for forested lands, is now a major issue in grazing lands around the world and that it has clear consequences for people, wildlife, and grazing lands. We address both the causes of fragmentation as well as the human responses to these changes, a phenomenon that is rarely addressed in scientific studies of fragmentation of the earth's ecosystems. Our body of work shows the critical role of access to large-scale landscape heterogeneity in sustaining pastoral livelihoods and ecosystem services (e.g., wildlife), particularly in the vast majority of the world's grazing lands where outside inputs are neither affordable nor available. The traditional view that exclusivity of land use promotes human well-being and sustains natural processes (e.g., the tragedy of the commons) may be deeply flawed in dry grazing lands, where such tenure regimes can result in declining productivity and degradation, just the type of outcome that exclusivity was designed to prevent. More aptly put, the well-known 'tragedy of the commons' may rather be a 'tragedy of enclosure' for most of the earth's grazing lands.

ACKNOWLEDGEMENTS

In writing this chapter, and this book, we gratefully follow in the intellectual footsteps of Jim Ellis, who created the Biocomplexity, Spatial Scale, and Fragmentation: Implications for Arid and Semi-arid Ecosystems (SCALE) project and secured its funding from the U.S. National Science Foundation (grant # DEB-0119618). We thank A. Ash, C. Stokes, P. Thornton, J. Lackett, M. Coughenour, C. Kerven, D. Nkedianye, R. Behnke, S. BurnSilver, J. Worden, and C. Wilson for stimulating discussions that influenced the ideas presented here. We thank P.G. Jones and P.K. Thornton for use of their high resolution P/PET map of Africa, and R.B. Bocne for providing Figure 1-2.

ENDNOTES

[1] We use the term 'grazing lands' as a more inclusive term than 'wildlife habitat' or 'pastoral rangelands' because we refer to both human and wildlife use of these lands. We also use the term 'ecosystem' to mean socio-ecological systems that include people.

[2] All calculations of grazing land were based on analysis of Figure 1-1. We defined grazing lands as those land cover categories (GLC 2003) without significant tree cover (thus we excluded forest/woodland) or cropland (we excluded all categories containing any cropland); these lands are dominated by shrubs and herbaceous plants. We also excluded urban areas (GRUMP 2004b) and lumped them with cropland. 'Intensive' and 'extensive' grazing lands were distinguished using human population density (GRUMP 2004a). We included protected areas (WDPA 2006) as part of extensive grazing lands because wildlife, and sometimes livestock, graze here. Livestock numbers (cattle, sheep and goats) were from FAO (2006), tropics were defined as all land area between the Tropics of Cancer and Capricorn. Grazing lands were distinguished using an aridity index (long-term mean precipitation divided by long-term potential evapo-transpiration between 1951-1980 (Deichmann and Eklundh 1991, Middleton and Thomas 1997, MEA 2005)), with hyper-arid: < 0.05, arid: 0.05-0.20, semi-arid: 0.20-0.50; sub-humid/humid: > 0.50; the African part of these maps is a more recent, higher resolution version created by P.G. Jones and P.K. Thornton from WorldClim datasets (Hijmans et al. 2005). Cold grazing lands are those with more than six months when average temperatures are below freezing (0^oC) and not more than three months when temperatures reach above 6^oC.

[3] Herders also move for a variety of social and political reasons, e.g., McCabe (2004), Gulliver (1975).

REFERENCES

Andrew, M. H. 1988. Grazing impact in relation to livestock watering points. Trends in Ecology & Evolution 3:336-339.

Archer, S. A. 1994. Woody plant encroachment into SW grasslands and savannas: rates, patterns and proximate causes. Pages 13-68 *In* M. Vavra, W. Laycock, and R. Pieper, editors. Ecological implications of livestock herbivory in the West. Society for Range Management, Denver, Colorado, USA.

Archer, S., D. S. Schimel, and E. A. Holland. 1995. Mechanism of shrubland expansion: land use, climate or CO_2. Climatic Change 29:91-99.

Asner, G. P., A. J. Elmore, L. P. Olander, R. E. Martin, and A. T. Harris. 2004. Grazing systems, ecosystem responses, and global change. Annual Review of Environment and Resources 29:261-299.

Bailey, D. W., J. E. Gross, E. A. Laca, L. R. Rittenhouse, M. B. Coughenour, D. M. Swift, and P. L. Sims. 1996. Mechanisms that result in large herbivore grazing distribution patterns. Journal of Range Management 49:386-400.

Banks, T. 2003. Property rights reform in rangeland China: On the road to the household ranch. World Development 31:2129-2142.

Banks, T., C. Richard, L. Ping, and Y. Zhaoli. 2003. Community-based grassland management in western China. Mountain Research and Development 23:132-140.

Behnke, R. H. 1983. Production rationales: the commercialization of subsistence pastoralism. Nomadic Peoples 14:3-27.

Behnke, R. H. 1985. Measuring the benefits of subsistence versus commercial livestock production in Africa. Agricultural Systems 16:109-135.

Behnke, R. H. 2003. Reconfiguring property rights and land use. Pages 75-107 *In* C. Kerven, editor. Prospects for pastoralism in Kazakstan and Turkmenistan: From state farms to private flocks. Routledge Curzon, New York, USA.

Behnke, R. H., I. Scoones, and C. Kerven. 1993. Range ecology at disequilibrium: new models of natural variability and pastoral adaptation in African savannas. Overseas Development Institute, London.

Beutler, M. K. 2003. Impact of South Dakota agriculture, 2002. South Dakota State University, Brookings, South Dakota, USA.

Blench, R. 2000. 'You can't go home again', extensive pastoral livestock systems: issues and options for the future. ODI/FAO, London, UK.

Blench, R. and F. Sommer. 1999. Understanding rangeland biodiversity. Working paper # 121 Overseas Development Institute, London, UK.

Bond, W. J., G. F. Midgley, and F. I. Woodward. 2003. The importance of low atmospheric CO^2 and fire in promoting the spread of grasslands and savannas. Global Change Biology 9:973-982.

Boone, R. B. and N. T. Hobbs. 2004. Lines around fragments: Effects of fencing on large herbivores. African Journal of Range & Forage Science 21:147-158.

Boone, R. B., S. B. BurnSilver, P. K. Thornton, J. S. Worden, and K. A. Galvin. 2005. Quantifying declines in livestock due to land subdivision. Rangeland Ecology & Management 58:523-532.

Bourlière, F. and M. Hadley. 1983. Present-day savannas: An overview. Pages 1-17 *In* F. Bourliere, editor. Tropical savannas. Ecosystems of the world 13. Elsevier, Amsterdam.

Brockington, D., J. Igoe, and K. Schmidt-Soltau. 2006. Conservation, human rights, and poverty reduction. Conservation Biology 20:250-252.

Bromley, D. W. 1992. Making the commons work. Institute for Contemporary Studies Press, San Francisco, California, USA.

Child, R. D., H. F. Heady, W. C. Hickey, R. A. Peterson, and R. D. Piper. 1984. Arid and semi-arid lands: Sustainable use and management in developing countries. Winrock International, Morrilton, Arkansas, USA.

CIE. 1997. Sustainable natural resource management in the rangelands. Centre for International Economics, Canberra & Sydney, Australia.

Conrad, V. 1941. The variability of precipitation. Monthly Weather Review 69:5-11.

Coughenour, M. B. 1991. Spatial components of plant-herbivore interactions in pastoral, ranching, and native ungulate ecosystems. Journal of Range Management 44:530-542.

Cumming, D. H., M. B. Fenton, I. L. Rautenbach, R. D. Taylor, G. S. Cumming, and M. S. Cumming. 1997. Elephants, woodlands and biodiversity in southern Africa. South African Journal of Science 93:231-236.

Darwin, C. R. 1909-1914. The voyage of the Beagle. Vol. XXIX. The Harvard Classics. P.F. Collier & Son, New York.

de Haan, C., H. Steinfeld, and H. Blackburn. 1997. Livestock and the environment: finding a balance. WRENmedia, Fressingfield, UK.

de Leeuw, J., M. N. Waweru, O. O. Okello, M. Maloba, P. Nguru, M. Y. Said, H. M. Aligula, I. M. A. Heitkonig, and R. S. Reid. 2001. Distribution and diversity of wildlife in northern Kenya in relation to livestock and permanent water points. Biological Conservation 100:297-306.

Deichmann, U. and L. Eklundh. 1991. Global digital data sets for land degradation studies: a GIS approach. UNEP/GEMS and GRID, Nairobi, Kenya.

Delgado, C., M. Rosegrant, H. Steinfeld, S. Ehui, and C. Courbois. 1999. Livestock to 2020: the next food revolution. IFPRI, FAO, and ILRI, Washington, D.C.

Desta, S., and D. L. Coppock. 2004. Pastoralism under pressure: Tracking system change in southern Ethiopia. Human Ecology 32:465-486.

Ellis, J. and D. M. Swift. 1988. Stability of African pastoral ecosystems: Alternative paradigms and implications for development. Journal of Range Management 41:450-459.

Ellis, J. E. and K. A. Galvin. 1994. Climate patterns and land-use practices in dry zones of Africa. BioScience 44:340-349.

FAO. 1999. 1998 Production Yearbook. Food and Agriculture Organization of the United Nations, Rome, Italy.

FAO. 2006. Livestock mapping project, http://www.fao.org/ag/AGAinfo/resources/en/glw/default.html, Rome, Italy.

Farmer, G. 1986. Rainfall variability in tropical Africa: some implications for policy. Land Use Policy 3:336-342.

Fernández-Giménez, M. E. 2002. Spatial and social boundaries and the paradox of pastoral land tenure: A case study from postsocialist Mongolia. Human Ecology 30:49-78.

Fratkin, E. 2004. Ariaal pastoralists of Kenya: Studying pastoralism, drought, and development in Africa's arid lands. Pearson Education, Boston, Massachusetts, USA.

Fratkin, E. M., E. A. Roth, and M. A. Nathan. 1999. When nomads settle: the effects of commoditization, nutritional change, and formal education on Ariaal and Rendille pastoralists. Current Anthropology 40:729-735.

Fratkin, E. and R. Mearns. 2003. Sustainability and pastoral livelihoods: Lessons from East African Maasai and Mongolia. Human Organization 62:112-122.

Fratkin, E. and E. A. Roth, editors. 2005. As pastoralists settle: Social, health, and economic consequences of pastoral sedentarization in Marsabit District, Kenya. Kluwer Academic Publishers, Dordrecht, Netherlands.

Fryxell, J. M., J. F. Wilmshurst, and A. R. E. Sinclair. 2004. Predictive models of movement by Serengeti grazers. Ecology 85:2429-2435.

Fujita, M., E. A. Roth, M. A. Nathan, and E. Fratkin. 2004. Sedentism, seasonality, and economic status: a multivariate analysis of maternal dietary and health statuses between pastoral and agricultural Ariaal and Rendille communities in northern Kenya. American Journal of Physical Anthropology 123:277-291.

Galaty, J. G. 1994. Rangeland tenure and pastoralism in Africa. Pages 185-204 *In* E. Fratkin, K. A. Galvin, and E. A. Roth, editors. African pastoralist systems: An integrated approach. Lynne Reiner Publishers, Boulder, Colorado, USA.

Galvin, K. A., P. K. Thornton, R. B. Boone, and J. Sunderland. 2004. Climate variability and impacts on east African livestock herders: the Maasai of Ngorongoro Conservation Area, Tanzania. African Journal of Range and Forage Science 21:183-189.

Geist, H. J. and E. F. Lambin. 2004. Dynamic causal patterns of desertification. BioScience 54:817-829.

Gill, M. 1999. Meat production in developing countries. Proceedings of The Nutrition Society 58:371-376.

Gill, R. A. and I. C. Burke. 1999. Ecosystem consequences of plant life form changes at three sites in the semiarid United States. Oecologia 121:551-563.

GLASOD. 1990. Global assessment of soil degradation. International Soil Reference and Information Centre, Wageningen, Netherlands, and United Nations Environment Program, Nairobi, Kenya.

GLC. 2003. Global Land Cover 2000 Database. Joint Research Centre, European Commission, http://www-gem.jrc.it/glc2000.

Groombridge, B. E. 1992. Global biodiversity: Status of the earth's living resources. Chapman & Hall, London, UK.

GRUMP. 2004a. Global Rural-Urban Mapping Project (GRUMP), Alpha Version: Population Grids. Socioeconomic Data and Applications Center (SEDAC), Columbia University. http://sedac.ciesin.columbia.edu/gpw. Center for International Earth Science Information Network (CIESIN), Columbia University; International Food Policy Research Institute (IFPRI); The World Bank; and Centro Internacional de Agricultura Tropical (CIAT). Palisades, NY.

GRUMP. 2004b. Global Rural-Urban Mapping Project (GRUMP), Alpha Version: Urban Extents. Socioeconomic Data and Applications Center (SEDAC), Columbia University. http://sedac.ciesin.columbia.edu/gpw. Center for International Earth Science Information Network (CIESIN), Columbia University; International Food Policy Research Institute (IFPRI); The World Bank; and Centro Internacional de Agricultura Tropical (CIAT). Palisades, NY.

Gulliver, P. H. 1975. Nomadic movements: causes and implications. Pages 369-386 *In* T. Monod, editor. Pastoralism in tropical Africa. Oxford University Press, Oxford.

Hein, L. and N. De Ridder. 2006. Desertification in the Sahel: a reinterpretation. Global Change Biology 12:751-758.

Hiernaux, P. 1998. Effects of grazing on plant species composition and spatial distribution in rangelands of the Sahel. Plant Ecology 138:191-202.

Hijmans, R. J., S. E. Cameron, J. L. Parra, P. G. Jones, and A. Jarvis. 2005. Very high resolution interpolated climate surfaces for global land areas. International Journal of Climatology 25:1965-1978.

Horowitz, M. M. and P. D. Little. 1987. African pastoralism and poverty: some implications for drought and famine. Pages 59-82 *In* M. Glantz, editor. Drought and famine in Africa: Denying drought a future. Cambridge University Press, Cambridge, UK.

Illius, A. W. and T. G. O'Connor. 1999. On the relevance of nonequilibrium concepts to arid and semiarid grazing systems. Ecological Applications 9:798-813.

Illius, A. W. and T. G. O'Connor. 2000. Resource heterogeneity and ungulate population dynamics. Oikos 89:283-294.

ILRI. 2006. Changing livestock landscapes: Drivers of change are creating a 'new livestock economy' that could spur pro-poor growth. Summary of a keynote address given by the International Livestock Research Institute (ILRI) at a meeting hosted by ILRI and the Indian Council for Agricultural Research (ICAR) in New Delhi 31 January–1 February 2006.

IPCC. 2000. Land use, land-use change, and forestry. Intergovernmental Panel on Climate Change, Cambridge University Press, Cambridge, UK.

James, C. D. J., J. Landsberg, and S. R. Morton. 1999. The provision of watering points in the Australian arid zone: a review of effects on biota. Journal of Arid Environments 41:87-121.

Janzen, J. 2005. Changing political regime and mobile livestock keeping in Mongolia. Geography Research Forum 25:62-82.

Kerven, C. 2003. Agrarian reform and privatisation in the wider Asian region: comparison with Central Asia. Pages 10-26 *In* C. Kerven, editor. Prospects for pastoralism in Kazakstan and Turkmenistan: From state farms to private flocks. Routledge Curzon, New York, USA.

Koocheki, A. and S. R. Gliessman. 2005. Pastoral nomadism, a sustainable system for grazing land management in arid areas. Journal of Sustainable Agriculture 25:113-131.

Leneman, J. M. and R. S. Reid. 2001. Pastoralism beyond the past. Development 44:85-89.

Little, P. D. 2003. Somalia: Economy without a State. International African Institute, James Currey, Indiana University Press, Btec Books, Oxford.

Lunch, C. 2003. Shepherds and the state. Pages 171-193 *In* C. Kerven, editor. Prospects for pastoralism in Kazakstan and Turkmenistan: From state to private flocks. Routledge Curzon.

Mabbutt, J. A. 1984. A new global assessment of the status and trends of desertification. Environmental Conservation 11:100-113.

Marshall, F. and E. Hildebrand. 2002. Cattle before crops: The beginnings of food production in Africa. Journal of World Prehistory 16:99-143.

McAllister, R. R. J., I. J. Gordon, M. A. Janssen, and N. Abel. 2006. Pastoralists responses to variation of rangeland resources in time and space. Ecological Applications 16:572-583.

McCabe, J. T. 2004. Cattle bring us to our enemies. University of Michigan Press, Ann Arbor.

McCabe, J. T., R. Dyson-Hudson, and J. Wienpahl. 1999. Nomadic movements. Pages 108-121 *In* M. A. Little and P. W. Leslie, editors. Turkana herders of the dry savanna. Oxford University Press, Oxford, England.

MEA. 2005. Ecosystems and human well-being: Current state and trends, volume 1. Millenium Ecosystem Assessment, Island Press, Washington, DC, USA.

Menaut, J.-C. 1983. The vegetation of African savannas. Pages 109-149 *In* F. Bourliere, editor. Tropical Savannas. Ecosystems of the World 13. Elsevier, Amsterdam.

Middleton, N. and D. Thomas. 1997. World atlas of desertification. Arnold, London, UK.

Misselhorn, A. A. 2005. What drives food security in southern Africa? A meta-analysis of household economy studies. Global Environmental Change – Human and Policy Dimensions 15:33-43.

Murdock, G. P. 1959. Africa - Peoples and their culture, history. McGraw Hill, New York.

Neupert, R. F. 1999. Population, nomadic pastoralism and the environment in the Mongolian Plateau. Population and Environment 20:413-441.

Niamir-Fuller, M. 1999. Managing mobility in African rangelands. FAO and Beijer International Institute of Ecological Economics, London.

Nicholson, S. E., C. J. Tucker, and M. B. Ba. 1998. Desertification, drought and surface vegetation: an example from the West African Sahel. Bulletin of the American Meteorological Society 79:815-830.

Ostrom, E. 1990. Governing the commons: The evolution of institutions for collective action. Cambridge University Press, Cambridge, UK.

Owen-Smith, N. 1987. Pleistocene extinctions - The pivotal role of megaherbivores. Paleobiology 13:351-362.

Potkanski, T. 1997. Pastoral economy, property rights and traditional mutual assistance mechanisms among the Ngorongoro and Salei Maasai of Tanzania. IIED Drylands Programme, London, UK.

Reid, R. S., P. K. Thornton, and R. L. Kruska. 2004a. Loss and fragmentation of habitat for pastoral people and wildlife in East Africa: Concepts and issues. South African Journal of Grass and Forage Science 21:171-181.

Reid, R. S., P. K. Thornton, G. J. McCrabb, R. L. Kruska, F. Atieno, and P. G. Jones. 2004b. Is it possible to mitigate greenhouse gas emissions in pastoral ecosystems of the tropics? Environment, Development and Sustainability 6:91-109.

Reid, R. S., S. Serneels, M. Nyabenge, and J. Hanson. 2005. The changing face of pastoral systems in grass-dominated ecosystems of East Africa. FAO, Rome, Italy.

Ritchie, M. E. and H. Olff. 1999. Spatial scaling laws yield a synthetic theory of biodiversity. Nature 400: 557-560.

Rohde, R. F., N. M. Moleele, M. Mphale, N. Allsopp, R. Chanda, M. T. Hoffman, L. Magole, and E. Young. 2006. Dynamics of grazing policy and practice: Environmental and social impacts in three communal areas of southern Africa. Environmental Science & Policy 9:302-316.

Rutten, M. 1992. Selling wealth to buy poverty: The process of individualisation of land ownership among the Maasai pastoralists of Kajiado District, Kenya, 1890-1990. Breitenbach Publishers, Saarbrucken, Germany.

Salzman, P. C. 1980. When nomads settle: Processes of sedentarization as adaptation and response. Praeger, New York.

Sato, S. 1997. How the East African pastoral nomads, especially the Rendille, respond to the encroaching market economy. African Studies Monographs 18:121-135.

Scholes, R. J. and S. R. Archer. 1997. Tree-grass interactions in savannas. Annual Review of Ecology and Systematics 28:517-544.

Scoones, I. 1991. Wetlands in Drylands - Key resources for agricultural and pastoral production in Africa. Ambio 20:366-371.

Scoones, I. 1995. Exploiting heterogeneity - Habitat use in cattle in dryland Zimbabwe. Journal of Arid Environments 29:221-237.

Senft, R. L., M. B. Coughenour, D. W. Bailey, R. W. Rittenhouse, O. E. Sala, and D. M. Swift. 1987. Large herbivore foraging and ecological hierarchies. BioScience 37:789-795.

Shinn, E. A., G. W. Smith, J. M. Prospero, P. Betzer, M. L. Hayes, V. Garrison, and R. T. Barber. 2000. African dust and the demise of Caribbean coral reefs. Geophysical Research Letters 27:3029-3032.

Skarpe, C. 1991. Impact of grazing in savanna ecosystems. Ambio 20:351-356.

Solbrig, O. T. 1996. The diversity of the savanna ecosystems. Pages 1-30 *In* O. T. Solbrig, E. Medina, and J. F. Silva, editors. Biodiversity and savanna ecosystem processes. Springer-Verlag, Berlin, Germany.

Spedding, C. R. W. 1971. Grassland Ecology. Oxford University Press, Oxford, UK.

Sullivan, S., and R. Rohde. 2002. On non-equilibrium in arid and semi-arid grazing systems. Journal of Biogeography 29:1595-1618.

Swift, J., and K. Hamilton. 2003. Household and food livelihood security. Pages 67-92 *In* S. Devereux and S. Maxwell, editors. Food Security in Sub-saharan Africa. ITDG Publishing, London, UK.

Tolsma, D. J., W. H. O. Ernst, and R. A. Verwey. 1987. Nutrients in soil and vegetation around two artifical waterpoints in eastern Botswana. Journal of Applied Ecology 24:991-1000.

Turner, M. D. 1999a. The role of social networks, indefinite boundaries and political bargaining in maintaining the ecological and economic resilience of the transhumance systems of Sudano-Sahelian West Africa. Pages 97–123 *In* M. Niamir-Fuller, editor. Managing mobility in African rangelands. Intermediate Technology Publications, London, UK.

Turner, M. D. 1999b. Spatial and temporal scaling of grazing impact on the species composition and productivity of Sahelian annual grasslands. Journal of Arid Environments 41:277-297.

Van Auken, O. W. 2000. Shrub invasions of North American semi-arid grasslands. Annual Review of Ecology and Systematics 31:197-215.

Verlinden, A., J. S. Perkins, M. Murray, and G. Masunga. 1998. How are people affecting the distribution of less migratory wildlife in the southern Kalahari of Botswana? A spatial analysis. Journal of Arid Environments 38:129-141.

Vetter, S. 2005. Rangelands at equilibrium and non-equilibrium: recent developments in the debate. Journal of Arid Environments 62:321-341.

Villard, M.-A. 2002. Habitat fragmentation: major conservation issue or intellectual attractor? Ecological Applications 12:319-320.

Wang, G. M., N. T. Hobbs, R. B. Boone, A. W. Illius, I. J. Gordon, J. E. Gross, and K. L. Hamlin. 2006. Spatial and temporal variability modify density dependence in populations of large herbivores. Ecology 87:95-102.

Warren, A. 2002. Land degradation is contextual. Land Degradation & Development 13:449-459.

WDPA. 2006. WDPA Consortium 2006 World Database on Protected Areas, http://www.unep-wcmc.org/wdpa/. UNEP-World Conservation Monitoring Centre (UNEP-WCMC), Cambridge, UK.

Western, D. 1982. The environment and ecology of pastoralists in arid savannas. Development and Change 13:183-211.

Western, D. and D. Maitumo. 2004. Woodland loss and restoration in a savanna park: a 20-year experiment. African Journal of Ecology 42:111-121.

Williams, D. M. 2002. Beyond great walls. Environment, identity, and development on the Chinese grasslands of Inner Mongolia. Stanford University Press, Stanford, CA.

Witt, G. B., J. Luly, and R. J. Fairfax. 2006. How the west was once: vegetation change in south-west Queensland from 1930 to 1995. Journal of Biogeography 33:1585-1596.

WRI. 2000. World Resources. World Resources Institute, Oxford University Press, New York.

Chapter 2

FRAGMENTATION OF ARID AND SEMI-ARID ECOSYSTEMS: IMPLICATIONS FOR PEOPLE AND ANIMALS

N. Thompson Hobbs[1,2], Robin S. Reid[3], Kathleen A. Galvin[1,4], and James E. Ellis*

[1]Natural Resource Ecology Laboratory, Colorado State University, Fort Collins, CO 80523-1499, USA; [2]Department of Forest, Rangeland, and Watershed Stewardship, Colorado State University, Fort Collins, CO 80523, USA; [3]International Livestock Research Institute, Nairobi, Kenya; [4]Department of Anthropology, Colorado State University, Fort Collins, CO 80523, USA

Human action has modified the earth in many ways, but one of the most pervasive effects of humans on the environment is dissection of natural systems into spatially isolated parts, a process generally known as fragmentation. Fragmentation of environments is not only caused by humans; dynamic natural processes like landslides, fires, and floods can create barriers that dissect natural systems. Understanding the consequences of human-caused and natural sources of fragmentation has been a fundamental challenge in ecology, a problem occupying theoretical and empirical workers for decades (see reviews of Usher 1987, Andren 1994, Collinge 1996, Turner 1996, Young et al. 1996, Harrison and Bruna 1999, Debinski and Holt 2000, Niemela 2001, Chalfoun et al. 2002, de Blois et al. 2002, Schmiegelow and Monkkonen 2002). Moreover, anthropologists and other social scientists have worked to understand the human forces that drive fragmentation of landscapes (Khazanov 1984, Little and Leslie 1999, Kerven 2003). Despite these efforts, understanding of the consequences of landscape fragmentation for human economies and social systems remains rudimentary.

* deceased

K. A. Galvin et al. (eds.), Fragmentation in Semi-Arid and Arid Landscapes: Consequences for Human and Natural Systems, 25–44.

This book is about the effects of fragmentation on arid and semi-arid ecosystems. More specifically, we describe how fragmentation influences people and animals, and in so doing, shapes human economies and ecological processes operating in rangelands throughout the world. Our central thesis is this: Socio-economic forces, particularly the privatization of land once used communally by pastoralists, have caused increasing exclusivity of use of the world's rangelands. Exclusivity of use, in turn, has restricted the movements of people, livestock, and wildlife across landscapes, thereby limiting their access to resources that vary over space. As a result of this limitation, people and animals have fewer options for responding to temporal variability in production of vegetation and availability of water, variability that characterizes arid and semi-arid lands. A diminished ability to compensate for temporal heterogeneity in vegetation and water by exploiting its spatial heterogeneity has interrupted ecological processes that sustain natural and human economies. Degradation of human welfare has followed, requiring substantial inputs of policy and capital to offset the effects of fragmentation. Our thesis contrasts sharply with the traditional view that exclusivity of use promotes human welfare and sustains natural processes by preventing the "tragedy of the commons." The traditional view holds that the sum of the productivities of privately-owned parcels is greater than the whole landscape productivity because of the incentive for land stewardship provided by property rights (Lund 2000). In contrast, our thesis is that in many systems, the sum of the productivities of land fragments may be less than the productivity of the unfragmented landscape.

In this chapter, we develop concepts supporting a more detailed treatment of our ideas, which will follow throughout the remainder of the book. We begin by talking about fragmentation, contrasting it with other human-caused changes in land use and land cover. We then discuss the related concepts of scale and heterogeneity. Next we describe a mechanism that mediates the way that land fragmentation influences people and animals in arid and semi-arid ecosystems. We subsequently develop the case that fragmentation of arid and semi-arid lands is a nearly universal outcome of modern systems of land tenure with important economic and social consequences. We close by outlining a conceptual model integrating these concepts and providing an overarching framework for the chapters that follow.

1. FRAGMENTATION, HABITAT LOSS, AND HABITAT MODIFICATION

Humans change landscapes in many ways and the variety of these changes has created confusion in terminology used to describe them. To avoid this confusion, we will clarify some terms as we will use them throughout this book. We begin with the concept of habitat loss, which will allow us subsequently to define habitat fragmentation. When human or natural disturbances convert land cover from one form to another, that conversion often changes the suitability of habitat for livestock and wildlife. For example, the conversion of grassland to intensive agriculture or to urban development changes an area of landscape that was once suitable for grazing animals to an area that is unsuitable. We will refer to changes like these as *habitat loss*—the conversion of landscape occurring in such a way that the *area* of habitat suitable for a species or community of animals is diminished. As the area of habitat declines, the patches or pieces of habitat that remain usually become more isolated from each other, because reduction in habitat area also expands the distances among the patches (Figure 2-1A).

Historically, the process of habitat loss was considered synonymous with habitat fragmentation. However, because these concepts were not distinct, the effects of reductions in habitat area were confounded with the effects of habitat isolation. The contemporary definition of fragmentation seeks to disentangle these effects; by contrast with habitat loss, the modern view of fragmentation refers to changes in relative isolation of habitat distinct from changes in their area (Figure 2-1B, C) (Fahrig 1997, 1998, 2002, Schmiegelow and Monkkonen 2002, Villard 2002, Ryall and Fahrig 2006, Betts et al. 2006). One of the most important, unresolved questions in contemporary studies of fragmentation asks: "How does habitat fragmentation influence the abundance of organisms apart from the effects of habitat loss?" (for review see Andren 1994, Fahrig 1997, Bender et al. 1998, Schmiegelow and Monkkonen 2002, Tscharntke et al. 2002, Ryall and Fahrig 2006, Betts et al. 2006). On rangelands, this question becomes "How does habitat fragmentation influence the number of people and animals that can be supported by a given landscape?"

Understanding this influence is important for two reasons. First, there are many cases where landscapes are fragmented without changes in habitat area, the clearest examples arising when landscapes are dissected by fences (Boone and Hobbs 2004) or roads (reviewed by Forman and Alexander 1998, Spellerberg 1998, Trombulak and Frissell 2000). In these cases, movement of people and animals can be restricted with negligible change in the total area of habitat available. We maintain that rangelands are the ideal laboratory to use to isolate fragmentation from loss because we think the most widespread change in rangelands is habitat fragmentation alone, without habitat loss. Second, it is possible that the effect of fragmentation can amplify the effect of habitat loss; that is, isolated patches of habitat may support smaller populations than patches of the same size that are not isolated.

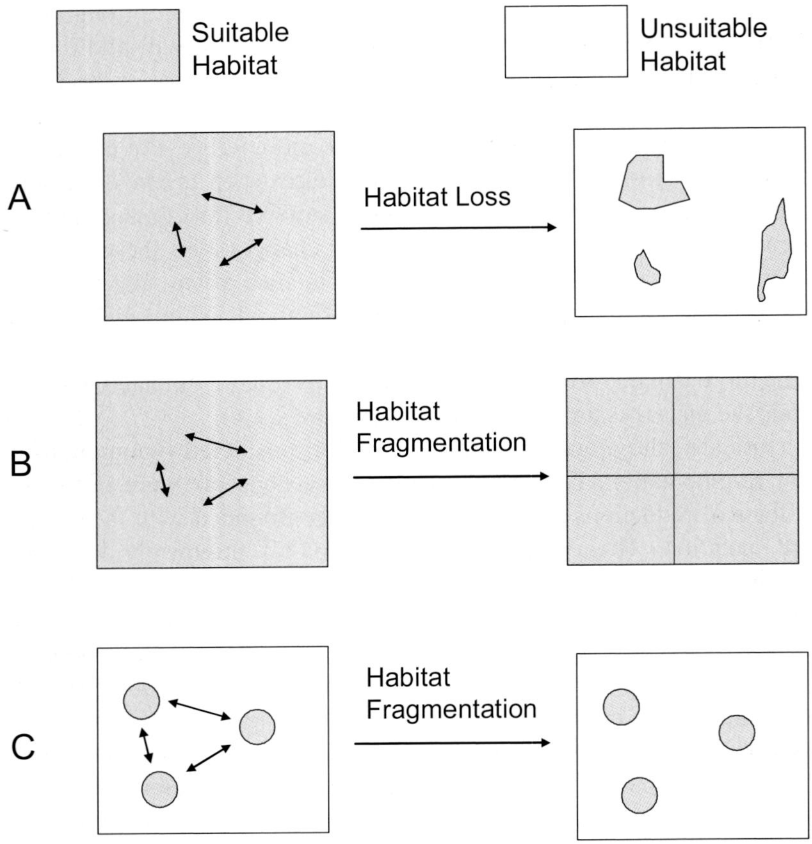

Figure 2-1. Shaded areas represent habitat that is suitable for a species; unshaded areas are unsuitable. Arrows represent hypothetical movements of animals. Habitat loss implies the conversion of habitat that is suitable to an unsuitable form (A). This conversion has two effects—the loss of habitat area and the isolation of suitable fragments imbedded in an unsuitable matrix (A). Fragmentation refers to isolation of habitat apart from any habitat loss. Fragmentation causes isolation by creating barriers to movement like fences or roads (B) or by interrupting corridors for movement through unsuitable habitat (C).

What this means is that a fragmented landscape can sustain lower densities of people and animals than an intact landscape of the same area and containing the same resources.

The preponderance of effort invested in understanding the ecological consequences of fragmentation has focused on populations and communities of plants and animals (see review of Debinski and Holt 2000). Examples of this focus include studies of fragmentation on species and genetic diversity, population stability, and extinction risk (reviewed by Saunders et al. 1991, Andren 1994, Turner 1996, Harrison and Bruna 1999, Fahrig 2002). However, the effects of fragmentation on populations may be mediated by effects on individuals. Below, we develop the idea that habitat fragmentation shapes interactions between people, animals, and landscapes, (e.g., Bowers et al. 1996, Ritchie 1998) and in so doing, determines the welfare and performance of their populations.

2. SCALE AND RESOURCE HETEROGENEITY

Virtually all ideas about the effects of fragmentation have overlooked the fact that isolation of habitat fragments can compress the scale of interaction between consumers and the ecological and social resources they require to survive and reproduce. We use scale in this book to mean spatial extent or area.[1] The scale of interaction between consumers and resources will become smaller whenever the area of habitat fragments is smaller than the area that could be used for foraging by people and animals on an intact landscape. Thus, our view holds that when a habitat becomes fragmented, a single intact set of interactions is transformed into multiple sets of interactions, each occurring over a smaller spatial scale than would occur in an unfragmented system (Figure 2-2). The area of habitat does not change, nor does the number of actors, but the spatial extent over which the interactions play out is compressed by dissection (Figure 2-2).

Because fragmentation implies a reduction in the spatial scale of ecological interactions (Figure 2-1), and because scale and heterogeneity are inextricably linked (Milne 1991, Levin 1992, Dolloff et al. 1997, Schneider 1998, Hobbs 2003), it is important to consider heterogeneity as an integral part of efforts to understand effects of fragmentation. Like fragmentation, heterogeneity is not crisply defined so we offer an operational definition here. In our view, heterogeneity contains three components: variety, pattern, and grain. Variety is what most people think of when talking about heterogeneity. We will use variety to characterize resources. Resources required by people and animals in grazing ecosystems can be assigned to categories. Examples of such categories might include vegetation types, elevation zones, plant functional groups, and water. We will assume that heterogeneous

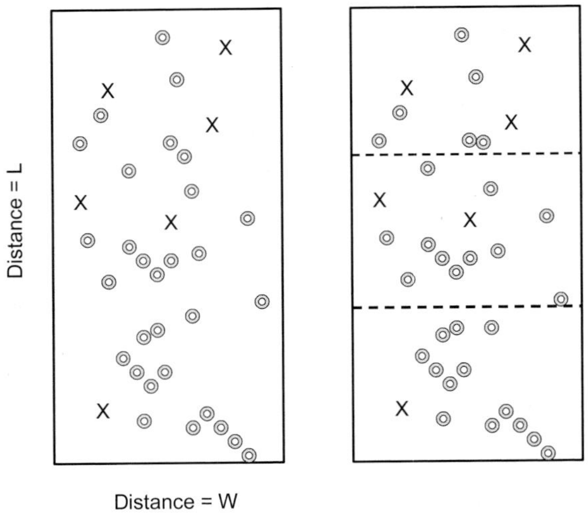

Figure 2-2. Fragmentation compresses the scale of ecological interactions. Presume that the x's are consumers and the o's are resources. In the unfragmented "habitat" on the left the scale of interaction is the square root of L*W and there are 6 x 30 = 180 potential interactions between consumers and resources. In contrast, the scale of interaction is the square root of 1/3 (L*W) in the fragmented habitat on the right and there are 18 + 22 + 13 = 53 potential consumer x resource interactions. Note that fragmentation meaningfully reduces the number of potential interactions without changing habitat area.

systems have a greater variety of these categories than homogeneous systems—a larger number of vegetation types, a greater range of elevation and aspect, and more diverse plant species and functional groups. The pattern component of heterogeneity arises as follows. We equate spatial heterogeneity with spatial dependence—the relationship between values of variables observed at different locations (Pastor et al. 1998, Adler et al. 2001). When spatial heterogeneity is low, spatial dependence is low and the average of a variable of interest distributed over a given area will closely match the average taken from a different area (Figure 2-3). When spatial heterogeneity is high, spatial dependence is also high, and the average value of a variable of interest in one area is not the same as an average taken at another area. It follows that significant spatial heterogeneity implies strong spatial dependence or patchiness. Spatial homogeneity implies spatial independence or the absence of pattern, which results when objects are randomly or uniformly distributed (Figure 2-3). A similar definition can be applied to heterogeneity in time—temporally heterogeneous systems are those where the value of a variable averaged over one time interval does not inform us about values in the future or past (Figure 2-3). So, for example, in a spatially heterogeneous system, the production of vegetation at one point in space conveys very little information about vegetation production at another point in space. In a temporally heterogeneous system, knowledge of the production of vegetation during one year does very little to allow you to predict the production of vegetation during another year.

We will use the term grain to bring together the concepts of scale, pattern, and variety of resources. In systems that are fine grained, the spatial pattern of resources is such that the full diversity of resource types are found at fine scales (Figure 2-4A). In a system where the pattern and variety of resources is coarse grained, large areas of landscape are required to include the same diversity (Figure 2-4B).

3. CRITICAL SCALES OF FRAGMENTATION

Because the concept of grain integrates scale, spatial pattern, and resource variety, it is particularly relevant to understanding the effects of fragmentation, and in particular, to defining critical scales where fragmentation may cause abrupt disruption of natural and human economies. These critical scales arise as follows. We presume there is a set of key resources (sensu Illius and O'Connor 1999) that is necessary to support growth, reproduction, and survival of people and animals exploiting rangelands. For example, in western North America, migratory ungulates require access to high elevation habitats that provide nutritious, productive forage during summer (Wallmo et al. 1977, Frank et al. 1998). However, because these areas are made inhospitable by accumulation of deep snow during winter, ungulates also require access to low elevation areas that accumulate less snow during the winter, areas that are too low in productivity to support populations year-round (Wallmo et al. 1977). These broad categories of winter vs. summer ranges may themselves contain critical components; for example, wind swept ridges that remain snow free, south facing slopes with early green up, forests that provides thermal cover, and soils that offer minerals at high concentration (Hobbs et al. 1981, Hobbs 1989, Frank and McNaughton 1992). In many tropical systems, there is a similar dependence of ungulates on dry and wet-season ranges and the resource heterogeneity that is nested within them (Frank and McNaughton 1992). If any of these critical resources are absent from the ranges of animals, then populations suffer reduced performance. This is because these resources are not substitutable, that is, no amount of increase in summer range can compensate for loss of winter

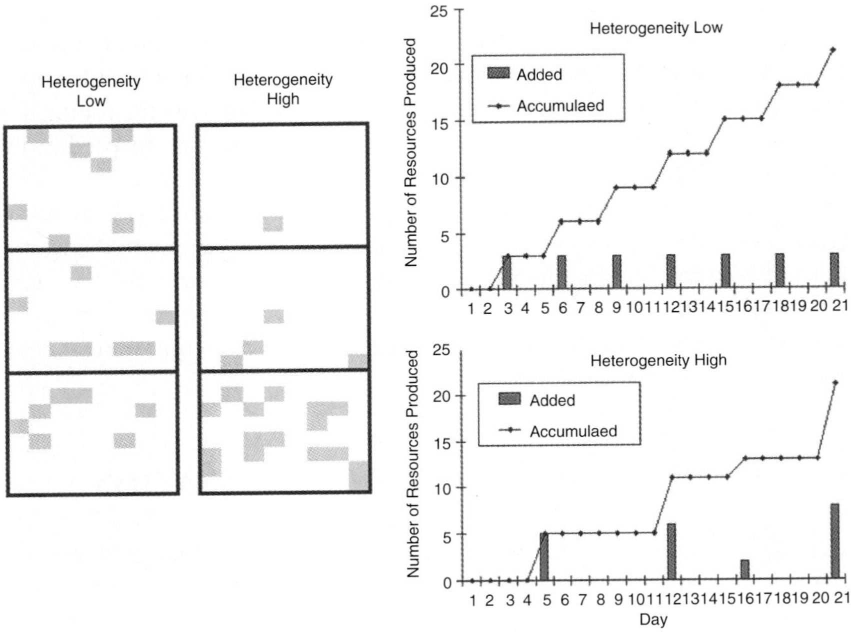

Figure 2-3. Assume the shaded areas are resources. Systems that are heterogeneous in space show high spatial dependence. In the rectangle on the left, the average number of resources per unit area is largely independent of location in each third of the rectangle's area. In the rectangle on the right the number of resources per unit area varies strongly with spatial location. Similar concepts apply to time. When temporal heterogeneity is low, it is possible to predict the future rate of resource production by sampling a brief time interval. When temporal heterogeneity is high, knowing the rate over one time interval does not inform what that rate will be over a future time interval.

range to development in temperate systems; expansion of wet season range cannot compensate for loss of dry season range to intensive agriculture in tropical systems. It follows from these ideas that there may be critical scales of fragmentation, scales where habitat fragments fail to include the full set of key resources (Figure 2-4C).

These critical scales are determined by resource grain. In habitats where heterogeneity in resources is fine grained in time and space, then habitat fragments may contain all key resources (Figure 2-4C), and as a result, we would not expect that the isolation of these fragments would affect access of people and animals to resource heterogeneity. However, when heterogeneity in resources is sufficiently coarse grained relative to fragment size, habitat fragmentation can reduce the variety of resources available to consumers below critical levels (Figure 2-4C). Such reduction might reasonably be

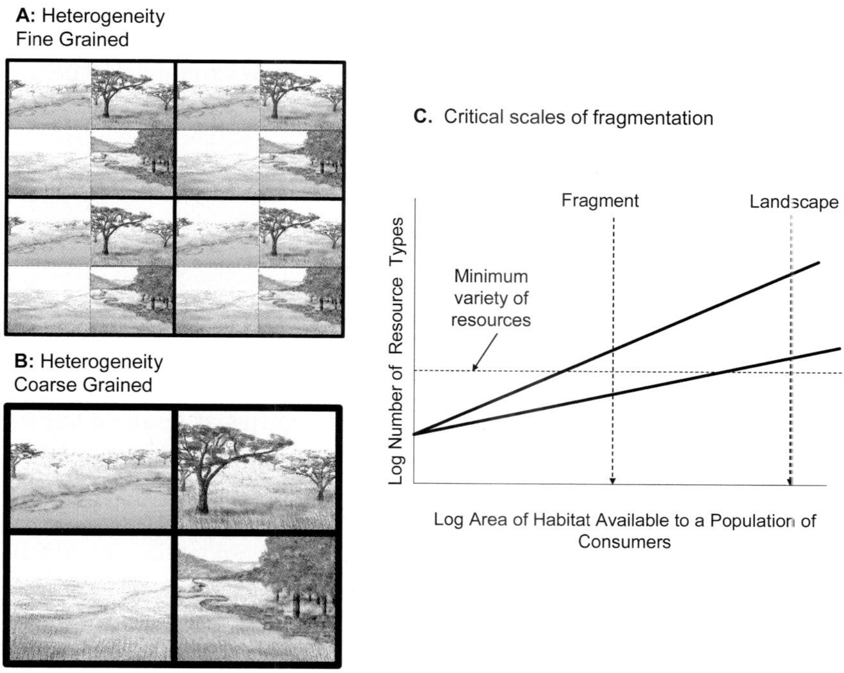

Figure 2-4. Solid lines represent boundaries of landscape units. A. When heterogeneity in resources is fine grained, then a landscape unit with a small area will contain all resource types. B. When heterogeneity is coarse grained, a large area is required to contain all resource types. Assume an intact landscape is fragmented into parts, with no loss of area. In this case, the solid lines now represent barriers to movement defining four habitat fragments. Each fragment contains a sub-population of consumers. Assume there is some minimum variety of resources that is required to support population growth. The area of habitat available to each sub-population is reduced by fragmentation even if the total amount of resource remains constant. If heterogeneity in resources is fine grained (upper solid line, panel C) then fragmentation will not reduce the variety of resources below the minimum level. However, if heterogeneity in resources is coarse grained relative to the scale of fragmentation (lower solid line, panel C), then fragmentation can prevent consumers from obtaining the range of resources they need to survive and reproduce, even if the total amount of resource remains unchanged.

expected to cause sudden interruption of ecological and social processes needed to sustain populations of people and animals. For example, the elimination of access to dry season ranges in the tropics will exacerbate mortality during drought years (Illius and O'Connor 2000). Lack of access to south-facing slopes on temperate winter ranges will prolong the period during which animals must draw down fat reserves (Hobbs 1989). In the next section, we describe general mechanisms that cause fragmentation of rangelands to harm the ability of people and animals to exploit heterogeneous resources.

4. MECHANISTIC EFFECTS OF FRAGMENTATION

Two mechanisms give rise to the effects of fragmentation on people and animals in arid and semi-arid ecosystems. These mechanisms translate effects of fragmentation on individual consumers into effects on population dynamics. We will refer to the first of these mechanisms as resource tracking. Temporal variability in production of vegetation and availability of water shapes virtually all ecological processes in arid and semi-arid lands (Ellis and Swift 1988, Illius et al. 1998, Illius and O'Connor 1999). This variability occurs over several temporal scales. Annual differences in precipitation among growing seasons create ten-fold differences in plant growth among years. Within years, there are seasonal differences in plant production and nutritional quality driven by precipitation and temperature. Within seasons, there is variation caused by the timing of plant growth and by consumption of plants by herbivores. These patterns of temporal variation in vegetation are composed of pulses of resources interrupted by resource shortages at scales of decades, years, and days.

The welfare of people and animals in arid and semi-arid rangelands depends in a fundamental way on coping with this variation. Consumers can cope with temporal variability in resources by exploiting their variation in space—in essence, temporal variability can be damped by selective use of resources that vary in quantity and quality over space. There are many examples of this selectivity. In western North America, elevation gradients created spatial variation in the timing of green-up of vegetation. Such variation in plant phenology is important for herbivores because young, rapidly growing plants are far more nutritious than plants that are mature. Ungulates exploit this spatial variation in plant quality by moving up the elevation gradient, matching their distribution with the highest quality forage, tracking a "green wave" of young, nutritious vegetation (Frank and McNaughton 1992). In a similar way, large herbivores in African savannas track variation in plant quality created by spatial heterogeneity in rainfall (Fryxell et al. 2005). This tracking is important because it expands the window of time during which animals can consume vegetation at peak nutriational quality. Fragmentation can restrict the movement needed to achieve this tracking, and hence, can compress the amount of time when peak quality forage is accessible, resulting in poorer quality diets. Populations may not be able to persist in temporally heterogeneous environments if they are not able to track resources that vary over space (Fryxell et al. 2005). Moreover, the inability to track temporally variable resources over space can create bottlenecks in resource consumption if animals are forced to forage in areas when resources are in short supply or are of poor quality (Ellis and Swift 1988, Illius and O'Connor 2000). The ability to avoid these temporal bottlenecks by selective use of space can have fundamental implications for consumer population dynamics (Ellis and Swift 1988, Illius and O'Connor 1999, 2000). These examples illustrate that access to heterogeneity is a requirement for dietary selectivity, and selectivity is fundamental to achieving high quality diets. Recent results suggest that resource heterogeneity can enhance the carrying capacities of habitats for large herbivores (Wang et al. 2006).

The second mechanism we will call resource trade-offs. There are no perfect food resources for animals; instead, virtually all resources involve some trade-off in their value (Rapport 1980). For example, the biomass of plant tissue is inversely related to its nutritional quality (Auclair and Rencz 1982, Breman and Wit 1983, Hendrickson 1988). Plants that contain essential nutrients at high concentrations may be indigestible or defended with secondary compounds (Robbins and Moen 1975, Hobbs et al. 1981, Belovsky 1981, Robbins et al. 1987). Resources that are low quality may be stable over time, while high quality ones are ephemeral. Consumers can balance these trade-offs by using a range of resource types (Rapport 1980, Hobbs et al. 1981, Belovsky 1984a, b, 1986, Bernays et al. 1997). This is not merely a matter of mixing diets with complementary nutrients. It also entails exploitation of heterogeneity in space to compensate for heterogeneity in time and requires the exploitation of rare resources by use of those that are common. However, coping with these multiple trade-offs requires access to more than one resource type. If these resources are found in different locations, and if habitat fragmentation prevents animals from moving among these locations, then we should expect that consumers in habitat fragments will suffer impaired nutrition.

5. SOURCES OF FRAGMENTATION FOR HUMAN AND NATURAL SYSTEMS

Movement-mediated connectivity among heterogeneous landscape units allows resource tracking and resource mixing, strategies that are crucial to the welfare of people and animals in arid and semi-arid ecosystems. However, human land use and land tenure systems tend to fragment these ecosystems into disconnected parcels. Fragmentation occurs with the imposition of a land tenure system, usually to facilitate protection or usurpation of some key portion of the ecosystem, to implement private property rights, promote economic intensification, enforce sedentarization of nomads, or to facilitate other policies or political agendas (Galaty and Johnson 1990, Perkins and Thomas 1993, Starrs 1998, Behnke 1999, Ellis 1999, Ellis and Lee 1999). Four idealized property systems (Table 2-1) provide the theoretical justifycation for different types of land tenure regimes. These idealized systems are distinguished by characteristic property-owning units and by the distinctive mechanisms intended to control rates of resource exploitation for each property type.

Table 2-1. Alternative types of property systems.

Tenure type	Owners	Putative regulatory mechanism
State property	State	Administrative control
Common property	Corporate groups	Collective restraint – 'stinting'
Private property	Individuals	Internalization of resource rents
Open access	No one	Low levels of resource demand

These theoretical forms of land tenure have been used to understand existing property rights regimes, and, more polemically, to create these systems by influencing policy. Each property type has been appropriated by one of the grand theories of political economy including capitalism, communism, and Euro-American notions of primitive political systems. For our purposes, it is noteworthy that fragmentation, justified in different ways in different political systems, is a near-universal feature of modern land tenure systems. Today's dominant concepts of land tenure developed and flourished in the relatively mesic environments of western Europe and eastern North America. The transfer of these mesic tenure systems to arid and semi-arid ecosystems has caused ecological damage and economic disruption (Ellis 1988, Williamson et al. 1988, Behnke 1994, Sneath 1998, Humphrey and Sneath 1999). Although benefits, such as ease of management and security of investment, may arise from fragmentation, other results are far from beneficial (Figure 2-1).

There are several sources of fragmentation of rangelands worldwide. Pastoral populations, while owning livestock privately, have communal or cooperative ownership of pastures and water. Generally subdivisions of nomadic societies, like 'sections' or territorial political groups, have rights of possession and use of pastures, use of wells, boundaries of routes of migration, etc. These rules of use are varied in time and place and are sometimes very complex. These rules of use allow the carrying out of socially expedient managerial functions. The social organizational level that is imbued with decision-making authority varies among groups. For example, in East Africa where climate variability is extreme, the decision- making body for movement of herds is often a group of households who possess the flexibility to respond to rapidly changing ecological conditions. In more mesic areas the community may make decisions about pasture use from one season to the next.

Communal ownership means that pastures are used (and thereby fragmented) through formal and informal indigenous institutions, that is, customary laws which determine both physical (e.g., fences) and institutional barriers (e.g., dry season grazing reserves) to pasture and water use. These same institutions also create opportunities for pasture and water use. Even when pasture is owned by the state, pastoralists have generally had rights to use pasture.

Water, sometimes the key and most limiting resource in pastoral systems, is controlled by the local group which could be a group of households or a territorial group. It is generally the case that the more effort put into building and maintaining wells, the more exclusionary the well is to users. Other wells, more easily accessed, are generally more open to use by a wider array of people and livestock. Rules and regulations over well and pasture use is dependent on environmental, sociopolitical, and climatic conditions. Thus, rules can change from year to year and season to season (Evans-Pritchard 1940, Gulliver 1955, Spencer 1965).

Modern political and economic imperatives favor fragmentation and the removal of connectivity of arid and semi-arid rangelands. Although benefits may arise from fragmentation, the dissection of rangelands into small, disconnected units can compromise ecosystem function and the viability of grazing systems by restricting movements and reducing access to ecosystem heterogeneity.

6. ECONOMIC DIMENSIONS OF ECOSYSTEM FRAGMENTATION

Neo-classical economic perspectives routinely undervalue ecosystem-natural capital resources and assume these can be perfectly substituted by economic inputs (Prugh et al. 1999). Thus, fragmentation and loss of access to heterogeneity are not perceived as negative aspects of development or land use, but rather as necessary steps toward intensification and economic growth. Economic inputs may be rewarded by higher regional carrying capacity and productivity per unit area, but in the past, the value of heterogeneity has not been costed properly. Only the economic side of the equation is considered; the ecological side and its value are ignored. However, ecosystem scientists and ecological-economic practitioners understand that complex systems are self-sustaining, whereas simplified (fragmented) ecosystems often require capital inputs, subsidies and/or management to be sustainable (Ellis and Peel 1995). Although ecosystem fragmentation is often justified as a means of economic intensification in the neo-classical framework, in fact, it costs money (fodder, infrastructure, etc.) to replace the access to natural capital lost through fragmentation (Prugh 1999). Land use patterns, driven by economic or political agendas, are unlikely to be perfectly superimposed on spatial complexity patterns. Where land tenure dictates a small-scale pattern of exploitation, economic inputs are needed to compensate for the natural capital lost to fragmentation. We hypothesize that inputs per unit area increase exponentially with fragmentation and decreasing scale (Figure 2-5). Alternately, scale expansion through consolidation adds greater complexity to the grazing orbit, reducing economic inputs until at some larger scale, the *minimum level of complexity* for unsubsidized exploitation is reached, and economic inputs approach zero (Figure 2-5, Figure 2-4C).

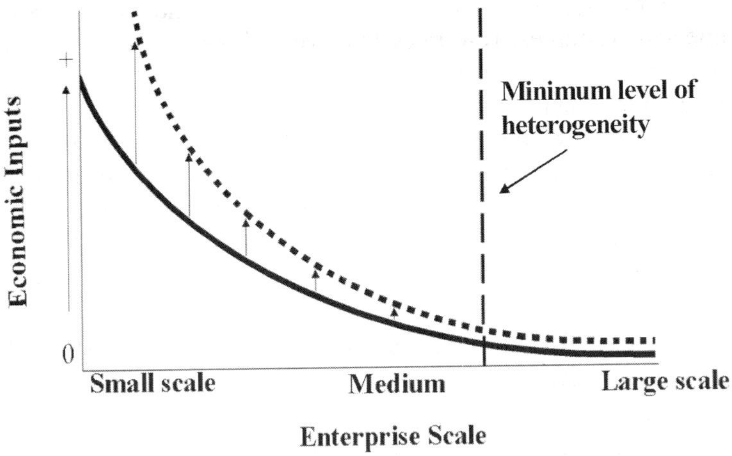

Figure 2-5. The need for economic inputs declines with increasing enterprise scale because larger enterprises encompass greater environmental heterogeneity (solid line). Under conditions of high climatic variability, greater inputs are required to offset effects of climate perturbations (dotted line). At the minimum heterogeneity level, inputs are not needed and are internalized by the enterprise.

A critical issue is to understand the trade-offs between loss of access to heterogeneity and the benefits of intensified land use, given different forms of economic substitution and a proper ecological-economic accounting of natural ecosystem values. The question then arises whether there are thresholds in increasing levels of spatial and temporal heterogeneity above which this compensation no longer occurs in any practical sense. To the best of our knowledge, this sort of economic assessment has not been conducted, although many of the building blocks to permit such an analysis are in place.

Human land tenure or land-use patterns, dictated by political or economic imperatives, are seldom superimposed on ecosystem spatial complexity patterns. Where land tenure dictates a sub-optimal scale of exploitation, economic inputs are required to compensate for the natural capital lost to fragmentation. Benefits derived from economic subsidies may or may not compensate for the loss of access to heterogeneity.

7. A CONCEPTUAL MODEL OF ECOSYSTEM FRAGMENTATION

The ideas we sketched above give rise to a conceptual model, a model we offer to create a useful organizing framework for the chapters that follow. Historically, the world's rangelands contained a diverse set of interconnected resources, resources that were linked by the movements of people and animals (Figure 2-6). A variety of proximate drivers, including changes in land tenure and livelihood strategies, have interrupted these movements, leading to fragmentation of rangelands worldwide. Ultimately, these proximate drivers respond to a context of underlying driving forces created by human population change, the natural resource base supporting the human population, and climate. Human economies and political institutions further shape the proximate causes of fragmentation. Roy Behnke (Chapter 13) describes these global drivers of fragmentation. Fragmentation interrupts the movement of people and animals among landscape units and in so doing, restricts their access to resources. Mike Coughenour (Chapter 3) describes responses of pastoralists and large herbivores to landscape heterogeneity in resources, and the consequences of these responses for animal nutrition and population dynamics. Restriction of movement creates short and long term consequences for human and natural economies. These are described by Boone et al. (Chapter 14) and Galvin (Chapter 15).

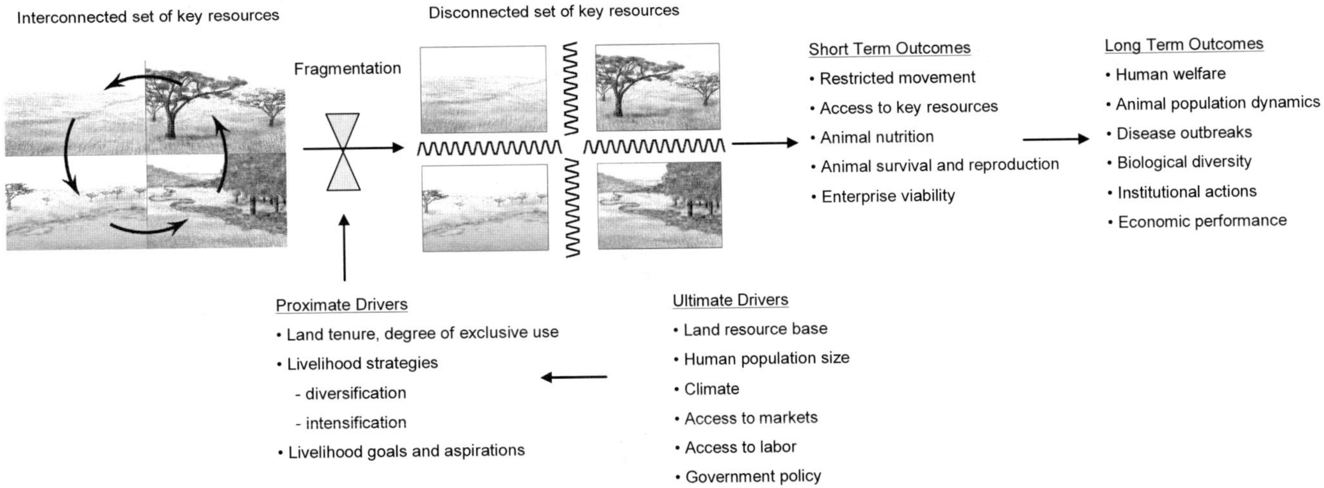

Figure 2-6. Conceptual model illustrating the central thesis of this book. Fragmentation of arid and semi-arid ecosystems is driven by forces operating over different temporal scales. Ultimately, the land resources base, climate and human economies drive proximal changes in land tenure, which cause land to be dissected into isolated parcels, transforming an intact set of resources into a disconnected set. Fragmentation exerts short term effects on the ability of people and animals to exploit spatial heterogeneity in resources, which eventually causes long term impacts on ecosystem function, economic performance, and human welfare.

ENDNOTES

[1] Used this way, scale is usually quantified as the square root of an area of landscape.

REFERENCES

Adler, P. B., D. A. Raff, and W. K. Lauenroth. 2001. The effect of grazing on the spatial heterogeneity of vegetation. Oecologia 128:465-479.

Andren, H. 1994. Effects of habitat fragmentation on birds and mammals in landscapes with different proportions of suitable habitat - a review. Oikos 71:355-366.

Auclair, A. N. D. and A. N. Rencz. 1982. Concentration, mass, and distribution of nutrients in a subarctic Picea mariana-Cladonia alpestris ecosystem. Canadian Journal of Forest Research 12:947-968.

Behnke, R. 1994. Natural resource management in pastoral Africa. Development Policy Review 12:5-27.

Behnke, R. 1999. Reconfiguring property rights in livestock production systems. Overseas Development Institute (ODI), London.

Belovsky, G. E. 1981. Food plant selection by a generalist herbivore: the moose. Ecology 62:1020-1030.

Belovsky, G. E. 1984a. Herbivore optimal foraging - a comparative test of 3 models. American Naturalist 124:97-115.

Belovsky, G. E. 1984b. Snowshoe hare optimal foraging and its implications for population-dynamics. Theoretical Population Biology 25:235-264.

Belovsky, G. E. 1986. Optimal foraging and community structure - implications for a guild of generalist grassland herbivores. Oecologia 70:35-52.

Bender, D. J., T. A. Contreras, and L. Fahrig. 1998. Habitat loss and population decline: A meta-analysis of the patch size effect. Ecology 79:517-533.

Bernays, E. A., J. E. Angel, and M. Augner. 1997. Foraging by a generalist grasshopper: The distance between food resources influences diet mixing and growth rate (Orthoptera: Acrididae). Journal of Insect Behavior 10:829-840.

Betts, M. G., G. J. Forbes, A. W. Diamond, and P. D. Taylor. 2006. Independent effects of fragmentation on forest songbirds: An organism-based approach. Ecological Applications 16:1076-1089.

Boone, R. B. and N. T. Hobbs. 2004. Lines around fragments: effects of fencing on large herbivores. African Journal of Range and Forage Science 21:79-90.

Bowers, M. A., K. Gregario, C. J. Brame, S. F. Matter, and J. L. Dooley. 1996. Use of space and habitats by meadow voles at the home range, patch and landscape scales. Oecologia 105:107-115.

Breman, H. and C. T. de Wit. 1983. Rangeland productivity and exploitation in the Sahel. Science 221:1341-1347.

Chalfoun, A. D., F. R. Thompson, and M. J. Ratnaswamy. 2002. Nest predators and fragmentation: a review and meta-analysis. Conservation Biology 16:306-318.

Collinge, S. K. 1996. Ecological consequences of habitat fragmentation: Implications for landscape architecture and planning. Landscape and Urban Planning 36:59-77.

de Blois, S., G. Domon, and A. Bouchard. 2002. Landscape issues in plant ecology. Ecography 25:244-256.

Debinski, D. M. and R. D. Holt. 2000. A survey and overview of habitat fragmentation experiments. Conservation Biology 14:342-355.

Dolloff, C. A., H. E. Jennings, and M. D. Owens. 1997. A comparison of basinwide and representative reach habitat survey techniques in three southern Appalachian watersheds. North American Journal of Fisheries Management 17:339-347.

Ellis, J. E. 1999. Extensive grazing systems: Persistence under political stress and environmental risk. Page 10 *In* Ruminations: Newsletter of the Global Livestock Collaborative Research Support Program.

Ellis, J. E. and D. M. Swift. 1988. Stability of African pastoral ecosystems: alternative paradigms and implications for development. Journal of Range Management 41:450-459.

Ellis, J. E. and M. Peel. 1995. Economies of spatial scale in dryland ecosystems. *In* Arid Zone Ecology Forum, Kimberly, South Africa.

Ellis, J. E. and R. Lee. 1999. Ecosystem dynamics and ecological perspectives on the collapse of the livestock sector in southeastern Kazakstan. Overseas Development Institute (ODI), London.

Evans-Pritchard, E.E. 1940. The Nuer. Oxford University Press, London.

Fahrig, L. 1997. Relative effects of habitat loss and fragmentation on population extinction. Journal of Wildlife Management 61:603-610.

Fahrig, L. 1998. When does fragmentation of breeding habitat affect population survival? Ecological Modelling 105:273-292.

Fahrig, L. 2002. Effect of habitat fragmentation on the extinction threshold: A synthesis. Ecological Applications 12:346-353.

Forman, R. T. T. and L. E. Alexander. 1998. Roads and their major ecological effects. Annual Review of Ecology and Systematics 29:207-231.

Frank, D. A. and S. J. McNaughton. 1992. The ecology of plants, large mammalian herbivores and drought in Yellowstone National Park. Ecology 73:2043-2058.

Frank, D. A., S. J. McNaughton, and B. F. Tracy. 1998. The ecology of the Earth's grazing ecosystems. Bioscience 48:513-521.

Fryxell, J. M., J. F. Wilmshurst, A. R. E. Sinclair, D. T. Haydon, R. D. Holt, and P. A. Abrams. 2005. Landscape scale, heterogeneity, and the viability of Serengeti grazers. Ecology Letters 8:328-335.

Galaty, J. and D. Johnson, editors. 1990. The world of pastoralism: Herding systems in comparative perspective. The Guilford Press, New York.

Gulliver, P. H. 1955. The Family Herds. A study of two pastoral tribes in East Africa. The Jie and Turkana. Routledge and Kegan Paul, London.

Harrison, S. and E. Bruna. 1999. Habitat fragmentation and large-scale conservation: what do we know for sure? Ecography 22:225-232.

Hendrickson, O. Q. 1988. Biomass and nutrients in regenerating woody vegetation following whole-tree and conventional harvest in a northern mixed forest. Canadian Journal of Forest Research 18:1427-1436.

Hobbs, N. T. 1989. Linking energy balance to survival in mule deer: development and test of a simulation model. Wildlife Monographs 101:1-39.

Hobbs, N. T. 2003. Challenges and opportunities for integrating ecological knowledge across scales. Forest Ecology & Management 181:222-238.

Hobbs, N. T., D. L. Baker, J. E. Ellis, and D. M. Swift. 1981. Composition and quality of elk winter diets in Colorado. Journal of Wildlife Management 45:156-171.

Humphrey, C. and D. Sneath. 1999. The end of nomadism? Society, state and the environment in inner Asia. The White Horse Press, Cambridge.

Illius, A. W., J. F. Derry, and I. J. Gordon. 1998. Evaluation of strategies for tracking climatic variation in semi-arid grazing systems. Agricultural Systems 57:381-398.

Illius, A. W. and T. G. O'Connor. 1999. On the relevance of nonequilibrium concepts to arid and semiarid grazing systems. Ecological Applications 9:798-813.

Illius, A. W. and T. G. O'Connor. 2000. Resource heterogeneity and ungulate population dynamics. Oikos 89:283-294.

Kerven, C., editor. 2003. Prospects for pastoralism in Kazakstan and Turkmenistan. Routledge Curzon, London.

Khazanov, A. M. 1984. Nomads and the outside world. Cambridge University Press, Cambridge.

Levin, S. A. 1992. The problem of pattern and scale in ecology. Ecology 73:1943-1967.

Little, M. A. and P. W. Leslie, editors. 1999. Turkana herders of the dry savanna: Ecology and biobehavioral response of nomads to an uncertain environment. Oxford University Press, Oxford.

Lund, C. 2000. African land tenure: Questioning basic assumptions. IIED Issue paper no 100.

Milne, B. T. 1991. Heterogeneity as a multiscale characteristic of landscapes. Pages 68-84 *In* J. Kolasa and S. T. A. Pickett, editors. Ecological heterogeneity. Springer-Verlag, New York.

Niemela, J. 2001. Carabid beetles (Coleoptera: Carabidae) and habitat fragmentation: a review. European Journal of Entomology 98:127-132.

Pastor, J., B. Dewey, R. Moen, D. J. Mladenoff, M. White, and Y. Cohen. 1998. Spatial patterns in the moose-forest-soil ecosystem on Isle Royale, Michigan, USA. Ecological Applications 8:411-424.

Perkins, J. S. and D. S. G. Thomas. 1993. Spreading deserts or spatially confined environmental impacts? Land degradation and cattle ranching in the Kalahari Desert of Botswana. Land Degradation and Rehabilitation 4:179-194.

Prugh, T., R. Constanza, J. Cumberland, H. Daly, R. Goodland, and R. Norgaard. 1999. Natural capital and human economic survival, 2nd edition. CRC Press, Bacon Raton, Florida.

Rapport, D. J. 1980. Optimal foraging for complementary resources. American Naturalist 116:324-346.

Ritchie, M. E. 1998. Scale-dependent foraging and patch choice in fractal environments. Evolutionary Ecology 12:309-330.

Robbins, C. T. and A. N. Moen. 1975. Composition and digestibility of several deciduous browses in the northeast. Journal of Wildlife Management 39:337-341.

Robbins, C. T., S. Mole, A. E. Hagerman, and T. A. Hanley. 1987. Role of tannins in defending plants against ruminants: reduction in dry matter digestion? Ecology 68:1606-1615.

Ryall, K. L. and L. Fahrig. 2006. Response of predators to loss and fragmentation of prey habitat: a review of theory. Ecology 87:1086-1093.

Saunders, D. A., R. J. Hobbs, and C. R. Margules. 1991. Biological consequences of ecosystem fragmentation: a review. Conservation Biology: the Journal of the Society for Conservation Biology 5:18-32.

Schmiegelow, F. K. A. and M. Monkkonen. 2002. Habitat loss and fragmentation in dynamic landscapes: avian perspectives from the boreal forest. Ecological Applications 12:375-389.

Schneider, D. C. 1998. Applied scaling theory. Pages 254-269 *In* D. L. Peterson and V. T. Parker, editors. Ecological scale: Theory and applications. Columbia University Press, New York.

Sneath, D. 1998. State policy and pasture degradation in inner Asia. Science 281:1147-1148.

Spellerberg, I. F. 1998. Ecological effects of roads and traffic: a literature review. Global Ecology and Biogeography Letters 7:317-333.

Spencer, P. 1965. The Samburu: A Study of Gerontocracy in a Nomadic Tribe. University of California Press, Berkeley.

Starrs, P. F. 1998. Let the cowboy ride: Cattle ranching in the American west. Johns Hopkins University Press, Baltimore.

Trombulak, S. C. and C. A. Frissell. 2000. Review of ecological effects of roads on terrestrial and aquatic communities. Conservation Biology 14:18-30.

Tscharntke, T., I. Steffan-Dewenter, A. Kruess, and C. Thies. 2002. Characteristics of insect populations on habitat fragments: A mini review. Ecological Research 17:229-239.

Turner, I. M. 1996. Species loss in fragments of tropical rain forest: A review of the evidence. Journal of Applied Ecology 33:200-209.

Usher, M. B. 1987. Effects of fragmentation on communities and populations: a review. Pages 103-121 *In* D. A. Saunders, G. W. Arnold, A. A. Bumbidge, and A. J. M. Hopkins, editors. Nature conservation: The role of remnants of native vegetation. Surrey Beatty & Sons, Chipping Norton, N.S.W. Australia.

Villard, M. A. 2002. Habitat fragmentation: major conservation issue or intellectual attractor? Ecological Applications 12:319-320.

Wallmo, O. C., L. H. Carpenter, W. L. Regelin, R. B. Gill, and D. L. Baker. 1977. Evaluation of deer habitat on a nutritional basis. Journal of Range Management 30:122-127.

Wang, G. M., N. T. Hobbs, R. B. Boone, A. W. Illius, I. J. Gordon, J. E. Gross, and K. L. Hamlin. 2006. Spatial and temporal variability modify density dependence in populations of large herbivores. Ecology 87:95-102.

Williamson, D., J. Williamson, and K. T. Ngwamotsoko. 1988. Wildebeest migration in the Kalahari. African Journal of Ecology 26:269-280.

Young, A., T. Boyle, and T. Brown. 1996. The population genetic consequences of habitat fragmentation for plants. Trends in Ecology & Evolution 11:413-418.

Chapter 3

CAUSES AND CONSEQUENCES OF HERBIVORE MOVEMENT IN LANDSCAPE ECOSYSTEMS

Michael B. Coughenour
Natural Resource Ecology Laboratory, Colorado State University, Fort Collins, CO 80523, USA

1. INTRODUCTION

Over the last century, ecosystems with large herbivores have been increasingly threatened by land conversion, land use intensification, resource extraction, and artificial barriers. These ecosystems are threatened by these changes, because many, if not most, of the expansive and spatially heterogeneous habitats that large herbivore species evolved in have been increasingly compressed, subdivided, fragmented, and homogenized, disrupting the movements of these species. There are many reasons why large herbivores are adapted to move over scales of meters to hundreds of kilometers; there are many consequences if they cannot, not only for the herbivores but for ecosystems.

In this chapter, I explore the causes and consequences of herbivore movement and disruptions of movement in ecosystems. I begin by describing the primary ways that large herbivore movements may be disrupted, each leading to different potential consequences. I elaborate the primary reasons why large herbivores move, primarily involving access to temporally varying distributions of forage, water, and habitat. I then examine the consequences of movement for herbivores, plants and soils, biodiversity, and ecosystem functioning. The ultimate goal of this exploration is to

K. A. Galvin et al. (eds.), Fragmentation in Semi-Arid and Arid Landscapes: Consequences for Human and Natural Systems, 45–91.

heighten appreciation for the importance of herbivore movement in ecosystems and the need for integrated, ecosystem-level assessments of the consequences of habitat loss, fragmentation, and sedentarization. Such assessments would provide a more informed basis for coping with the effects of past and ongoing land conversion and designing more sustainable and effective grazing and browsing ecosystems for the future.

2. MOVEMENT DISRUPTIONS

Three major categories of movement disruption are pertinent for large herbivore ecosystems: habitat loss, habitat fragmentation, and sedentaryzation. Each has different causes and different consequences (Reid et al. 2004). Habitat loss is the permanent loss of access to land surface area. Fragmentation refers to the breaking apart and isolation of a habitat into unconnected pieces. Sedentarization is a pervasive disruption of pastoral movement patterns worldwide, usually resulting from various forms of modernization and development. It is possible for the three types of disruption to occur simultaneously. Habitat loss frequently results in fragmentation, for example.

There are many ways for habitat loss to occur, and many ways it can affect herbivore movements. 1) Habitat may be lost by converting vegetation from forage species to crops or other non-forage species. 2) Habitat may be lost by cutting off access, for example, by surrounding it with fences. Although fencing is often a cause of fragmentation, if the fenced area is depopulated and if the fencing is effective in preventing recolonization, the result may be habitat loss, or habitat loss plus fragmentation. 3) A population may be compressed into a reduced area. 4) Fencing a core habitat may result in the loss of a dispersal sink. In some cases, a core habitat may become full, at which point the population produces a surplus of animals that either die or emigrate into marginal habitats. The marginal habitat may support the dispersers, but be incapable of supporting net population growth (Owen-Smith 1981). 5) Habitat may be lost because a critical resource is removed. For example, a critical water point may become inaccessible or dysfunctional, resulting in the loss of use of the surrounding area.

Similarly, there are different varieties of fragmentation. 1) A previously continuous habitat may be subdivided into small, discontinuous and disjoint pieces by fencing or other intentional restrictions on access. The individual fragments are still viable habitats at least in the short term. Over longer periods isolated populations may be at risk of extinction. 2) A habitat may be fragmented without necessarily subdividing it by the placement of artificial boundaries such as roads. 3) An area adjacent to suitable habitat may be suitable for movement but not for foraging. This area might be made untransversable due to changes in land cover or the presence of humans (e.g., Berger 2004). As a result, the suitable habitat may become a fragment by cutting off its linkage with other habitats.

A combination of habitat loss and habitat fragmentation is probably the most common form of movement disruption. The conversion and loss of habitat often creates isolated or poorly connected patches within a matrix of newly unsuitable habitat. The loss component results in decreased forage and inasmuch as forage is limiting, decreased herbivore population size. The fragmentation component results in isolated populations inhabiting the remaining fragmented areas. A lack of movement between the isolated sub-populations has the additional effect of increasing risks of sub-population extinction (Hill and Caswell 1999, Fahrig and Nuttle 2005), and of decreasing options for herbivores to move in response to localized disturbances or climatic variations. Thus, the combined outcome of these two disruptions on herbivore population size is at least additive, if not multiplicative.

In some cases, land cover change could either fragment or remove habitat, depending upon the viability of the remaining habitats. For example, cutting off a migratory pathway may be a form of habitat loss or it may be a form of fragmentation, depending upon the viability of the excised area as a habitat for a non-migratory sub-population. If the excised portion of land cannot support a viable sub-population, it is lost habitat. If it can support a viable sub-population, it is a fragment.

Sedentarization does not have to involve either fragmentation or habitat loss, but it could. Nomadic, semi-nomadic, and transhumant pastoralists in many parts of the world have become increasingly less mobile as a consequence of development and modernization schemes and the increasing prevalence of developed infrastructure, such as, roads, schools, water sources, and medical facilities (Darling and Farvar 1972, Widstrand 1975, Garcia-Ruiz and Lasanta-Martinez 1990, Turner 1993, El-Shorbagy 1998, Finan and Al Haratani 1998, Sneath 1998, Niamir-Fuller and Turner 1999, Humphrey and Sneath 1999, Alimaev 2003, Robinson and Milner-Guilland 2003, Fratkin and Roth 2005, Turner et al. 2005, Zhaoli et al. 2005). Sedentarized pastoralists may retain access to traditional grazing areas while basing their operations out of a more permanent dwelling due to modern forms of transportation that permit access to remote areas more quickly. However, sedentarization may involve habitat loss when the sedentarized pastoralists lose access to habitats that would have previously been reached through movement. Sedentarization is often associated with subdivision, and thus fragmentation. The combined effects of more concentrated herbivory around settlements or households and loss of the ability to opportunistically move could lead to negative consequences for pastoralists as well as rangelands.

3. REASONS FOR MOVING

3.1 Forage

Even if forage is uniformly distributed, herbivores must move to forage. Movement is required simply because forage has a spatial dimension. While trivial, this is a useful starting point for a conceptual model. A second step in building this model would be to represent the way the necessary foraging area varies with forage productivity. Clearly, the size of the foraging area must be larger in drier and less productive environments. As there is less forage per hectare, the number of hectares must rise in compensation. Consequently, spatial extent becomes more of an issue in dry climates. A third step is to consider how the size of the required area varies with herbivore body size. Larger-bodied herbivores require larger foraging areas because they require more forage. Thus, it is well known that home range size scales with body size due to increased energetic requirements (McNab 1963, Owen-Smith 1988). We could then consider the consequences of forming herds or social groups for the amount of land that each individual must cover. Large herbivores exhibit every conceivable group size, from solitary, to herds of many thousands. The area of land required for a group of animals scales directly with the number of animals in the group. If N animals live in a group, then each animal in that group will need to cover N times the area that it would have to cover if it were a solitary individual. Larger herds must move over larger areas. One possible example of this is the observation that larger pastoral herds travel faster and spend a larger fraction of their time moving than small herds (Copolillo 2000). Thus, individual level mobility must be expanded to accommodate social grouping and herding.

The most straightforward way to calculate the area required by herbivores, or conversely, the number of herbivores that can be supported by a given land area, is to sum up the total mass of forage (kg). If forage is heterogeneously distributed, the total amount of forage can be assessed using appropriate sampling procedures, for example, by taking random samples of forage biomass across the landscape or within strata of the landscape. The result could be described in terms of a probability distribution for the landscape or for each stratum. The required area or home range for a group of animals could then be calculated by integrating over the distribution, or equivalently, just taking the average. If strata are used, the averages for the strata could be combined in an area-weighted average for the landscape. Forage quality could be considered by not including items that are of insufficient quality or are unpalatable (e.g., Hobbs and Swift 1985).

However, the problem is not as simple as determining total forage biomass. First, because forage intake rate declines at low forage biomass densities (Hudson and Watkins 1986, Wilmshurst et al. 1999a), the "value" of the forage is not proportional to forage biomass. Patches with low densities of forage may not meet the energetic demands of the herbivores, despite the fact that there is forage present. Second, the rate of forage intake may be non-linearly related to forage biomass (Spalinger and Hobbs 1992, Farnsworth and Illius 1998, Owen-Smith 2002, Hobbs et al. 2003). Consequently, a different result would be obtained by explicitly considering how forage intake rate varies across a heterogeneous landscape than by calculating forage intake rate based upon the average forage biomass of the landscape. Third, spatial heterogeneity and patch geometry may influence foraging efficiency (Beecham and Farnsworth 1998, Hobbs et al. 2003). The first way that heterogeneity affects foraging efficiency is through its influence on movement time, which is affected by the spatial arrangement of the patches on the landscape and the movement pattern of the herbivores among the patches. Herbivores make decisions about which patches to use, and the decisions may not necessarily be optimal (Bailey et al. 1996). A patch may not be used because of the time required to travel to it, or to find it. Herbivores may decide to use a lower quality patch that is close, than a better quality patch that is far. Thus, the net value of a heterogeneous habitat must be calculated by integrating the costs and benefits of foraging over space. The value could be determined by integrating over the continuous distribution mathematically, or by using a spatially explicit movement and foraging model.

There is a substantial literature on foraging theory that predicts the optimal ways for herbivores to move or invest time while foraging in a patchy or heterogeneous landscape (Pyke 1984, Bailey et al. 1996, Owen-Smith 2002, Fortin et al. 2003, Searle et al. 2005, 2006). These models assume that herbivores will move among patches to maximize net benefits to the individual, for example by using the marginal value theorem (Charnov 1976, Fryxell et al. 2004), rules of thumb (Ward and Salz 1994, Bailey et al. 1996), or rules that consider digestive constraints, social behaviors, and foraging environments (Searle et al. 2006). For large herbivores, benefits may be defined in terms of energy intake, accounting for the decline in energy intake due to reduced digestibility with increased forage biomass (Wilmshurst et al. 1999a, b, Fryxell et al. 2004). Alternatively, fitness has been defined as the product of body mass and survivorship from predation (Morales et al. 2005). For any given time interval, the model must allocate foraging effort among patches across the landscape, calculate forage intake rate during the interval, and multiply the intake rate by the time length of the interval. This can be done by simply redistributing the herbivore population in piece-wise time intervals in relation to habitat suitability, forage distribution, or energy intake rate based on a functional response and forage digestible energy content (Coughenour 1993, Boone et al. 2002, Weisberg et al. 2006). Or, it can be done by spatially explicit modeling of individual movements and foraging behaviors (Siniff and Jessen 1969, Taylor and Taylor 1977, Turner et al. 1993, Moen et al. 1997, Beecham and Farnsworth 1998, Farnsworth and Beecham 1999, Morales et al. 2005), or by mass flows of population segments between neighboring locations (Fryxell et al. 2004, 2005). The evolution of adaptive movement behaviors in response to particular spatio-temporal landscapes has recently been simulated using genetic algorithms (Morales et al. 2005, Boone et al. 2006). Another approach represents individual movements in response to dynamic habitat suitability index maps, which are in turn generated by a spatially explicit ecosystem model (Rupp 2005). Many of these studies have been based on theoretical landscapes. However, models which predict forage-based herbivore movements from actual resource distributions are usually more useful for natural resource management and planning (e.g., Coughenour 2002, 2005, Boone et al. 2002, Weisberg et al. 2002, Rupp 2005).

Characterizing the distribution of forage is more complex than mapping total biomass. Forage quality, often quantified in terms of digestibility and protein content, varies among species and tissues, giving it a spatial distribution. Forage can be characterized with a frequency distribution of food items or of biomass amounts in different quality categories (Hobbs and Swift 1985), or it can be characterized spatially. Temporal variations in forage quality are an issue in environments with pronounced cold or dry seasons. As grasses senesce and enter dormancy, energy and protein is translocated belowground and aboveground forage quality declines. Other mineral nutrients such as phosphorous, calcium, and sodium may also be important (McNaughton 1988, 1990, Murray 1995). Despite the importance of forage quality and mineral content, few assessments have considered how spatio-temporal variations in these attributes influence habitat quality, herbivore habitat selection and movement, or the consequences of movement disruption. What is required, essentially, are a set of maps depicting how forage nutritional quality varies over time, essentially a dynamic 'nutrient map'. If the dynamic nutritional qualities of forage species are known, the set of maps could be created from a map of plant species composition. Other forage quality characteristics can also be important and their distributions could be mapped. Secondary compounds such as tannins and phenolics influence digestibility (Owen-Smith 1982, Cooper et al. 1988). Spinescence is an effective physical defense, deterring herbivory (Cooper and Owen-Smith 1986). Browse species escape herbivory simply by growing out of reach, and canopy height provides a third dimension for ungulate niche partitioning (du Toit 2003). In a spatially explicit model, these attributes would be linked to the spatial distributions of plant species. The distributions of these variables would then alter the spatial distributions of herbivores and plant-animal interactions.

Habitat selection and diet selection are closely interrelated. Herbivores move to select their diets in response to the spatial distributions of dietary items. Conversely, spatial location constrains dietary composition. If a herbivore is located within a certain patch or landscape, it must select a diet from the forage species present in that area. A realistic movement model would therefore allow herbivores to select habitats and movements in response to the distribution of preferred food items. For example, a habitat suitability index used to drive herbivore distributions might incorporate forage abundances scaled by dietary preference weights. Conversely, a realistic diet selection model must consider the location of the herbivore and the dietary items that are available to choose from at that location.

Two conclusions can be drawn from these observations. First, movement plays a fundamental role in determining the ability of a landscape to support herbivores, even in this simple case of a single, heterogeneously distributed resource. The time required for movement, the rate of movement, and the energetic cost of movement could all be variables in a model that predicts the consequences of a movement disruption for a herbivore population. Second, the spatial arrangement of the resources matters. Two landscapes with the same total quantity of forage may support different numbers of herbivores because the forage is distributed differently. Consequently, the spatial arrangement of forage is relevant for predicting the outcome of movement disruptions, and predicting which portions of a landscape are most important to conserve.

3.2 Water

Water is a major determinant of large herbivore movements in arid and semi-arid environments; it is often the primary determinant of the area that is available for foraging, and it influences the distribution of herbivory and the effects of herbivory on plants, soils, and ecosystems. Water plays a central role in fragmentation and habitat loss inasmuch as it influences herbivores and their interactions with other components of ecosystems. Changes in water distribution have considerable potential for disrupting herbivore movements. Loss of access to water sources could contribute to fragmentation and habitat loss. Conversely, provisioning of water may result in a loss of heterogeneity, and unsustainable herbivore distributions and densities.

Herbivore movements and spatial distributions are jointly affected by water requirements and capabilities of moving to water. Water requirements generally scale with body size (du Toit 2002, Brown 2006), however some species are independent of surface water. Browsers or mixed feeders are more likely to be water independent (du Toit 2002), obtaining the bulk of their water from forage. Species adapted to arid environments often have physiological adaptations to reduce sweating, store water, recycle water more efficiently, or reduce water losses in feces and urine (Cain et al. 2006). Differential species requirements affect their interactions. In northern Kenya, livestock were concentrated closer to water while wildlife were often farther away, thus resulting in an inverse correlation in their distributions (de Leeuw et al. 2001)

What is important for movement is the required frequency of watering, and the distance from water that can be tolerated based upon the combination of watering frequency, rate of travel, and time available for travel. For example, horses must be watered daily if eating dry vegetation and every 2-3 days if vegetation is green; cattle must be watered every 1-2 days, sheep and goats every 2-3 days, and camels every 4-5 days (Heady and Child 1994, Holechek et al. 2004). Watering frequency may vary seasonally, for example wildebeest, zebra, and impala water twice as often in the dry season as in the wet season (Gaylard et al. 2003). A certain number of hours per day might be spent walking to water, with varying degrees of trade-off with time spent foraging. Depending on the environment, foraging and traveling to water may occur simultaneously. Traveling rate scales with body size (Cumming and Cumming 2003), but is affected by topography. From these facts, required distances to water could be estimated. The distances may change seasonally or in response to the fraction of green versus dry biomass in the diet.

The influence of water on herbivore distributions can be easily modeled by combining a distance to water map with a function describing decreasing probability of use with increasing distance to water. However, in the case of wildlife, the development of distance to water maps may not be a straightforward task. Different species prefer different water sources (Jarman and Mmari 1971, Gaylard et al. 2003), with respect to vegetation cover, visibility, topography, water depth, and rivers versus pools or troughs. Thus, water maps may need to be developed for each species, and the probability of habitat use may be affected by the type of water source as well as distance to water. Further complicating the problem is the heterogeneity of water points along stream channels. Distance from a stream bank may mean little, as there are definite access points, places where water pools, and places having the right combination of cover and topography. Finally, relative preferences may change with availability. As water becomes scarce, preferences for certain types of water might be expected to decrease.

Herbivore movements may be constrained by the supply rates of water sources and by water quality. Wells or boreholes may only have a certain recharge rate (in liters per day). Consequently, watering points may have a limited capacity to support herbivores. Access to a watering point could be physically limited. If only a certain number of herbivores can access a watering point at a time, this will constrain the number that can be supported by that source per day. Water quality, particularly in terms of salinity, mineral content, or alkalinity, could determine whether a water source is usable. Tolerance of mineral water varies among species (Wolanski and Gereta 2001, Gaylard et al. 2003). Water quality may vary with season due to evaporation and stagnation (Wolanski and Gereta 2001). Few assessments or models have attempted to take these constraints into account. In an ecosystem model for Turkana, Kenya (Coughenour 1992, 1993), maps of water discharge rates were used to limit the density of herbivores in a given location. Separate distance to water and water discharge maps were employed for mineral and fresh water sources, and for wet season vs. dry season sources. Pastoralists informed us that camels were tolerant of mineral water, while other livestock species were not.

It is well known that water distribution influences the distribution of grazing impacts on rangelands. Grazing impacts increase closer to the water point, thus creating "piospheres", or zones of influence around individual water points (Lange 1969, Foran 1980, Andrew 1988, James et al. 1998, Thrash 2000, Brits et al. 2002). Localized areas of degradation around water points are often referred to as "sacrifice areas", as heavy grazing and trampling impacts near water points are an inevitable outcome of providing access to surrounding forage. Bell (1973) noted that when animals have to travel far to water, they graze while moving out to water, but follow trails back, making trails deeper and longer, inducing erosion and draining water from the rangeland. Lange (1969) found that track and dung densities decreased linearly from water. Others have found that rangeland condition varies with distance from water according to a logistic curve, increasing exponentially at first but asymptotically further away (Graetz and Ludwig 1978, Thrash 1993). Forage production may decrease gradually with distance from water, but the proportion of production utilized may be greatest at intermediate distances (Adler and Hall 2005). Adler and Hall attempted to predict production and utilization gradients from an individual-based movement model coupled with a simple plant growth model. They succeeded in producing gradients, but unrealistically steep gradients in grass biomass developed close to water. They suggested other factors such as knowledge of forage biomass further away or social processes may also drive animal movements near water sources. Piospheres do not necessarily develop around water sources if movements are strongly affected by other factors. For example, piospheres did not develop in Rukwa, Tanzania because pastoralists returned their livestock to their settlements each night for predator protection (Copolillo 2000).

Water development is often used to make fuller use of the landscape, to improve animal distributions and grazing impacts, and to provide adequate water during drought. However, water development can have unintended consequences. Negative impacts on vegetation can become more widespread. For example, much of the Sahel was opened up to year-long pastoral access when boreholes were created throughout the region. This had a marked effect on rangeland condition, possibly leading to desertification in some areas (Sinclair and Fryxell 1985). In Saudi Arabia, water was a major constraint on pastoral grazing (Finan and Al Haratani 1998). This constraint ensured rangeland sustainability by reducing grazing offtake on distant ranges, which then functioned as a source of seeds for range regeneration elsewhere. Now, motorized transport has removed that constraint, leading to degradation.

Unexpected consequences for herbivores are also common (Owen-Smith 1996). For example, water development may result in the depletion of forage that would otherwise be critical for survival during drought (Walker et al. 1987). Thus, while water development may increase herbivore numbers in the short term, it may lead to fewer numbers in the long term (Owen-Smith 1996). Water development may cause water-dependent species to increase at the expense of rarer, water-independent species. An interesting situation arose in Kruger National Park, South Africa; areas that previously received little use by most herbivores due to lack of water became available due to borehole development. Increased forage availability and reduced mortality during drought then led to increased herbivore densities, which then caused predator numbers to increase (Gaylard et al. 2003). Unfortunately, rare antelope species such as roan antelope previously thrived in these water-free areas due to their weaker dependence on water and because there were few predators. The increased availability of prey and predators around water sources subsequently caused roan antelope to decline (Owen-Smith 1996, Harrington et al. 1999).

3.3 Temporally varying forage distributions

While herbivores move to distribute themselves in relation to resource distributions, the more interesting and compelling reason to move is to respond to changing resource distributions. In a static environment, foraging movements at multiple scales are required to procure resources on spatially heterogeneous landscapes. If the resource distributions are invariant, herbivores can learn the distributions and make foraging decisions based upon reliable knowledge. In a predictably seasonal environment, foraging is more complex because it requires memory of the way forage distributions change. Herbivores can clearly manage that, as evidenced by the many examples of seasonal migrations found in nature. However, in an environment where changes are unpredictable, foraging is considerably more complex because it involves continuous searching and relearning. In the extreme, memory and learning either become impossible, or of no use.

The most common cause of temporal variations in forage distributions is probably spatially heterogeneous precipitation. There are many different scales and types of rainfall heterogeneity and it is important to characterize these patterns when assessing potential consequences of habitat loss or fragmentation. Temporal variations may be influenced by gradients or by patchiness. Rainfall may be more patchy in environments with convective thundershower activity. Time and space scales of patchiness are interrelated. Finer scale patchiness occurs at finer temporal scales. On a daily basis rainfall may be highly patchy, but averaged over a month, rainfall patterns may be smoother, giving rise to discernable patterns at larger spatial scales. Gradients, or larger scale variations, are important in many systems, particularly when the gradient spans a short enough distance for herbivores to respond to. The drier end of a gradient is often characterized by shorter growing seasons, particularly in regions where most precipitation is concentrated during a wet season. In the Serengeti, for example, the growing season is shorter on the shortgrass plains (McNaughton 1985). During the growing season plant growth rates are high, foraging conditions are optimal, and predators are easy to detect. Migratory herbivores must make use of this area while it is available or it is simply a wasted resource. During the dry season, migrants are forced to move to the wet end of the gradient. Movement pathways that optimize access to areas with recent rainfall and green biomass are consistent with the actual migratory pathways of the wildebeest (Boone et al. 2006).

There are other causes of temporally varying forage distributions. The depletion of forage by herbivory is fundamental. After an area has been grazed, herbivores must move. Patchy fires may cause an initial loss, but subsequent regrowth of higher quality forage attracts herbivores (Wilsey 1996, Archibald et al. 2005). Temporally varying snow cover has a large influence in many systems. For example, the size of the elk winter range in Yellowstone varies within and among years, depending upon the distribution of snow depth. When snow is deep, the winter range shrinks because snow depths are shallower at lower elevations and the area that is at a sufficiently low elevation decreases (Houston 1982, Coughenour and Singer 1996b). This introduces a density independent variation in recruitment and mortality among years (Coughenour and Singer 1996a). As snow melts off at progressively higher elevations, elk follow a green wave of plant growth as the range expands to its full summer extent (Frank and McNaughton 1992).

The net outcome of movement responses to changing forage distributions is increased stability of forage intake. In other words, herbivores buffer temporal variations in forage distributions by moving. This seems like an obvious outcome, particularly where herbivores are clearly migrating in response to changes in forage distributions. In less predictable environments, however, movement is particularly important. To cope with uncertainty, herbivores must be flexible and opportunistic. They must have a range of options for responding to the changing forage distribution (Perevolotsky 1987, Coughenour 1991, Scoones 1995, McAllister et al. 2006). Opportunistic movements in spatially variable environments not only buffer variations in forage, they also reduce uncertainty by decreasing the probability that an animal will find itself in a forage deficit situation. This was a key finding of research on the Turkana nomadic pastoralists (McCabe 1983, 2004, McCabe et al. 1988, Ellis and Swift 1988). Some of the first spatially explicit modeling studies that were carried out in the course of that research demonstrated the importance of opportunistic movement for pastoral viability (Coughenour 1989, 1991). Similar results have been obtained by modeling responses of nomadic gazelle to spatially variable rainfall in the Serengeti (Fryxell et al. 2005). While this form of movement is often referred to as nomadic, movements are not simply random wanderings. More accurately, such movements are opportunistic, but purposeful and calculated responses to changing forage distributions. Movements are devised to minimize variability in forage intake, based upon experience and the best information that can be obtained. Pastoralists can obtain information about distant pastures from scouts and through word of mouth. Wildlife, in contrast, may not have access to information about forage in distant locations. In the Fryxell et al. (2004) model of Thompson's gazelles, for example, the most realistic movement rule was based upon local rather than global scale matching of redistribution and energy intake rate. Field observers have often anecdotally noted that herbivores such as wildebeest or oryx suddenly move great distances in the direction of distant rainfall, but this author is unaware of any firm evidence.

Frequency of movement is affected by the abundance and patchiness of forage, which changes seasonally. Forage may be more patchily distributed in seasons of intermediate dryness, and movements may consequently be more frequent. McCabe (2004) found this to be the case with Turkana pastoralists in Kenya. During very good wet seasons, movements were less frequent. During normal and poor wet seasons, movement frequency increased, probably in response to patchy forage. In normal and bad dry seasons, movements were less frequent, possibly because the benefits of moving decrease compared to the costs. In very dry conditions, movement frequency increased again because remaining forage was very patchy and it was more critical to access forage to prevent mortality.

Commonly, herbivore populations are limited by the amount of forage that is available on a limited portion of the landscape during winter, the dry season, or a drought. We found this to be true in Turkana, Kenya. During dry seasons and droughts, livestock populations were limited to forage in locations that were little used during wet seasons, and further from water but still within a close enough distance to water for animals to travel to at the necessary frequency (Swift et al. 1996). In many pastoral grazing systems, there is a 'dry season grazing reserve'. These are often areas that are less desirable to use during the growing season for some reason, such as overly warm temperatures at lower elevations, presence of parasitic insects (e.g., ticks, biting flies), long distances to water, or difficult topography. Dry season areas are often more productive and have longer growing seasons than the wet season areas. In this case, the wet season area (with the shorter growing season) is utilized while it is available, freeing the dry season area from use during a portion of the year. Population size is still limited by the forage on the dry season area, thus explaining why the population does not increase to fill up both wet and dry season grazing areas during the wet season. It also explains why a small fraction of the forage in the whole system is consumed and why the herbivore population is smaller than might be expected given the apparent forage surplus (Coughenour et al. 1985, Ellis and Swift 1988). Limiting areas of this sort have been termed "key resource areas" (Scoones 1995, Illius and O'Connor 1999, 2000). These situations have been modeled by representing the time-varying distribution of forage on the landscape, other constraints on herbivore distributions, and the resultant forage limitations on herbivore population growth (Coughenour 1992, 1993, Illius and O'Connor 2000, Weisberg et al. 2006).

Key resource areas were central to Illius and O'Connor's (1999) criticisms of the concept of non-equilibrial plant-herbivore systems in variable environments. Ellis and Swift (1988) proposed that when precipitation and forage are highly variable, herbivore populations are unable to equilibrate with forage. After a drought induced die-off, there is a subsequent time lag for recovery of the herbivore population. Before the population can reach an equilibrium with forage, there is another die-off. Illius and O'Connor (1999, 2000) argued that there is, in fact, density dependence based upon competition for key resource areas during dry seasons and droughts. Either way, the livestock population can never grow large enough to fully exploit the forage base. Thus, it has been asserted that overgrazing is unlikely in drought-prone environments (Behnke and Schoones 1993, Schoones 1994). Degradation may or may not occur in key resource areas, depending upon the spatial distribution and timing of key resource availability. If there are a few large key areas, herbivores would be expected to congregate on them during droughts, imposing heavy grazing pressure. However, if the plants go into dormancy, either as annual seeds or as perennial roots, the heavy grazing may have little impact. Resource availability in the key resource area should limit herbivore populations so if the key resource area is degraded, herbivore populations will decline in response. On the other hand, if the key resource areas consist of many widely scattered, small patches, herbivore populations will be less likely to find them and less likely to aggregate upon them in excessive densities, lessening the probability of degradation. The extreme case is where forage is homogeneously distributed and there are no key resource areas. Instead, as the forage senesces and is grazed down, herbivores face some probability of starvation that is independent of their density and more related to the frequency distribution of energy reserves in the population, genetic variation, and exposure and susceptibility to weather and disease. The same fraction of the population perishes, irrespective of the number of animals. Ecosystems exhibit a wide range of variation in the relative importance of key resources and density dependent vs. independent controls on population regulation, depending upon the spatial configuration of resources on the landscape.

There are two points to be made here. One is that herbivores must be able to opportunistically move to access critical resource areas when precipitation has a high degree of temporal variability. The other is that herbivore population dynamics are emergent outcomes of movements and the spatio-temporal distributions of forage on the landscape.

3.4 Temporally varying water distributions

Water requirements and supplies may be seasonally variable. Because water requirements increase with temperature, dry matter forage intake, and lactation, water requirements often vary seasonally. Grazers may be independent of water in the wet season and dependent in the dry season. The distribution of water may vary seasonally, as surface water sources evaporate in the dry season and are replenished in the wet season. For example, water collects in numerous surface depressions during the wet season on the Serengeti shortgrass plains. However, these sources dry up and become increasingly saline due to evaporation, making them unsuitable or nonexistent during the dry season (Wolanski et al. 1999, Wolanski and Gereta 2001). These authors hypothesized that the Serengeti wildebeest migration is largely driven by seasonally varying distributions of available surface water. However, the spatio-temporal distribution of water is confounded with the distribution of rainfall and green biomass (Pennycuick 1975, Wilmshurst et al. 1999b, Boone et al. 2006), making it difficult to attribute movements to one factor or the other. In Amboseli National Park, Kenya, herbivores concentrate near areas with surface water in the dry season, but disperse more widely in the wet season (Western 1975). Similarly, pastoralists often concentrate around a limited number of perennial water points during the dry season, and disperse widely during the wet season as livestock increasingly obtain water from ephemeral surface water sources and green forage (e.g., Sinclair and Fryxell 1985, Turner 1998). The location of dry season water sources in the Rukwa Valley of Tanzania influenced pastoral land use year-round, with increased grazing near water in the dry season, and increased grazing away from water in the wet season (Copollilo 2000). In Turkana, by contrast, during the wet season there are ample water sources and they are well distributed. These include surface water sources such as ephemeral streams and ponds. Ample water coupled with increased forage productivity permits pastoralists to congregate in favored locations. During the dry season, surface water and some wells dry up, forage production ceases, and standing forage is depleted, leaving a limited number of far-flung water points and associated foraging areas. Thus, the seasonal pattern is one of contraction in the wet season and dispersion in the dry season.

Seasonal variations in water provisioning have been recommended as a method of balancing uses of wet and dry season ranges (Owen-Smith 1996). Dispersal to a more expansive wet season area is important for allowing plants near dry season concentration areas to regrow. If artificial water points are opened on the wet season range during the dry season, this may make more forage available, increasing herbivore populations, and subsequently imposing increased grazing pressure in the wet season areas. Thus, the balance of grazing pressure between wet season and dry season areas is determined by seasonal water availability. A portion of the artificial water sources can be closed during the dry season to reduce grazing pressure in the wet season area. Because herbivores spend more time on the dry season range, it was recommended to control water availability so as to achieve a 2:1 ratio of wet season to dry season areas (Owen-Smith 1996).

3.5 Habitat quality and other factors that constrain movement

The consequences of habitat loss and fragmentation cannot be assessed unless the suitability of habitats can be mapped based upon a suite of relevant variables. Movements in response to forage and water are constrained by other habitat factors, some of which are temporally variable. Habitat utilization or preference is usually characterized in terms of a probability of use (Boyce et al. 2002, Manly et al. 2002). This probability of use can be thought of as an index of habitat quality. In a statistical sense, there is a higher probability of finding an animal in a higher quality habitat. Importantly, habitat quality is a continuous, rather than a dichotomous variable. While many studies of the effects of fragmentation on wildlife populations, mostly using metapopulation modeling, have characterized the landscape as a set of favorable habitat patches within a matrix of unfavorable habitat, this is an oversimplification (Weigand et al. 2005, Fahrig and Nuttle 2005). The effects of habitat loss and fragmentation are likely to be proportional to habitat quality, because by definition, higher quality habitats support more animals. Fragmentation or loss of high quality habitat will clearly have more serious consequences. In this section, I briefly review approaches for modeling and assessing habitat utilization, and other factors besides forage and water that must often be considered to study the importance of movement and the consequences of its disruption.

Habitat models have been used since the 1970s to assess the effects of land use on wildlife (Berry 1986). Habitat suitability index (HSI) models were developed for many species by the US Fish and Wildlife Service as part of their habitat evaluation procedures (Fish and Wildlife Service 1980, 1981). The approach involves combining multiple habitat factors into a single index of relative suitability. Statistical approaches have included multiple regression (Marzluff 1986) and multivariate statistics (Brennan et al. 1986). The volume edited by Verner et al. (1986) provided numerous examples of the state of the art at that time, including several examples of linking habitat models with simulation models of vegetation dynamics. Employing a dynamic HSI in spatial ecosystem models has led to many useful analyses (Coughenour 2005, Weisberg et al. 2006). Combining an HSI model with fluid dispersion models resulted in realistic patterns of marine organism invasions (Inglis et al. 2006). A simple example of a dynamic application was the use of a dynamic HSI function of snow depth and forage to estimate ungulate carrying capacity (Coughenour and Singer 1996a).

Resource selection functions (RSFs) are being widely used to model animal movements and distributions (Manly et al. 2002, Boyce et al. 2002, Keating and Cherry 2004, Anderson et al. 2005). Basically, resource selection functions predict the probability of use among available areas using logistic regression (Manly et al. 2002, Boyce et al. 2002, Keating and Cherry 2004). A limitation of this approach is that it must differentiate available from unavailable habitat (Buskirk and Millspaugh 2003), which may be affected by a variety of factors such as topography. A more recent approach uses resource utilization functions (RUFs) (Marzluff et al. 2004, Millspaugh et al. 2006). A spatially smoothed utilization distribution (UD), a probability density surface, is created using kernel density estimators. Then this surface is regressed against habitat variables. Although RSFs and RUFs have been primarily applied to predict year-long or seasonal distributions, there is no reason why they cannot be used to model dynamic distributions or movements either using a redistribution approach, or individual-based movement modeling.

Another recent set of modeling approaches called ecological niche models (ENMs) has been developed to predict habitat suitability, probability of occurrence, and species ranges. These include generalized regression modeling (Lehmann et al. 2002), genetic algorithms (Stockwell and Peters 1999, Peterson and Vieglas 2001), a multivariate statistical approach called ecological niche factor analysis (Hirzel et al. 2002), the product of Gaussian-shaped curves for multiple habitat variables (Brown et al. 1995), and an approach based upon Mahalanobis D^2 (Rotenberry et al. 2006).

Climate may directly influence herbivore distributions. Mammalian herbivores are limited to thermally viable environments. Cold temperatures, wind, and snow contribute to winter-summer movements in many temperate zone ecosystems, particularly in mountainous landscapes. While deep snow affects forage availability, it also impedes movement. Snow depth has well known influences on elk, deer, and bison distributions, defining winter ranges and summer ranges in western North America. Similarly, vertically transhumant movements of pastoralists occur throughout the world in response to winter weather conditions at higher elevations, including Spain (Ruiz and Ruiz 1986, Garcia-Gonzalez et al. 1990, Garcia-Ruiz and Lasanta-Martinex 1990), the Tibetan Plateau (Cincotta et al. 1992), and Iran (Beck 1991). North-south movements over longer distances are also common. Historically, pastoralists in Mongolia and Kazakstan migrated long distances southwards in the fall to avoid severe winter weather (Fernandez-Gimenez and Allen-Diaz 1999, Robinson and Milner-Guilland 2003, Kerven 2004, Kerven et al. 2004). Conversely, overly warm conditions could induce movements to summer ranges. For example, high temperatures in the southern deserts of Kazakstan could have contributed to northerly movements in the spring (Robinson and Milner-Guilland 2003).

Woody cover is an important habitat feature for many large herbivores. Shade is necessary in many tropical environments where direct beam solar radiation may lead to excessively high body temperatures. Cover is often necessary for shelter from wind and precipitation. Cover may also be important for temperate zone ungulates in winter, because it provides a less stressful infrared (thermal) environment during the night (Porter et al. 2002). Cover may decrease predator detection and provide escape refugia. Woody plants can also have negative effects on habitat suitability. Bison, for example, prefer open habitats. Thickets of shrubs or shrub-height trees may be difficult or impossible to move in. Downed timber in previously burned forests can also impede movement.

Insects and disease are less commonly documented but often important drivers of herbivore movement. Herbivores may move to avoid biting flies or mosquitoes. For example, caribou form groups and travel to habitats providing relief from insects in summer (Fancy et al. 1989). Livestock seek wind-exposed areas to avoid insects in the Pyrenees (Garcia-Gonzalez et al. 1990). One reason elk and bison may prefer higher elevations in the summer is to avoid biting flies (Meagher 1973, Houston 1982). Tick avoidance is important for pastoralists in many areas. For example, in Turkana, pastoralists avoid longer grass areas during the growing season to avoid ticks and the diseases they carry. These areas then serve as dry season or drought grazing reserves. In Ngorongoro Conservation Area, Tanzania, Maasai pastoralists avoid certain areas due to ticks and diseases (McCabe 1995, Boone et al. 2002). They also avoid areas inhabited by calving wildebeest due to the transmission of Malignant Catarrhal Fever. Tse-tse fly and trypanosomiasis influence livestock distributions throughout Africa (Jordan 1986, Ormerod 1986, Reid et al. 1997).

Herbivore movements may be significantly modified by predators and their spatial distributions. It has often been theorized that herbivores live in herds in order to reduce the risk of predation (Hamilton 1971, Bertram 1978, Fryxell 1995). In turn, herding behavior influences patterns of movement and distribution for reasons discussed above. Similarly, risk of predation may influence the way animals move. For example, migratory wildebeest may travel in lines because the risk of predation is lower for individual animals (Bertram 1979). Herbivores may avoid areas of the landscape with higher predator densities, with subsequent consequences for plants. For example, it is possible that wolves are altering elk density distributions in Yellowstone National Park, resulting in reduced herbivory on willows and aspen in some locations (Ripple and Beschta 2004, Fortin et al. 2005, Creel et al. 2005). Elk were less likely to forage and more likely to relocate in areas with high wolf use (Frair et al. 2005). Herbivores may avoid areas where predators are more effective. Vegetation cover and fine scale physical features influence predator effectiveness (Hopcraft et al. 2005). Herbivores may consequently avoid areas with tall grass, woody cover, or other suitable ambush sites.

Conversely, herbivore movements modify the effects of predators on their prey. If herbivores are successful in escaping predation through movement or altered spatial distributions, it follows that their population dynamics will be less affected by predation. In the extreme, migratory movements of herbivores enable them to largely escape predators, in the sense that their populations are no longer regulated by predation (Fryxell et al. 1988, Fryxell and Sinclair 1988, Sinclair 2003). This is a result of the fact that predators are non-migratory, and their densities are consequently limited by the length of time prey are in their hunting ranges. Movements of pastoralists are often constrained by the presence of enemies (Niamir-Fuller 1999, McCabe 2004). Pastoralists throughout Africa and elsewhere have engaged in intertribal warfare and livestock raiding for centuries. In Turkana, different tribal sections have different enemies, depending on who is living on their border. In South Turkana, the Pokot tribe was the principle enemy, however there were also bandits known as Ngoroko who posed threats in certain areas and at certain times. A large fraction of South Turkana was avoided due to the threats of Pokot raiding. These areas were mainly used when the grass was finished elsewhere. The effect of security risk had to be considered in modeling the Turkana ecosystem, as it had significant consequences for forage availability (Coughenour 1992). A 'force' effect on herbivore distributions was used to influence habitat suitability as a function of distance from the Pokot border.

3.6 Socio-economic factors

Movements of pastoralists and their herds are the outcomes of socio-economic as well as biophysical factors (Turner 1993, McCabe 1994, 2004, Copolillo 2000, Boone et al. 2002, Thornton et al. 2006, Baker and Hoffman 2006, Galvin et al. 2006). Livestock distributions are determined by the management of daily and seasonal grazing movements, and this management is, in turn, affected by changing patterns of access to pastures, labor resources, and livestock wealth. Consequently, grazing movements are tied to changes in the broader political economy (Turner 1993). Turner criticized the view that the relationship between stocking rate and forage production should form the basis for managing African rangelands as being overly simplistic, and exclusive of socio-economic processes.

The influence of socio-economic factors is especially apparent in day-to-day movement decisions (Baker and Hoffman 2006). These authors found that in the aggregate, the overall goal of movement was to manage environmental variability, but on a daily basis, factors such as needs to share labor, and maintain gardens or businesses often took precedence. Pastoralists remained sedentary due to the costs of moving, poor health, and the need to tend their homes. Pastoralists moved to acquire or provide shared labor, and for a variety of personal and social reasons. Baker and Hoffman (2006) noted that socio-economic factors are becoming more important because of restricted land access and more diversified income sources.

Interactions between biophysical and socio-economic processes demand interdisciplinary approaches to understanding pastoral ecosystems (Coughenour et al. 1985, McCabe 2004, Baker and Hoffman 2006). Understanding pastoral movements requires ecologists and social scientists to work together. Increasingly, a systems level approach is being taken in which ecological, household decision-making, and land-use change models are interlinked (Boone et al. 2002, Thornton et al. 2003, 2006, Galvin et al. 2006).

4. THE EFFECTS OF MOVEMENT AND THE CONSEQUENCES OF DISRUPTION

4.1 Plant communities and ecosystems

Herbivore movement patterns affect plant communities and ecosystems as a consequence of direct and indirect effects on plants, other above- and belowground consumers, predators, and nutrient cycles. Herbivores exert numerous effects on ecosystems, and many of the effects involve herbivore movement (Ruess and McNaughton 1987, Seagle et al. 1992, Hobbs 1996, 2006, Frank et al. 1998, du Toit and Cumming 1999, Pastor et al. 1999, 2006, Sinclair 2003, Frank 2006, Suominen and Danell 2006). Through ecological engineering (Jones et al. 1994) or landscaping (Sinclair 2003), large herbivores may act as keystone species that determine diversity for the rest of the system (Collins et al. 1998, Sinclair 2003). Some of these effects involve generation of spatial heterogeneity, which feeds back to herbivores through their movements. A few cascading effects are of particular importance and clearly related to herbivore movement. These involve heterogeneity, biodiversity, spatial food webs and nutrient cycles, and the integration of spatially separated patches at the ecosystem level of organization. The linkages between herbivore movement, disturbance regimes, biodiversity, and ecosystem functioning require further elucidation. An increased understanding of these linkages would be useful for predicting ecosystem-level outcomes of disrupted movements.

Multiscaled herbivore movement and foraging increases the spatial heterogeneity of vegetation (Senft et al. 1987, Hobbs 1996, 2006, Adler et al. 2001), with subsequent effects on biodiversity and ecosystem functioning. Spatial heterogeneity contributes to coexistence and biodiversity of plant species (Tilman 1994, Lehman and Tilman 1997, Pacala and Levin 1997). A diversity of forage qualities promotes coexistence of herbivores of different body sizes (Prins and Olff 1998). If the patterns generated by herbivory are multiscaled, they could contribute to consumer diversity by increasing the variety of vegetation patch sizes, which in turn, facilitates coexistence of herbivores with different body sizes (Ritchie and Olff 1999). If the spatial heterogeneity that is generated by mobile herbivores alters biodiversity, then it also affects ecosystem functioning (Schulze and Mooney 1993, Loreau et al. 2002).

Herbivore movement is central to the process of patch dynamics (Pickett et al. 1989, 2003). Herbivores may graze or browse patches of vegetation, moving onto other ones while vegetation in previously visited patches regenerates. The result is a shifting mosaic of patches in different stages of regrowth or succession on the landscape (e.g., Fuhlendorf and Engle 2004). Plant species with different life history strategies may occupy disturbed versus recovered patches. As a result, plant species diversity at the landscape scale is increased. Increased plant species diversity contributes to increased herbivore diversity, due to differences in their diets. Ruminant herbivores with different body sizes should also select habitats and vegetation patches with different biomass densities and forage qualities (Wilmshurst et al. 2000). For this process of patch and biodiversity generation to be sustainable, herbivore movements must operate at a sufficiently large spatial scale. The rate of patch creation must not exceed the rate of patch regeneration.

Similarly, when the rate of plant or plant population growth is low, diversity will be highest when the frequency or intensity of disturbance is low (Huston 1979, 1994). When plant growth rate is high, diversity should be maximized at a high frequency or intensity of disturbance. At intermediate plant growth rates, diversity is maximal with intermediate disturbance frequencies. The outcomes are determined by whether the plants are able to recover from biomass mortality fast enough to keep up with disturbance, and whether competitive exclusion among plants occurs due to rapid plant growth relative to disturbance frequency. As noted above, the frequency and intensity of herbivory are combined outcomes of the rate of movement and the local density of the herbivores. Consequently, herbivore movement patterns have the potential to affect plant species diversity in different ways, depending upon interactions between the resulting disturbance regime and rates of plant regrowth following defoliation.

Certain frequencies and intensities of herbivory and thus movement are required to initiate and maintain grazing lawns. Lawns are created by frequent, close grazing, which alters the morphology of grazing-adapted graminoids towards higher leaf:stem ratios, denser tillers, and more prostrate leaf angles (Vesey-Fitzgerald 1960, McNaughton 1984). Nutrient cycling is enhanced, light penetration is increased, there is less rainfall interception by herbage and litter, and soil temperatures may be warmer. As a result, lawns may be more productive than ungrazed swards (McNaughton 1979, Coughenour 1984, Coughenour et al. 1984, Ruess and McNaughton 1987, Seagle et al. 1992), and more suitable for smaller-bodied herbivores. Lawns may be initiated by intense grazing or fire, and they may be sustained within a matrix of less palatable taller grasses. A positive feedback cycle is established as the lawn continues to attract herbivores. A certain visitation frequency and spatial distribution of herbivory is required to maintain grazing lawns. For example, fire alters the distribution of grazing pressure, with subsequent effects on the distribution of sward structure and the balance between grazing-tolerant and intolerant grass species and their distributions on the landscape (Archibald et al. 2004, 2005). Fires draw herbivores off existing grazing lawns and herbivory becomes more diffusely distributed, resulting in the conversion of lawns back to tall grass. Over the long-term, if fires are frequent and large enough, then grazing-induced lawns, and grazing-adapted lawn grass species, could disappear (Archibald et al. 2005).

Herbivore movement is involved in grazing succession (Vesey-Fitzgerald 1960, Bell 1971), in which large-bodied herbivores convert taller grass swards with low leaf:stem ratios to shorter swards with higher leaf:stem ratios that are subsequently utilized by smaller-bodied herbivores. In so doing, larger-bodied species can facilitate energy flow to smaller-bodied species (McNaughton 1976). The smaller-bodied species selectively forage for high-quality tissues, reducing the quality of the patch, which forces less selective larger-bodied species to move on to a new patch (Murray and Illius 2000, du Toit 2003). Movement is integral to this interactive process because the large-bodied species must be able to move in response to changing distributions of swards with high biomass swards and sufficient quality, while the small-bodied species must be able to move in response to the changing distribution of swards with low biomass but high quality. Similarly, elephants convert woodlands to grasslands or more open savannas, creating habitats that are more suitable for grazers rather than browsers (Dublin 1995, Whyte et al. 2003, Western and Maitumo 2004). Thus, movement may be central to the maintenance of a diversity of ungulates with varied body sizes and digestive physiologies.

Herbivore movement may alter ecosystem functioning by laterally transferring nutrients across the landscape and by creating zones of nutrient enrichment or 'hotspots'. In the arid Turkana ecosystem of northern Kenya, we noted that herbivores can transport nutrients from upland interfluvial areas to areas along ephemeral streams which support large Acacia trees. This occurred as there was an increased amount of dung deposition under the trees where the animals sought shade during the heat of the day, or where pastoralists often corralled their livestock overnight. We also found that there was a considerable concentration of nutrients in the temporary livestock corrals built by the pastoralists in new locations every 10-20 days (Ellis et al. 1985, Reid and Ellis 1995). These corrals were not only enriched in nutrients for many years afterwards, they also served as sites for tree regeneration. Similarly, sites of ancient pastoral settlements and corrals in Kenya and South Africa were converted from nutrient poor savanna types to nutrient rich savanna types, comprised of characteristic tree species (Blackmore et al. 1990, Young et al. 1995), or herbaceous species (Augustine 2003), and elevated forage nutrient concentrations (Augustine 2004). Herbivores are subsequently attracted to these hotspots (Young et al. 1995, Augustine 2004). If animals graze on one patch or landscape and ruminate or bed on another, nutrients will be transported from the grazing area to the ruminating/bedding area. In Rocky Mountain National Park, nitrogen was moved away from willow communities where elk obtained considerable forage, and towards conifers where they bedded (Schoenecker et al. 2004). Livestock created soil nutrient concentration gradients radially from water points and encampments in the Sahel due to the fact that a large fraction of resting, ruminating, and suckling time is spent at or near these locations (Turner 1998). It is also possible for migrating animals to transport nutrients between seasonal ranges. If animals gain weight on one range and lose it on another, nutrients are transported through the gain of lean body nitrogen in one area and the excretion of nitrogen due to lean body metabolism in the other (Hobbs 1996). For example, elk transport nitrogen by gaining weight on summer range and losing weight on winter range (e.g., Singer and Schoenecker 2003, Schoenecker et al. 2004). Deer transport nitrogen from croplands to forests (Seagle 2003). If animals die disproportionately on one seasonal range, the carcasses will provide a net nutrient input to that area (Holdo et al. 2006).

Through the movements of herbivores, spatially separated patches on a landscape become integrated into a single functioning system and the patches start to affect each other indirectly. For example, herbivores may be drawn to one area because of higher forage quality or quantity. The area they are drawn from is thus affected by the area they are drawn to, in terms of the level of herbivory it experiences. Suppose a diverse landscape contains partially overlapping habitats for two different herbivore species. The presence of patches in the landscape which support species A thus influences the number of herbivores of species B through competition in the overlap area. This feeds back to the number of species A, which then affects plants in patches utilized by species A, but not B. Patches may be connected through spatially distributed food webs (Polis et al. 1997, 2004 , Knight et al. 2005). If forage in area A supports a mobile herbivore population that also uses another area B inhabited by a predator population, then the predator population in area B becomes connected to the forage in area A through herbivore movements. The prey that are eaten by predators in area B are also affected by the forage in area A. These types of spatially distributed indirect interactions are likely to be quite pervasive in ecosystems. Consequently, they represent critical elements of ecosystem functioning.

Many of the movement effects discussed above are likely to increase ecosystem resilience. Ecosystem resilience focuses on persistence, adaptiveness, variability, and unpredictability and it is measured in terms of the magnitude of disturbance that a system can absorb before it changes its structure and functioning (Holling and Gunderson 2002). Ecosystems, particularly those in arid and semi-arid climate zones, experience variability and uncertainty. Rather than stabilizing about an equilibrium, they adapt to change. Movement is a key element of adaptability and ecosystem resilience in regions with spatially variable climate (Walker and Abel 2002). The importance of opportunistic movements for herbivores was discussed above. At the ecosystem level, movement promotes resilience by giving herbivores access to diverse land systems that offer a range of opportunities in time and space (Walker and Abel 2002). Herbivore movements increase resilience by providing opportunities for plant regrowth and by creating mosaics of patches with varied functions. Patch dynamics result in meta-stability or persistence at large scales, as opposed to the transient dynamics that occur at local scales (Wu and Loucks 1995). Movement-induced biodiversity contributes to ecosystem resilience by increasing the variety of ways that different species can respond to change. The increased variety essentially insures that a suitable range of functions will be available to respond to changing environmental conditions (Yachi and Loreau 1999, Loreau et al. 2002). Fragmentation and habitat loss reduces resilience by diminishing movement-based adaptations to variability and disturbance, by reducing opportunities for plant regrowth, and by increasing homogeneity.

Access to landscape diversity may contribute to resilience by supporting an increased number of trophic energy flow pathways. Different pathways originate in different parts of the landscape and propagate through different populations of herbivores and consumers. Different pathways may come into play under different environmental conditions. For example, multiple species of livestock make use of varied forage resources on different parts of the landscape in Turkana, Kenya (Coughenour et al. 1985, Coppock et al. 1986). Livestock species include pure grazers (cattle), mixed feeders (sheep and goats), and browsers (camels). The grazers exploit large but short-lived pulses of herbaceous primary production following rain. Browsers exploit woody plant resources that remain green longer or store nutrients in stems, thus attenuating the rainfall pulse into the dry season (Swift et al. 1996). The diversity of herbaceous and woody plants is linked to landscape diversity (Coughenour et al. 1990, Coughenour and Ellis 1993, Patten and Ellis 1995).

Water redistribution on the landscape supported woody plants in water concentration zones. Herbaceous plants were more abundant on runoff areas, thin soils, and areas with high enough rainfall to support a sufficient fuel load to carry a fire. Herbivore movement enabled access to the diversity of the landscape. As a result, the flow of energy to pastoralists, and the ecosystem, was stabilized despite the extreme temporal variability of rainfall.

Through these processes and interactions, herbivore movements effectively integrate landscape sub-elements into a landscape meta-ecosystem (Loreau et al. 2002, Lovett et al. 2006). Ecosystem stability and dynamics are affected by interactions and feedbacks among heterogeneous animal movements and the heterogeneity of resource distributions on the landscape (Pastor et al. 1997). Because spatial heterogeneity, complexity, and diversity are critical elements for sustained ecosystem functioning, and humans have pervasive influences on natural disturbance regimes, natural resource management requires the explicit recognition of these linkages and well-defined operational goals for sustaining them (Christensen 1997, Rogers 1997, 2003). Management of ecosystems with large herbivores requires knowledge of linkages among movement, heterogeneity, and ecosystem functioning, and the ability to predict the effects of human activities on landscape meta-ecosystems.

4.2 Spatially distributed plant-herbivore interactions

A question of interest is whether habitat loss and fragmentation can indirectly affect plants and soils on the remaining viable habitat. There are many situations where this could occur, but they all involve a change in grazing pressure on the remaining habitat. A possible example of this is the Yellowstone northern elk range (Pengelly 1963, Houston 1982, Coughenour and Singer 1991, 1996a, Singer et al. 1998, Boyce 1998, National Research Council 2001). When the park was established, the northern boundary cut off portions of the winter range. The park is at a higher elevation than the surrounding area and receives more snowfall. Land outside the park at lower elevations was largely converted to domestic livestock grazing and irrigated hayfields, and it was heavily hunted. Some have argued that this led to an increased concentration of elk on the remaining winter range inside the park because it removed winter range that would have previously been preferred due to lower snow cover. As a result, some argued that elk were artificially confined and as a result, were overgrazing the remaining winter range. This is debatable because the remaining winter range proved to be a viable winter habitat and elk numbers rose considerably after culling was stopped in 1968.

The idea that confinement or compression of herbivore populations can lead to overabundance and ecosystem degradation has been at the center of much discussion regarding herbivore population management (Myers 1973, Jewell et al. 1981, Owen-Smith 1983, Dublin 1995, Whyte et al. 2003, Western and Maitumo 2004). The compression hypothesis is often invoked to explain herbivore overabundance and to justify management interventions to reduce population sizes. An interesting example is the conversion of vegetation by elephants in Tsavo National Park, Kenya (Myers 1973, Parker 1983). Elephant densities increased inside the park in response to increasing human populations in surrounding areas. Between 1957 and 1972 elephants reduced most of the woodlands to grassland. A severe drought in 1970-71 resulted in a massive elephant die off, putatively in part because they had nowhere else to go. Since 1970, woody vegetation has partly recovered, in response to the reduced numbers of elephants. Elephant numbers were kept low by illegal poaching in the 1980s. However, as a result of increased protection since the 1990s, elephant numbers could grow again and history could repeat itself (Leuthold 1996).

As pointed out by Owen-Smith (1981, 1983) and Pulliam (1988), landscapes might consist of source and sink areas for dispersing herbivores. The source areas produce surplus animals that disperse to the less favorable sink areas where mortality exceeds recruitment. When reserves are created, they are usually created on areas containing the best habitat. These areas could have functioned as source areas originally. When surrounding areas are made inaccessible due to fencing or human land use, sink areas often become inaccessible to herbivores. Consequently, the source area becomes overpopulated. This raises the question as to whether herbivores would disperse before reaching food-limited carrying capacity on the source area. If they do not, then herbivore numbers would grow to a point where forage production and offtake are in dynamic equilibrium. The same dynamic equilibrium would be attained even after the sink area has been excised. Or in the case of a system with frequent density independent mortality caused by droughts or severe winters, populations should still be limited by those events inside the reserve. These questions are raised to illustrate the complexity of predicting the consequences of disrupted movement for herbivores, plants, and soils.

In Yellowstone, there is evidence that bison have expanded their ranges (Meagher 1989) and that range expansion occurred when bison reached high local densities (Taper et al. 2000, Meagher et al. 2003). It was concluded that the bison expanded their ranges as a result of spatial density dependence and their normal capabilities to develop travel routes through snow (Bruggeman et al. 2006). Nutritional modeling showed that they expand their ranges when they begin to experience moderate nutritional deficits, and before the deficits become large enough to result in significantly increased mortality, decreased recruitment, or significant deterioration of the vegetation and soils (Coughenour 2005). The bison are essentially exhibiting density dependent dispersal, attempting to expand their ranges as a proactive behavior to prevent starvation. This is significant because it suggests that herbivore populations might regulate local densities through density dependent dispersal below food-limited carrying capacity as traditionally defined. If a herbivore population normally disperses in response to declining forage availability, and the dispersal is into sink areas where population growth rates are negative, then densities in the source area will always be less than food-limited carrying capacity. If dispersal is impaired, for example by habitat loss and fragmentation around reserves, population regulation could be achieved through food limitation, but densities would be elevated. Thus, processes within the reserve would be influenced by humans outside of the reserve. This would compromise the suitability of the reserve to serve as a benchmark for assessing the consequences of human activity (Sinclair 1998).

As noted elsewhere (May and Beddington 1981, Coughenour 1991), concepts of patch dynamics in predator-prey interactions are relevant to interactions between large herbivores and plants. A number of theoretical and experimental studies have shown that predator-prey systems are stabilized by movements of predators among patches of prey (Huffaker 1958, Hilborn 1975, Gurney and Nisbet 1978, Abrams 2000). This principle also applies to two-species metapopulation systems, where predators and prey exist in spatially subdivided habitats (Fahrig and Nuttle 2005). Although most of these studies have involved predators and prey, the same principles apply to spatially distributed systems of consumers and regenerating resources. If consumers move to locations with high resource density, patches will act as refugia where resources can recover while consumers are absent. Consumer and resource densities are maximized at intermediate consumer movement rates relative to the resource regeneration rate (Abrams 2000). If movement rate is too rapid, there will not be a sufficient time lag for resource regeneration in the patches from which consumers have departed. If movement rate is too slow, the consumers do not disperse fast enough to prevent resource overutilization, consumer numbers decrease, and the system is destabilized.

While the preceding studies mainly involved the population dynamics of the consumer, the idea of time lags and resource regeneration can be applied to plant-herbivore systems with mobile herbivores. Rotation grazing systems have long been used to provide time for plant regrowth (Bell 1973, Heady and Child 1994, Holechek et al. 2004). However, the effectiveness of grazing systems for improving range health, forage productivity, or livestock productivity has been brought into question (Heitschmidt and Taylor 1991, Hart et al. 1993, Quirk 2002). The effects of different paddock sizes on the grazing regime experienced by plants, in terms of frequency and intensity of defoliation, and plant responses, was thoroughly examined by Noy-Meir (1976). With a given number of animals, creating smaller paddocks means animals must rotate through the paddocks more quickly. Also, creating larger subherds demands faster rotation. Decreasing paddock size with relatively large subherds comprises a 'short-duration' grazing system with high intensities of defoliation followed by the necessary periods of regrowth. Different movement patterns result in different intensities and frequencies of defoliation, and different combinations of intensity and frequency have different consequences for plants. The effects of various combinations of grazing intensity and frequency on plants and soils can be explored through simulation modeling (e.g., Coughenour et al. 1984). If non-forage resources are poorly distributed, there may be advantages to fencing and rotating animals among paddocks to correct mismatches between herbivore density and forage production. However, if critical non-forage resources are well distributed (e.g., water, salt, topography), there is no reason to believe livestock, or herders working without fences or at smaller scales than fenced areas, will not distribute grazing in proportion to forage productivity through movement. In addition to movements at scales of paddocks, it is important to recognize that herbivores develop their own movement patterns at scales smaller than paddocks. Thus, it is misleading to conclude that planned rotation grazing systems are ineffective from experiments conducted on landscapes where there was no need for fencing and rotation in the first place.

4.3 Herbivores and pastoralists

If forage is the sole limiting factor, it might seem like a trivial exercise to calculate how much forage and thus how many animals will be lost, provided there are spatial data on forage production across the landscape. However, complications arise due to diet and habitat selection because of the differential herbivore responses to plant species and portions of the landscape. Methods have been developed to factor diet and range site selection into calculations of sustainable stocking rates (Holecheck et al. 2004). A similar approach has been used to calculate wildlife population sizes at food-limited carrying capacity (Coughenour and Singer 1996b). Herbivores and herbivory are spatially distributed based upon habitat use patterns or preferences, and forage intake is then distributed among plants based upon dietary patterns or preferences.

Another complication is temporal variability. Seasonal variations can be taken into account by calculating forage on a monthly basis or by determining forage availability in the most limiting months (Heady and Child 1994, Coughenour and Singer 1996b). Interannual variations are more difficult to factor in. Suppose, for example, that there are periodic year-long droughts. To calculate the consequences of habitat loss, it would be necessary to know how much forage is available in the habitat that is lost and in the remaining habitat in drought and non-drought years, as well as the degree of use of these habitats by the herbivores. If the lost habitat contains key resources that are used during drought years, habitat loss would lead to disproportionately large consequences.

The effects of habitat loss and fragmentation on vertebrate population dynamics have been extensively researched and modeled. This is a central topic of conservation biogeography (Lomolino et al. 2006, Groom et al. 2006) and metapopulation dynamics (Hanski 1995, Fahrig 2002, Engen et al. 2002, Fahrig and Nuttle 2005). It is beyond the aim of this chapter to review these topics. However, it is important to point out that landscape configuration and patch connectivity exert important effects on population dynamics through their effects on movements between habitat patches. There is an increased risk of local extinction in fragments that support small populations or are poorly connected to other fragments. The sizes of populations that can be supported by individual fragments depend not only on fragment size, but also habitat quality. Fragmentation may cut off movement corridors, and habitat loss may create overly narrow corridors or corridors on poor habitat or difficult topography, thus precluding dispersal and replenishment of locally declining populations.

The degree to which animal production is affected by habitat area depends upon the degree of synchrony among patches (Ash and Stafford-Smith 1996, Ash et al. 2004). If patches vary synchronously, there is little advantage for herbivores to have access to larger areas, because movement cannot exploit heterogeneity to buffer temporal variation. If patches vary asynchronously, movement among patches is advantageous, and larger areas provide more opportunities to find favorable patches at different times. Consequently, a reduction in habitat area or fragmentation, or confining livestock to smaller paddocks, would reduce animal productivity and population sizes in asynchronously varying landscape mosaics.

It is a relatively straightforward process for herbivores to be able to match foraging effort with forage availability by moving in an opportunistic manner. However, when the resource base involves more than just forage the task of choosing a movement pathway becomes more complex. An example would be a case where the animals require forage, water, and cover and the spatial distributions of the three resources vary independently. Trade-offs arise between gaining access to water vs. having cover, for example. As the problem becomes more complex, movement solutions will become more fragile and more sensitive to disruption.

Interferences with water access may affect herbivores directly and indirectly. During droughts, lack of access to water may result in dehydration and mortality. When water intake declines, forage consumption declines (Holechek et al. 2004). Increased energetic costs of traveling to alternative water points may lead to poorer body condition or weight loss, which in turn could affect mortality and recruitment rates. Increased time required for traveling also means that less time is available for foraging. Weakened animals could become increasingly susceptible to disease. These indirect effects could result in reduced or negative weight gains, lower recruitment, and elevated mortality.

4.4 Coping with disrupted movement

As a way to cope with the effects of subdivision on forage availability in climatically variable environments, schemes have been developed to allow pastoralists to reciprocally access each other's land. Such grazing cooperatives are being considered as a way to cope with subdivision of group ranches in southern Kenya (Boone 2005, BurnSilver and Mwangi 2006). In other cases, herders are no longer able to exploit natural resource variability through movement as they once did, so there is a need now to establish formal and informal mechanisms for negotiated access to grazing resources (Niamir-Fuller 1999). Commercial and other arrangements such as sub-leasing and agistment are used in subdivided areas of Australia (Ash et al. 2004, Davies and Sell 2005, McAllister et al. 2006). Internet-based agistment services are available to connect pastoralists and land owners. The reciprocation effectively simulates the opportunistic movements that have been cut off by subdivision. However, reciprocity is not a straightforward solution when subdivided patches are of differing qualities. In that case, owners of poor quality patches will more often require assistance of owners of better quality patches than the reverse. Although owners of poor quality patches may be granted access, at some point the good patches will be fully exploited. At or before that point, the owners of the better patches must deny further access. They may be in the minority, however, and thus be in a poor position to deny such access. Possibly, the majority would then demand that the better patches be converted to common resource areas.

The outcomes of such interactions would be affected by the predictability of patch quality during drought. If a patch is usually viable during drought, then that patch would be able to be self-sufficient. Herbivore populations would increase to sizes that can be supported by these patches during droughts. Predictably good patches would be more likely to become key components of defended territories. In wildlife systems, it would be advantageous for herbivores, males with harems, bands, or other social groups, to defend such territories (Crook et al. 1976). Outcomes would also depend upon the relative numbers of herbivores having access versus not having access. In pastoral systems a family or a group could try to defend a territory containing predictable drought reserves. If there are sufficient numbers of pastoralists who are excluded from this territory, they could ally themselves to challenge them. On the other hand, if a relatively large fraction of pastoralists have access to the reliable patches the disadvantaged pastoralists would be less likely mount a sufficient challenge.

In contrast, if there are few or no reliable drought reserves, or if viable patches shift locations due to patchy rainfall, for example, then fixed territories will be disadvantageous and pastoralists cannot be restricted to individual land parcels. The benefits of such open access will accrue equally to all of the herbivores or pastoralists when averaged over time. Pastoralists who are limited to individual fragments would be losers more often than not. But they would have bartering power in years when their fragment is viable. In exchange for granting access to others in one year, the owner is able to barter for access to others' plots in other years. Thus, reciprocity of land use should evolve because at some time or another it is advantageous to every individual in the population (Perevolotsky 1987). While this evolutionary process would adhere to the requirement for natural selection to act at the level of the individual, it also becomes advantageous for individuals to form groups with shared norms of behavior. Social processes give rise to rules which prevent cheating, thus preventing a tragedy of the commons (McCabe 1983, 2004). This situation probably exemplifies many pastoral land tenure systems. It would explain why many pastoral systems have communal lands in the sense of reciprocal use rights.

A key component of pastoralism is effective herding, which increases the match of herbivore distributions with forage distributions. In West Africa, agricultural development has led to reduced pastoral mobility and a diversion of labor away from herding (Turner et al. 2005). The reduced time available for herding and consequent reduced effectiveness of movement has, in turn, lowered the amount of forage encountered by livestock, and subsequently livestock productivity. These authors suggest that a way to cope with the labor reduction is for agro-pastoralists to pool their animals into larger herds, thus decreasing the labor needed per head of livestock.

5. CONCLUSIONS

Herbivore movement is an integral component of many arid and semi-arid ecosystems, and it is essential to their structure and functioning. Habitat loss, fragmentation, and sedentarization disrupt herbivore movements, thus threatening ecosystems. Boundaries have been drawn, land use has changed, wildlife have been confined, and pastoralists no longer move as they once did. Continued land conversions are inevitable. Economic development in arid and semi-arid environments will influence herbivore movements by providing increased infrastructure, water provisioning, and sedentarization. As spatially extensive grazing ecosystems are subdivided and pastoralists are sedentarized, there is an increasing likelihood that rangeland ecosystems will be degraded by unsustainable grazing regimes, and pastoral systems will become increasingly vulnerable. Loss of wildlife habitat and confinement of large herbivore populations to relatively small, unconnected reserves not only threatens the wildlife species, it also threatens entire ecosystems.

Past and present land-use changes have taken place based upon surprisingly little appreciation for the importance of movement for ecosystems with large herbivores. Ecosystem responses to habitat loss and fragmentation cannot be expected to be linear. In other words, a loss of 50% of the habitat does not translate into a simple 50% loss of herbivores, due to the manifold effects of movement on plants, herbivores, and ecosystem processes. Conversely, increasing the spatial extent of a habitat by removing fences or creating corridors will help in some but not all circumstances because it is not just spatial extent that matters. The heterogeneity and quality of the habitat must also be considered. It may be more effective to add a small but heterogeneous parcel than a large but homogeneous one. In order to effectively cope with the habitat changes that have already occurred, the importance of the lost or pre-fragmented habitat for plants, herbivores, biodiversity, and ecosystem functioning must be evaluated. Then, strategies and management interventions might be developed to cope with the loss of movement at all levels. Coping strategies could simulate the consequences of undisturbed movements. For example, through managed movements, find ways to increase opportunistic movements, moving animals among members of a grazing cooperative, emulating lost dispersal sinks by removing or translocating animals, or destocking and restocking to simulate migration. Supplementary forage may substitute for the loss of a critical habitat, such as a seasonal range or a key resource area. However, supplementary forage may also lead to increased sedentarization and overly dense herbivore populations. Water development may be used to better distribute grazing pressure, however overly dense water development could lead to sacrificed landscapes instead of sacrifice zones within sustainable landscapes.

The causes and consequences of changed herbivore numbers or spatial distributions involve complex interactions among plants, soils, other animal populations, and humans. Consequently, integrated approaches are required to assess the direct and indirect interactions between herbivores and their environments. The roles of herbivores as mobile agents of change in ecosystems must be explicitly included in conceptual and quantitative models of highly managed ecosystems as well as wildlife reserves. Consequences of movement disruption could be assessed and effective coping mechanisms could be designed by using spatially explicit, integrated assessments that describe and quantify the roles of herbivores and herbivore movements in ecosystems (Boone et al. 2002, Owen-Smith 2002, Weisberg et al. 2002, 2006). Spatially explicit, integrated assessments ultimately must include humans as mobile decision-makers, managers of mobile herbivores, and integral components of ecosystems (Thornton et al. 2003, 2006, Coughenour 2004, Galvin et al. 2006).

ACKNOWLEDGEMENTS

I would like to thank Kathy Galvin, Tom Hobbs, and Jim Ellis for the opportunity to write this, Iain Gordon for reading the manuscript and providing very valuable suggestions, and Jill Lackett for diligent editorial assistance.

REFERENCES

Abrams, P.A. 2000. The impact of habitat selection on the spatial heterogeneity of resources in varying environments. Ecology 81:2902-2913.

Adler, P.B. and S. Hall. 1995. The development of forage production and utilization gradients around livestock watering points. Landscape Ecology 20:319-333.

Adler, P.B., D.A. Raff, and W.K. Lauenroth. 2001. The effect of grazing on the spatial heterogeneity of vegetation. Oecologia 128:465-479.

Alimaev, I.I. 2003. Transhumant ecosystems: fluctuations in seasonal pasture productivity. Pages 31-51 *In* C. Kerven, editor. Prospects for pastoralism in Kazakstan and Turkmenistan: From state farms to private flocks. Routledge Curzon, London.

Anderson, D.P., M.G. Turner, J.D. Forester, J. Zhu, M.S. Boyce, and H. Beyer. 2005. Scale-dependent summer resource selection by reintroduced elk in Wisconsin, USA. Journal of Wildlife Management 69:298-310.

Andrew, M.H. 1988. Grazing impact in relation to livestock watering points. Trends in Ecology and Evolution 3:336-339.

Archibald, S. and W.J. Bond. 2004. Grazer movements: spatial and temporal responses to burning in a tall-grass African savanna. International Journal of Wildland Fire 13:377-385.

Archibald, S., W.J. Bond, W.D. Stock, and D.H.K. Fairbanks. 2005. Shaping the landscape: fire-grazer interactions in an African savanna. Ecological Applications 15:96-109.

Ash, A.J. and D.M. Stafford-Smith. 1996. Evaluating stocking impacts in rangelands: animals don't practice what we preach. Rangelands Journal 18:216-243.

Ash, A., J. Gross, and M. Stafford-Smith. 2004. Scale, heterogeneity and secondary production in tropical rangelands. African Journal of Range and Forage Science 21:137-145.

Augustine, D.J. 2003. Long-term, livestock-mediated redistribution of nitrogen and phosphorus in an East African savanna. Journal of Applied Ecology 40:137-149.

Augustine, D.J. 2004. Influence of cattle management on habitat selection by impala on central Kenyan rangeland. Journal of Wildlife Management 68:916-923.

Bailey, D.W., J.E. Gross, E.A. Laca, L.R. Rittenhouse, M.B. Coughenour, D.M. Swift, and P.L. Sims. 1996. Mechanisms that result in large herbivore grazing distribution patterns. Journal of Range Management 49:386-400.

Baker, L.E. and M. T. Hoffman. 2006. Managing variability: herding strategies in communal rangelands of semiarid Namaqualand, South Africa. Human Ecology 34:765-784.

Beck, L. 1991. Nomad: A year in the life of a Qashqa'i tribesman in Iran. University of California Press, Berkeley.

Beecham, J.A. and K.D. Farnsworth. 1998. Animal foraging from an individual perspective: an object orientated model. Ecological Modelling 113:141-156.

Behnke, R. and I. Scoones. 1993. Rethinking range ecology: implications for rangeland management in Africa. Pages 1-20 *In* R.H. Behnke, I. Scoones, C. Kerven, editors. Range ecology at disequilibrium. Overseas Development Institute, London.

Bell, H. 1973. Rangeland management for livestock production. University of Oklahoma Press, Norman.

Bell, R.H.V. 1971. A grazing ecosystem in the Serengeti. Scientific American 225:86-93.

Berger, J. 2004. The last mile: how to sustain long-distance migration in mammals. Conservation Biology 18:320-331.

Berry, K.H. 1986. Introduction: development, testing, and application of wildlife-habitat models. Pages 3-4 *In* Verner, J., M.L. Morrison, C.J. Ralph, editors. Wildlife 2000: Modeling habitat relationships of terrestrial vertebrates. University of Wisconsin Press, Madison.

Bertram, B.C.R. 1978. Living in groups: predators and prey. Pages 64-96 *In* J.R. Krebs, N.B. Davies, editors. Behavioral ecology. Blackwell Scientific Publications, Oxford.

Bertram, B.C.R. 1979. Serengeti predators and their social systems. Pages 221-248 *In* A.R.E. Sinclair, M. Norton-Griffiths, editors. Serengeti: Dynamics of an ecosystem. University of Chicago Press, Chicago.

Blackmore, A.C., M.T. Mentis, and R.J. Scholes. 1990. The origin and extent of nutrient-enriched patches within a nutrient-poor savanna in South Africa. Journal of Biogeography 17:463-470.

Boone, R.B. 2005. Quantifying changes in vegetation in shrinking grazing areas in Africa. Conservation and Society 3:150-173.

Boone, R.B., M.B. Coughenour, K.A.Galvin, and J.E. Ellis. 2002. Addressing management questions for Ngorongoro Conservation Area, Tanzania, Using the Savanna Modeling System. African Journal of Ecology 40:138-158.

Boone, R.B., S.J. Thirgood, and J.G.C. Hopcraft. 2006. Serengeti wildebeest migratory patterns modeled from rainfall and new vegetation growth. Ecology 87:1987-1994.

Boyce, M.S. 1998. Ecological-process management and ungulates: Yellowstone's conservation paradigm. Wildlife Society Bulletin 26:391-398.

Boyce, M.S., P.R. Vernier, S.E. Nielsen, and F.K.A. Schmiegelow. 2002. Evaluating resource selection functions. Ecological Modelling 157:281-300.

Brennan, L.A., W.M. Block, and R.J. Gutierrez. 1986. The use of multivariate statistics for developing habitat suitability index models. Pages 177-182 *In* Verner, J., M.L. Morrison,

C.J. Ralph, editors. Wildlife 2000: Modeling habitat relationships of terrestrial vertebrates. University of Wisconsin Press, Madison.

Brits, J., M.W. van Rooyen, and N. Van Rooyen. 2002. Ecological impact of large herbivores on the woody vegetation at selected watering points on the eastern basaltic soils in the Kruger National Park. African Journal of Ecology 40:53-60.

Brown, J.H., D.W. Hehlman, and G.C. Stevens. 1995. Spatial variation in abundance. Ecology 76:2028-2043.

Brown, L. 2006. Livestock watering factsheet. British Columbia Ministry of Agriculture and Lands, Resource Management Branch, Abbotsford, British Columbia.

Bruggeman, J.E., R.A. Garrott, D.D. Bjornlie, P.J. White, F.G.R. Watson, and J. Borkowski. 2006. Temporal variability in winter travel patterns of Yellowstone bison: the effects of road grooming. Ecological Applications 16:1539-1554.

BurnSilver, S. and E. Mwangi. 2006. Beyond group ranch subdivision: collective action for livestock mobility, ecological viability and livelihoods. Paper presented at: Pastoralism and Poverty Reduction in East Africa: A Policy Research Conference. June 27-28, Nairobi, Kenya. Unpublished MS. Natural Resource Ecology Laboratory, Colorado State University.

Buskirk, S.W. and J.J. Millspaugh. 2003. Defining availability and selecting currencies of use: key steps in modeling resource selection. Pages 1-11 *In* S. Hurzurbazar, editor. Resource Selection Methods and Applications. Proceedings of the First International Conference on Resource Selection. Omnipress, Madison, Wisconsin.

Cain, J.W. III., P.R. Krausman, S.S. Rosenstock, and J.C. Turner. 2006. Mechanisms of thermoregulation and water balance in desert ungulates. Wildlife Society Bulletin 34:570-581.

Charnov, E.L. 1976. Optimal foraging, the marginal value theorem. Theoretical Population Biology 9:129-136.

Christensen, N. 1997. Managing for heterogeneity and complexity on dynamic landscapes. Pages 167-186 *In* S.T.A. Pickett, R.S. Ostfeld, M. Shachak, G.E. Liken, editors. The ecological basis of conservation: Heterogeneity, ecosystems and biodiversity. Chapman and Hall, New York.

Cincotta, R.P., Z. Yanquing, and Z. Zingmin. 1992. Transhumant alpine pastoralism in northeastern Qinghai Province. Nomadic Peoples 30:3-25.

Collins, S.L., A.K. Knapp, J.M. Briggs, J.M. Blair and E.M. Steinauer. 1998. Modulation of diversity by grazing and mowing in native tallgrass prairie. Science 280:745-747.

Cooper, S. and N. Owen-Smith. 1986. Effects of plant spinescence on large mammalian herbivores. Oecologia 68:446-455.

Cooper, S., N. Owen-Smith, and J.P. Bryant. 1988. Foliage acceptability to browsing ruminants in relation to seasonal changes in the leaf chemistry of woody plants in a South African savanna. Oecologia 75:336-342.

Coppock, D.L., J.E Ellis, and D.M. Swift. 1986. Livestock feeding ecology and resource utilization in a nomadic pastoral ecosystem. Journal of Applied Ecology 23:373-383.

Coppolillo, P. 2000. The landscape ecology of pastoral herding: spatial analysis of land use and livestock production in East Africa. Human Ecology 28:527-560.

Coughenour, M. B. 1984. A mechanistic simulation analysis of water use, leaf angles, and grazing in East African graminoids. Ecological Modelling 26:203-220.

Coughenour, M. B. 1989. Ecosystem processes integrated by nomadic pastoralists. Fourth Annual Symposium, International Association of Landscape Ecologists US Chapter. Ft. Collins, Colorado. March 1989.

Coughenour, M. B. 1991. Spatial components of plant-herbivore interactions in pastoral, ranching, and native ungulate ecosystems. Journal of Range Management 44:530-542.

Coughenour, M.B. 1992. Spatial modeling and landscape characterization of an African pastoral ecosystem: a prototype model and its potential use for monitoring drought. Pages 787-810 *In* D.H. McKenzie, D.E. Hyatt, V.J. McDonald, editors. Ecological Indicators Vol. I. Elsevier Applied Science, London and New York.

Coughenour, M.B. 1993. The SAVANNA Landscape Model - Documentation and Users Guide. Natural Resource Ecology Laboratory, Colorado State University, Ft. Collins, CO.

Coughenour, M.B. 2002. Elk in the Rocky Mountain National Park ecosystem - A model-based assessment. Final Report to USGS Biological Resources Division, Ft. Collins, Colorado and U.S. National Park Service, Rocky Mountain. National Park. 125 pp. 116 Figs. (Externally reviewed).

Coughenour, M. 2004. The Ellis paradigm - humans, herbivores and rangeland systems. African Journal of Range and Forage Science 21:123-132.

Coughenour, M.B. 2005. Bison and elk in Yellowstone National Park - Linking ecosystem, animal nutrition, and population processes. Final Report to U.S. Geological Survey, Biological Resources Division, Bozeman, MT. Natural Resource Ecology Laboratory, Colorado State University, Fort Collins.

Coughenour, M.B., S.J. McNaughton, and L.L. Wallace. 1984. Simulation study of Serengeti perennial graminoid responses to defoliation. Ecological Modelling 26:177-201.

Coughenour, M.B., J.E. Ellis, D.M. Swift, D.L. Coppock, K. Galvin, J.T. McCabe, and T.C. Hart. 1985. Energy extraction and use in a nomadic pastoral ecosystem. Science 230:619-624.

Coughenour, M.B., D.L. Coppock, J.E. Ellis, and M. Rowland. 1990. Herbaceous forage variability in an arid pastoral region of Kenya: Importance of topographic and rainfall gradients. Journal of Arid Environments 19:147-159.

Coughenour, M.B. and J.E. Ellis. 1993. Landscape and climatic control of woody vegetation in a dry tropical ecosystem: Turkana District, Kenya. Journal of Biogeography 20:107-122.

Coughenour, M.B. and F.J. Singer. 1991. The concept of overgrazing and it's application to Yellowstone's northern range. Pages 209-230 *In* R. Keiter and M. Boyce, editors. The Greater Yellowstone ecosystem: Redefining America's wilderness heritage. Yale University Press, New Haven, CT.

Coughenour, M.B. and F.J. Singer. 1996a. Elk population processes in Yellowstone National Park under the policy of natural regulation. Ecological Applications 6:573-593.

Coughenour, M.B. and F.J. Singer. 1996b. Yellowstone elk population responses to fire - a comparison of landscape carrying capacity and spatial-dynamic ecosystem modeling approaches. Pages 169-180 *In* J. Greenlee, editor. The ecological implications of fire in Greater Yellowstone. International Association of Wildland Fire. Fairfield, WA.

Creel, S., J. Winnie Jr., B. Maxwell, K. Hamlin, and M. Creel. 2005. Elk alter habitat selection as an antipredator response to wolves. Ecology 86:3387-3397.

Crook, J.H., J.E. Ellis, and J.D. Goss-Custard. 1976. Mammalian social systems: structure and function. Animal Behavior 24:261-274.

Cumming, D.H.M. and G.S. Cumming. 2003. Ungulate community structure and ecological processes: body size, hoof area and trampling in African savannas. Oecologia 134:560-568.

Darling, F.F. and M.T. Farvar. 1972. Ecological consequences of sedentarization of nomads. Pages 671-683 *In* M.T. Farvar and J.P. Milton, editors. The careless technology: Ecology and international development. The Natural History Press, Garden City, NY.

Davies, L. and I. Sell. 2005. Agistment guidelines. New South Wales Department of Primary Industries. http://www.agric.nsw.gov.au/reader/drought-strategies/m16.htm

De Leeuw, J., M.N. Waweru, O.O. Okello, M. Maloba, P. Nguru, M.Y. Said, H.M. Aligula, I.M.A. Heitkonig, and R.S. Reid. 2001. Distribution and diversity of wildlife in northern

Kenya in relation to livestock and permanent water points. Biological Conservation 100:297-306.

Dublin, H. 1995. Vegetation dynamics in the Serengeti-Mara ecosystem: the role of elephants, fire, and other factors. Pages 71-90 *In* A.R.E. Sinclair and P. Arcese, editors. Serengeti II: Dynamics, management, and conservation of an ecosystem. University of Chicago Press, Chicago.

Du Toit, J.G. 2002. Water requirements. Pages 98-99 *In* J. du P. Botha, editor. Game ranch management, Fourth Edition. Van Schaik Publishers, Pretoria.

Du Toit, J.T. 2003. Large herbivores and savanna heterogeneity. Pages 292-309 *In* J.T. Du Toit, K.H. Rogers, and H. Biggs, editors. The Kruger experience: Ecology and management of savanna heterogeneity. Island Press, Washington.

Du Toit, J.T. and D.H.M. Cumming. 1999. Functional significance of ungulate diversity in African savannas and the ecological implications of the spread of pastoralism. Biodiversity and Conservation 8:1643-1661.

Ellis, J.E., D.S. Schimel, M.B. Coughenour, T.C. Hart, J.G. Wyant, and S. Lewis. 1985. Enhancement of tree establishment by pastoral nomads in an arid tropical ecosystem. Unpublished manuscript. Natural Resource Ecology Laboratory, Colorado State University.

Ellis, J.E. and D.M. Swift. 1988. Stability of African pastoral ecosystems: alternative paradigms and implications for development. Journal of Range Management 41:450-459.

El-Shorbagy, M.A. 1998. Impact of development programmes on deterioration of rangeland resources in some African and Middle Eastern countries. Pages 45-70 *In* V.R. Squires, A. E. Sidahmed, editors. Drylands: Sustainable use of rangelands into the twenty-first century. International Fund for Agricultural Development, Rome.

Engen, S., R. Lande, and B.E. Saether. 2002. Migration and spatiotemporal variation in population dynamics in a heterogeneous environment. Ecology 83:570-579.

Fahrig, L. 2002. The effect of habitat fragmentation on the extinction threshold: a synthesis. Ecological Applications 12:346-353.

Fahrig, L. and W. K. Nuttle. 2005. Population ecology in spatially heterogeneous environments. Pages 95-118 *In* G.M. Lovett, C.G. Jones, M.G. Turner, K.C. Weathers, editors. Ecosystem function in heterogeneous landscapes. Springer Science, New York.

Fancy, S.G., L.F. Pank, K.R. Whitten, and W.L. Regelin. 1989. Seasonal movements of caribou in arctic Alaska as determined by satellite. Canadian Journal of Zoology 67:644-650.

Farnsworth, K.D. and A.W. Illius. 1998. Optimal diet choice for large herbivores: an extended contingency model. Functional Ecology 12:74-81.

Farnsworth, K.D. and J.A. Beecham. 1999. How do grazers achieve their distribution? A continuum of models from random diffusion to the ideal free distribution using biased random walks. American Naturalist 153:509-526.

Fernandez-Gimenez, M. and B. Allen-Diaz. 1999. Testing a non-equilibrium model of rangeland vegetation dynamics in Mongolia. Journal of Applied Ecology 36:871-885

Finan, T.J. and E.R. Al Haratani. 1998. Modern Bedouins: the transformation of traditional nomad society in the Al-Tayasiyah region of Saudi Arabia. Pages 345-368 *In* V.R. Squires, A. E. Sidahmed, editors. Drylands: Sustainable use of rangelands into the twenty-first century. International Fund for Agricultural Development, Rome.

Fish and Wildlife Service. 1980. Habitat evaluation procedures. Ecological services manual 102. U.S. Department of Interior, Fish and Wildlife Service, Division of Ecological Services. Government Printing Office, Washington, D.C.

Fish and Wildlife Service. 1981. Standards for the development of suitability index models. Ecological services manual 103. U.S. Department of Interior, Fish and Wildlife Service, Division of Ecological Services. Government Printing Office, Washington, D.C.

Foran, B.D. 1980. Change in range condition with distance from watering points and its implications for field survey. Australian Rangeland Journal 2:59-66.

Fortin, D., J.M. Fryxell, L.O. Brodovich, and D. Frandsen. 2003. Foraging ecology of bison at the landscape and plant community levels: the applicability of energy maximization principles. Oecologia 134:219-227.

Fortin, D., H.L. Beyer, M.S. Boyce, D.W. Smith, T. Duchesne, and J.S. Mao. 2005. Wolves influence elk movements: behavior shapes a trophic cascade in Yellowstone National Park. Ecology 86:1320-1330.

Frair, J.L., E.H. Merrill, D.R. Visscher, D. Fortin, H.L. Beyer, and J.M. Morales. 2005. Scales of movement by elk (Cervus elaphus) in response to heterogeneity in forage resource and predation risk. Landscape Ecology 20:273-287.

Frank, D.A. 2006. Large herbivores in heterogeneous grassland ecosystems. Pages 326-347 *In* K. Danell, R. Bergstrom, P. Duncan, and J. Pastor, editors. Large herbivore ecology, ecosystem dynamics and conservation. Cambridge University Press, Cambridge.

Frank, D.A., and S.J. McNaughton. 1992. The ecology of plants, large mammalian herbivores, and drought in Yellowstone National Park. Ecology 73:2043-2058.

Frank, D.A., S.J. McNaughton, and B.F. Tracy. 1998. The ecology of the earth's grazing ecosystems. Bioscience 48:513-521.

Fratkin, E. and E.A. Roth, editors. 2005. As pastoralists settle: Social, health and economic consequences of pastoral sedentarization in Marsabit District, Kenya. Kluwer Academic and Plenum Publishers, New York.

Fryxell, J. 1995. Aggregation and migration by grazing ungulates in relation to resources and predators. Pages 257-273 *In* A.R.E. Sinclair and P. Arcese, editors. Serengeti II: Dynamics, management, and conservation of an ecosystem. University of Chicago Press, Chicago.

Fryxell, J.M. and A.R.E. Sinclair. 1988. Causes and consequences of migration by large herbivores. Trends in Ecology and Evolution 3:237-241.

Fryxell, J.M., J. Geever, and A.R.E. Sinclair. 1988. Why are migratory ungulates so abundant? American Naturalist 131:781-198.

Fryxell, J.M., J.F. Wilmshurst, and A.R.E. Sinclair. 2004. Predictive models of movement by Serengeti grazers. Ecology 85:2429-2435.

Fryxell, J.M., J.F. Wilmshurst, A.R.E. Sinclair, D.T. Haydon, R.D. Holt, and P.A. Abrams. 2005. Landscape scale, heterogeneity, and the viability of Serengeti grazers. Ecology Letters 8:328-335.

Fuhlendorf, S.D. and D.M. Engle. 2004. Application of the fire-grazing interaction to restore a shifting mosaic on tallgrass prairie. Journal of Applied Ecology 41:604-614.

Galvin, K.A., P.K. Thornton, J.R. de Pinho, J. Sunderland, and R.B. Boone. 2006. Integrated modeling and its potential for resolving conflicts between conservation and people in the rangelands of East Africa. Human Ecology 134:155-183.

Garcia-Gonzalez, R., R. Hidalgo, and C. Montserrat. 1990. Patterns of livestock use in time and space in the summer ranges of the western Pyrenees: a case study in the Aragon Valley. Mountain Research and Development 10:241-255.

Garcia-Ruiz, J.M. and T. Lasanta-Martinez. 1990. Land-use changes in the Spanish Pyrenees. Mountain Research and Development 10:267-2798.

Gaylard, A., N. Owen-Smith, and J. Redfern. 2003. Surface water availability: implications for heterogeneity and ecosystem processes. Pages 171-188 *In* J.T. DuToit, K.H. Rogers, and H. Biggs, editors. The Kruger experience: Ecology and management of savanna heterogeneity. Island Press, Washington.

Graetz, R.D. and J.A. Ludwig. 1978. A method for the analysis of piosphere data applicable to range assessment. Australian Rangeland Journal 1:126-136.

Groom, M.J., G. K. Meefe, and C.R. Carroll, editors. 2006. Principles of conservation biology. Third Edition. Sinauer Associates, Inc. Sunderland, Massachusetts.

Gurney, W.S.C. and R.M. Nisbet. 1978. Predator-prey fluctuations in patchy environments. Journal of Animal Ecology 47:85-102.

Hamilton, W.D. 1971. Geometry for the selfish herd. Journal of Theoretical Biology 31:295-311.

Hanski, I. 1995. A practical model of metapopulation dynamics. Journal of Animal Ecology 63:151-162.

Harrington, R., N. Owen-Smith, P.C. Viljoen, H.C. Piggs, D.R. Mason, and P. Funston. 1999. Establishing the cause of the roan antelope decline in the Kruger National Park, South Africa. Biological Conservation 90:69-78.

Hart, R.H., J. Bissio, M.J. Samuel, and J.W. Waggoner Jr. 1993. Grazing systems, pasture size, and cattle grazing behavior, distribution and gains. Journal of Range Management 46:81-87.

Heady, H.E. and R.D. Child. 1994. Rangeland ecology and management. Westview Press. Boulder, Colorado.

Heitschmidt, R. and C.A. Taylor. 1991. Livestock production. Pages 161-177 *In* R.K. Heitschmidt, and J.W. Stuth, editors. Grazing management: An ecological perspective. Timber Press, Portland.

Hilborn, R. 1975. The effect of spatial heterogeneity on the persistence of predator-prey interactions. Theoretical Population Biology 8:346-355.

Hill, M.F. and H. Caswell. 1999. Habitat fragmentation and extinction thresholds on fractal landscapes. Ecology Letters 2:121-127.

Hirzel, A.H., J. Hauser, D. Chessel, and N. Perrin. 2002. Ecological niche factor analysis: how to compute habitat suitability maps without absence data? Ecology 83:2027-2035.

Hobbs, N.T. 1996. Modification of ecosystems by ungulates. Journal of Wildlife Management 60:2397-2402.

Hobbs, N.T. 2006. Large herbivores as sources of disturbance in ecosystems Pages 261-288 *In* K. Danell, R. Bergstrom, P. Duncan, and J. Pastor, editors. Large herbivore ecology, ecosystem dynamics and conservation. Cambridge University Press, Cambridge.

Hobbs, N.T. and D.M. Swift. 1985. Estimates of habitat carrying capacity incorporating explicit nutritional constraints. Journal of Wildlife Management 49:814-822.

Hobbs, N.T., J.E. Gross, L.A. Shipley, D.E. Spalinger, and B.A. Wunder. 2003. Herbivore functional response in heterogeneous environments: A contest among models. Ecology 84:666-681.

Holdo, R.M., R.D. Holt, M.B. Coughenour, and M.E. Ritchie. 2006. Plant productivity and soil nitrogen as a function of grazing, migration, and fire in an African savanna. Journal of Ecology 95:115-128.

Holecheck, J.L., R.D. Pieper, and C.H. Herbel. 2004. Range management: Principles and practices. Pearson Education Inc. Upper Saddle River, New Jersey.

Holling, C.S. and L. H. Gunderson. 2002. Resilience and adaptive cycles. Pages 25-62 *In* L.H. Gunderson and C.S. Holling, editors. Panarchy: Understanding transformations in human and natural systems. Island Press, Washington D.C.

Hopcraft, J.G.C., A.R.E. Sinclair, and C. Packer. 2005. Planning for success: Serengeti lions seek prey accessibility rather than abundance. Journal of Animal Ecology 74:559-566.

Houston, D.B. 1982. The northern Yellowstone elk: ecology and management. MacMillan Press, New York.

Hudson, R.J. and W.G. Watkins. 1986. Foraging rates of wapiti on green ad cured pastures. Canadian Journal of Zoology 64:1705-1708.

Huffaker, C.B. 1958. Experimental studies on predation, dispersion factors and predatory-prey oscillations. Hilgardia 27:343-383.

Humphrey, C. and D. Sneath. 1999. The end of nomadism? Society, state and the environment in Inner Asia. Duke University Press, Durham, North Carolina.

Huston, M. 1979. A general hypothesis of species diversity. American Naturalist 113:81-101.

Huston, M.A. 1994. Biological diversity: The coexistence of species on changing landscapes. Cambridge University Press, Cambridge.

Illius, A. and T. O'Connor. 1999. The relevance of non-equilibrium concepts to arid and semi-arid grazing systems. Ecological Applications 9:798-813.

Illius, A.S. and T.G. O'Connor. 2000. Resource heterogeneity and ungulate population dynamics. Oikos 89:283-294.

Inglis, G.J., H. Hurren, J. Oldman, and R. Haskew. 2006. Using habitat suitability index and particle dispersion models for early detection of marine invaders. Ecological Applications 16:1377-1390.

James, C.D., J. Landsberg, and S.R. Morton. 1998. Provision of watering points in the Australian arid zone: a review of effects on biota. Journal of Arid Environments 41: 87-121.

Jarman, P.J. and P.E. Mmari. 1971. Selection of drinking places by large mammals in the Serengeti woodlands. East African Wildlife Journal 9:158-161.

Jewell, P.A., S. Holt, and D. Hart, editors. 1981. Problems in management of locally abundant wild mammals. Academic Press, New York.

Jones CG, J.H. Lawton, and M. Shachak. 1994. Organisms as ecosystem engineers. Oikos 69: 373-386

Jordan, A.M. 1986. Trypanosomiasis control and African rural development. Longman, London.

Keating, K.A. and S. Cherry. 2004. Use and interpretation of logistic regression in habitat-selection studies. Journal of Wildlife Management 68:774-789.

Kerven, C. 2004. The influence of cold temperatures and snowstorms on rangelands and livestock in northern Asia. Pages 41-55 *In* S. Vetter, editor. Rangelands at equilibrium and non-equilibrium: recent developments in the debate around rangeland ecology and management. Program for Land and Agrarian Studies, School of Government, University of the Western Cape. Capetown, South Africa.

Kerven, C., I. I. Alimaev, R. Behnke, G. Davidson, L. Franchois, N. Malmakov, E. Mathijs, A. Smailov, S. Temirbekov, and I. Wright. 2004. Retraction and expansion of flock mobility in Central Asia: costs and consequences. African Journal of Range and Forage Science 21:159-169.

Knight, T.M., M.W. McCoy, J. M. Chase, K.A. McCoy, and R.D. Holt. 2005. Trophic cascades across ecosystems. Nature 437:880-883.

Lange, R.T. 1969. The piosphere, sheep track and dung patterns. Journal of Range Management 22:396-400.

Lehman, C. and D. Tilman. 1997. Competition in spatial habitats. Pages 185-203 *In* D. Tilman and P. Karieva, editors. Spatial ecology: The role of space in population dynamics and interspecific interactions. Princeton University Press, New Jersey.

Lehmann, A., J. McC. Overton, and J.R. Leathwick. 2002. GRASP: generalized regression analysis and spatial prediction. Ecological Modelling 157:189-207.

Leuthold, W. 1996. Recovery of woody vegetation in Tsavo National Park, Kenya, 1970-1994. African Journal of Ecology 34:101-112.

Lomolinio, M.V., B.R. Riddle, and J.H. Brown. 2006. Biogeography. Third Edition. Sinauer Associates, Inc. Sunderland, Massachusetts.

Loreau, M., S. Naeem, and P. Inchausti. 2002. Biodiversity and ecosystem functioning. Oxford University Press, Oxford.

Lovett, G.M., C.G. Jones, M.G. Turner, and K.C. Weathers. 2006. Conceptual frameworks: plan for a half-built house. Pages 463-470 *In* G.M. Lovett, C.G. Jones, M.G. Turner, and K.C. Weathers, editors. Ecosystem function in heterogeneous landscapes. Springer Science, New York.

Manly, B.F., L.L. McDonald, D.L. Thomas, T.L. McDonald, and W.P. Erickson. 2002. Resource selection by animals: statistical design and analysis for field studies. Kluwer Academic, Norell, Massachusetts.

Marzluff, J.M. 1986. Assumptions and design of regression experiments: the importance of lack-of-fit testing. Pages 165-170 *In* Verner, J., M.L. Morrison, and C. J. Ralph, editors. Wildlife 2000: Modeling habitat relationships of terrestrial vertebrates. University of Wisconsin Press, Madison.

Marzluff, J.M., J. Millspaugh, P. Hurvitz, and M.A. Handcock. 2004. Relating resources to a probabilistic measure of space use: forest fragments and Steller's jays. Ecology 85:1411-1427.

May, R.M., and J.R. Beddington. 1981. Notes on some topics in theoretical ecology, in relation to the management of locally abundant populations of mammals. Pages 205-216 *In* P.A. Jewell, S. Holt, and D. Hart, editors. Problems in management of locally abundant wild mammals. Academic Press, New York.

McAllister, R.J., I.J. Gordon, M.A. Janssen, and N. Abel. 2006. Pastoralists' responses to variation of rangeland resources in time and space. Ecological Applications 16:572-583.

McCabe, J.T. 1983. Land use among the pastoral Turkana. Rural African 15/16:109-126.

McCabe, J.T. 1994. Mobility and land use among African pastoralists: old conceptual problems and new interpretations. Pages 69-89 *In* E. Fratkin, K.A. Galvin, and E.A. Roth, editors. African pastoralist systems: An integrated approach. Lyne Reinner Publishers, Boulder, London.

McCabe, J.T. 1995. Wildebeest/Maasai interaction in the Ngorongoro Conservation Area of Tanzania. Final report submitted to the National Geographic Society. Grant 4953-93.

McCabe, J.T. 2004. Cattle bring us to our enemies: Turkana ecology, politics, and raiding in a disequilibrium system. University of Michigan Press, Ann Arbor.

McCabe, J.T., R. Dyson-Hudson, P.W. Leslie, P.H. Fry, N. Dyson-Hudson, and J. Wienpahl. 1988. Movement and migration as pastoral responses to limited and unpredictable resources. Pages 727-734 *In* E.E. Whitehead, C.F. Hutchinson, B.N. Timmermann, and R.C. Varady, editors. Arid lands - Today and tomorrow: Proceedings of an International Research and Development Conference. Westview Press. Boulder, Colorado.

McNab, B.K. 1963. Bioenergetics and the determination of home range size. American Naturalist 97:133-140.

McNaughton, S.J. 1976. Serengeti migratory wildebeest: facilitation of energy flow by grazing. Science 191:92-93.

McNaughton, S.J. 1979. Grazing as an optimization process: grass-ungulate relationships in the Serengeti. American Naturalist 113:691-703.

McNaughton, S.J. 1984. Grazing lawns: animals in herds, plant form and coevolution. American Naturalist 124:863-886.

McNaughton, S.J. 1985. Ecology of a grazing ecosystem: the Serengeti. Ecological Monographs 55:259-294.

McNaughton, S.J. 1988. Mineral nutrition and spatial concentrations of African ungulates. Nature 334:343-345.

McNaughton, S.J. 1990. Mineral nutrition and seasonal movements of African migratory ungulates. Nature 345:613-615.

Meagher, M.M. 1973. The bison of Yellowstone National Park. National Park Service Scientific Monograph Series Number One. U.S. Dept. of Interior, National Park Service. U.S. Government Printing Office, Washington, D.C.

Meagher, M.M. 1989. Range expansion by bison of Yellowstone National Park. Journal of Mammalogy 70:670-675.

Meagher, M., M. Taper, and C. L. Jerde. 2003. Recent changes in population distribution: the Pelican bison and the domino effect. Pages 135-147 *In* R. J. Anderson and D Harmon, editors. Yellowstone Lake: Hotbed of chaos or reservoir of resilience? Proceedings of the 6th Biennial Scientific Conference on the Greater Yellowstone Ecosystem. A joint publication of the Yellowstone Center for Resources and the George Wright Society.

Millspaugh, J.J., R.M. Nielson, L. McDonald, J.M. Marzlff, R.A. Gitzen, C.D. Rittenhouse, M.W. Hubbard, and S.L. Sheriff. 2006. Analysis of resource selection using utilization distributions. Journal of Wildlife Management 70:384-395.

Moen, R., J. Pastor, and Y. Cohen. 1997. A spatially explicit model of moose foraging and energetics. Ecology 78:505-521.

Morales, J.M., D. Fortin, J.L. Frair, and E.H. Merill. 2005. Adaptive models for large herbivore movements in heterogeneous landscapes. Landscape Ecology 20:301-306.

Murray, M. 1995. Specific nutrient requirements and migration of wildebeest. Pages 231-256 *In* A.R.E. Sinclair and P. Arcese, editors. Serengeti II: Dynamics, management, and conservation of an ecosystem. University of Chicago Press, Chicago.

Murray, M. and A.W. Illius. 2000. Vegetation modification and resource competition in grazing ungulates. Oikos 89:501-508.

Myers, N. 1973. Tsavo National Park, Kenya and its elephants: an interim appraisal. Biological Conservation 5:123-132.

National Research Council. 2001. Ecological dynamics on Yellowstone's northern range. National Academy Press. Washington, D.C.

Niamir-Fuller, M. 1999. Conflict management and mobility among pastoralists in Karamoja, Uganda. Pages 149-183 *In* M. Niamir-Fuller, editors. Managing mobility in African rangelands: The legitimization of transhumance. Intermediate Technology Publications, London.

Niamir-Fuller, M. and M.T. Turner. 1999. A review of recent literature on pastoralism and transhumance in Africa. Pages 18-46 *In* M. Niamir-Fuller, editor. Managing mobility in African rangelands. Intermediate Technology Publications Ltd., London.

Noy-Meir, I. 1976. Rotational grazing in a continuously growing pasture: a simple model. Agricultural Systems 1:87-112.

Ormerod, W.E. 1986. A critical study of the policy of tsetse eradication. Land Use Policy 3:85-89.

Owen-Smith, N. 1981. The white rhino overpopulation problem and a proposed solution. Pages 129-150 *In* P.A. Jewell, S. Holt, and D. Hart, editors. Problems in management of locally abundant wild mammals. Academic Press, New York.

Owen-Smith, N. 1982. Factors influencing the consumption of plant products by large herbivores. Pages 359-404 *In* B.J. Huntley and B.H. Walker, editors. Ecology of tropical savannas. Springer-Verlag, Berlin.

Owen-Smith, N. 1983. Dispersal and the dynamics of large herbivore populations in enclosed areas: implications for management. Pages 127-143 *In* R.N. Owen-Smith, editor. Management of large mammals in African conservation areas. Haum Educational Publishers, Pretoria.

Owen-Smith, N. 1988. Megaherbivores: The influence of very large body size on ecology. Cambridge University Press, Cambridge.

Owen-Smith, N. 1996. Ecological guidelines for waterpoints in extensive protected areas. South African Journal of Wildlife Research 26:107-112.

Owen-Smith, R.N. 2002. Adaptive herbivore ecology: From resources to populations in variable environments. Cambridge Studies in Ecology, Cambridge University Press, Cambridge.

Pacala, S. and S. Levin. 1997. Biologically generated spatial pattern and the coexistence of competing species. Pages 204-232 *In* D. Tilman and P. Karieva, editors. Spatial ecology: The role of space in population dynamics and interspecific interactions. Princeton University Press, New Jersey.

Parker, I.C.S. 1983. The Tsavo story: an ecological case history. Pages 37-50 *In* R.N. Owen-Smith, editor. Management of large mammals in African conservation areas. Haum, Pretoria.

Pastor, J., R. Moen, and Y. Cohen. 1997. Spatial heterogeneities, carrying capacity, and feedbacks in animal-landscape interactions. Journal of Mammalogy 78:1040-1052.

Pastor, J., Y. Cohen, and R. Moen. 1999. Generation of spatial patterns in boreal forest landscapes. Ecosystems 2:439-450.

Pastor, J., Y. Cohen, and N.T. Hobbs. 2006. The roles of large herbivores in ecosystem nutrient cycles. Pages 289-325 *In* K. Danell, R. Bergstrom, P. Duncan, and J. Pastor, editors. Large herbivore ecology, ecosystem dynamics and conservation. Cambridge University Press, Cambridge.

Patten, R.S. and J.E. Ellis. 1995. Patterns of species and community distributions related to environmental gradients in an arid tropical ecosystem. Vegetatio 117:69-79.

Pengelly, W.L. 1963. Thunder on the Yellowstone. Naturalist 14:18-25.

Pennycuick, L. 1975. Movements of the migratory wildebeest population in the Serengeti area between 1960 and 1973. East African Wildlife Journal 13:65-87.

Perevolotsky, A. 1987. Territoriality and resource sharing among the Bedouin of southern Sinai: a socio-ecological interpretation. Journal of Arid Environments 13:153-161.

Peterson, A.T. and D.A. Vieglas. 2001. Predicting species invasions using ecological niche modeling. BioScience 51:363-371.

Pickett, S.T.A., A.J. Kolasa, J. Armesto, and S.L. Collins. 1989. The ecological concept of disturbance and its expression at various hierarchical levels. Oikos 54:129-136.

Pickett, S.T.A., M.L. Cadenasso, and T.L. Bunting. 2003. Biotic and abiotic variability as key determinants of savanna heterogeneity at multiple spatiotemporal scales. Pages 22-40 *In* J.T. Du Toit, K.H. Rogers, and H. Biggs, editors. The Kruger experience: Ecology and management of savanna heterogeneity. Island Press, Washington.

Polis, G.A., W.B. Anderson, and R.D. Holt. 1997. Towards an integration of landscape and food web ecology: the dynamics of spatially subsidized food webs. Annual Review of Ecology and Systematics 28: 289-316.

Polis, G.A, M.E. Power, and G.R. Huxel, editors. 2004. Food webs at the landscape level. University of Chicago Press, Chicago.

Porter, W.P., J.L. Sabo, C.R. Tracy, O.J. Reichman, and N. Ramankutty. 2002. Physiology on a landscape scale: plant-animal interactions. Integrative and Comparative Biology 42:431-453.

Prins, H.H.T. and H. Olff. 1998. Species-richness of African grazer assemblages: towards a functional explanation. Pages 449-490 *In* D. Newberrry, H.H.T. Prins, and G. Brown, editors. Dynamics of tropical communities. Blackwell Scientific Publications, Oxford.

Pulliam, H.R. 1988. Sources, sinks, and population regulation. American Naturalist 1132:652-661.

Pyke, G.H. 1984. Optimal foraging theory: a critical review. Annual Review of Ecology and Systematics 15:523-575.

Quirk, M. 2002. Managing grazing. Pages 131-146 *In* A.C. Grice and K.C. Hodgkinson, editors. Global rangelands: Progress and prospects. CABI Publishing Oxon, U.K. and New York.

Reid, R.S. and J.E. Ellis. 1995. Livestock-mediated tree regeneration: impacts of pastoralists on dry tropical woodlands. Ecological Applications 5:978-992.

Reid, R.S., C.J. Wilson, R.L. Kruska, and W. Mulatu. 1997. Impacts of tsetse control and land-use on vegetative structure and tree species composition in south-western Ethiopia. Journal of Applied Ecology 34:731-747.

Reid, R.S., P.K. Thornton, and R.L. Kruska. 2004. Loss and fragmentation of habitat for pastoral people and wildlife in East Africa: concepts and issues. African Journal of Range and Forage Science 21:171-181.

Ripple, W.J. and R. L. Beschta. 2004. Wolves and the ecology of fear: can predation risk structure ecosystems? Bioscience 54:755-766.

Ritchie, M.E. and H. Olff. 1999. Spatial scaling laws yield a synthetic theory of biodiversity. Nature 400:557-560.

Robinson, S. and K.J. Milner-Guilland. 2003. Contraction in livestock mobility resulting from state farm reorganization. Pages 128-145 *In* C. Kerven, editor. Prospects for pastoralism in Kazakstan and Turkmenistan. Routledge Curzon, London.

Rogers, K.H. 1997. Operationalizing ecology under a new paradigm: an African perspective. Pages 60-77 *In* S.T.A. Pickett, R.S. Ostfeld, M. Shachak, and G.E. Liken, editors. The ecological basis of conservation: Heterogeneity, ecosystems and biodiversity. Chapman and Hall, New York.

Rogers, K.H. 2003. Adopting a heterogeneity paradigm: implications for management of protected savannas. Pages 41-58 *In* J.T. Du Toit, K.H. Rogers, and H. Biggs, editors. The Kruger experience: Ecology and management of savanna heterogeneity. Island Press, Washington.

Rotenberry, J.T., K.L. Preston, and S.T. Knick. 2006. GIS-based niche modeling for mapping species' habitat. Ecology 87:12458-1464.

Ruess, R.W. and S.J. McNaughton. 1987. Grazing and the dynamics of nutrient and energy regulated microbial processes in the Serengeti grasslands. Oikos 49:101-110.

Ruiz, M. and J.P. Ruiz. 1986. Ecological history of transhumance in Spain. Biological Conservation 37:73-86.

Rupp, S. 2005. Ecological impacts of the Cerro Grande Fire: Predicting elk movements and distribution patterns in response to vegetative recovery through simulation modeling. PhD. Dissertation. Texas Technological University, Lubbick.

Schoenecker, K.A., F. J. Singer, L.C. Zeigenfuss, D. Binkley, and R. S.C. Meneses. 2004. Effects of elk herbivory on vegetation and nitrogen processes. Journal of Wildlife Management 68:837-849.

Schulze, E.-D. and H.A. Mooney, editors. 1993. Biodiversity and ecosystem function. Ecological Studies 99. Springer-Verlag, Berlin.

Scoones, I. 1994. New directions in pastoral development in Africa. Pages 1-36 *In* I. Scoones, editor. Living with uncertainty. Intermediate Technology Publications, London.

Scoones, I. 1995. Exploiting heterogeneity: habitat use by cattle in dryland Zimbabwe. Journal of Arid Environments 29:221-237.

Seagle, S. 2003. Can ungulates foraging in a multiple-use landscape alter forest nitrogen budgets. Oikos 103:230-234.

Seagle, S.W., S.J. McNaughton, and R.W. Ruess. 1992. Simulated effects of grazing on soil nitrogen and mineralization in contrasting Serengeti grasslands. Ecology 73:1105-1123.

Searle, K.R., N.T. Hobbs, and L.A. Shipley. 2005. Should I stay or should I go? Patch departure decisions by herbivores at multiple scales. Oikos 111:417-424.

Searle, K.R., T. Vandervelde, N.T. Hobbs, L.A. Shipley, and B. A. Wunder. 2006. Spatial context influences patch residence time in foraging hierarchies. Oecologia 148:710-719.

Senft, R. L., M. B. Coughenour, D. W. Bailey, L. R. Rittenhouse, O. E. Sala, and D. M. Swift. 1987. Large herbivore foraging and ecological hierarchies. Bioscience 37:789-799.

Sinclair, A.R.E. 1998. Natural regulation of ecosystems in protected areas as ecological baselines. Wildlife Society Bulletin 26:399-409.

Sinclair, A.R.E. 2003. Mammal population regulation, keystone processes and ecosystem dynamics. Philosophical Transactions of the Royal Society of London. B 358:1729-1740.

Sinclair, A.R.E. and J.M. Fryxell 1985. The Sahel of Africa: ecology of a disaster. Canadian Journal of Zoology 63:987-994.

Singer, F.J., D.M. Swift, M.B. Coughenour, and J. Varley. 1998. Thunder on the Yellowstone revisited: An assessment of natural regulation management of native ungulates, 1968-93. Wildlife Society Bulletin 26:375-390.

Singer, F.J. and K.A. Schoenecker. 2003. Do ungulates accelerate or decelerate nitrogen cycling? Forest Ecology and Management 181:189-204.

Siniff, D.B. and C.R. Jessen. 1969. A simulation model of animal movement patterns. Advanced Ecology Research 6:185-219.

Sneath, D. 1998. State policy and pasture degradation in Inner Asia. Science 281:1147-1148.

Spalinger, D.E. and N.T. Hobbs. 1992. Mechanisms of foraging in mammalian herbivores: new models of functional response. American Naturalist 140:325-348.

Stockwell, D.R.B. and A.T. Peters. 1999. The GARP modeling system: problems and solutions to automated spatial prediction. International Journal of Geographical Information Science 13: 143-158.

Suominen, O. and K. Danell. 2006. Effects of large herbivores on other fauna. Pages 383-412 *In* K. Danell, R. Bergstrom, P. Duncan, and J. Pastor, editors. Large herbivore ecology, ecosystem dynamics and conservation. Cambridge University Press, Cambridge.

Swift, D.M., M.B. Coughenour, and M. Atsedu. 1996. Arid and semiarid ecosystems. Pages 243-272 *In* T.R. McClanahan and T.P. Young, editors. East African ecosystems and their conservation. Oxford University Press, New York.

Taper, M.L., M. Meagher, and C.L. Jerde. 2000. The phenology of space: spatial aspects of bison density dependence in Yellowstone National Park. Final Report, U.S. Geological Service. Biological Resources Division. Bozeman, Montana State University.

Taylor, L.R. and R.A.J. Taylor. 1977. Aggregation, migration and population mechanics. Nature 265:415-420.

Thornton, P.K., K.A. Galvin, and R.B. Boone. 2003. An agro-pastoral household model for the rangelands of East Africa. Agricultural Systems 76:601-622.

Thornton, P.K., S.B. BurnSilver, R.B. Boone, and K.A. Galvin. 2006. Modeling the impacts of group ranch subdivision on agro-pastoral households in Kajiado, Kenya. Agricultural Systems 87:331-356.

Thrash, I. 1993. Impact of large herbivores at artificial watering compared to that at natural watering points in Kruger National Park, South Africa. Journal of Arid Environments 38:315-324.

Thrash, I. 2000. Determinants of the extent of indigenous large herbivore impact on herbaceous vegetation at watering points in the north-eastern lowveld, South Africa. Journal of Arid Environments 44:61-72.

Tilman, D. 1994. Competition and biodiversity in spatially structured habitats. Ecology 75:2-16.

Turner, M. 1993. Overstocking the range: a critical analysis of the environmental science of Sahelian pastoralism. Economic Geography 69:402-421.

Turner, M.D. 1998. Long term effects of daily grazing orbits on nutrient availability in Sahelian West Africa: 1. Gradients in the chemical composition of rangeland soils and vegetation. Journal of Biogeography 25:669-682.

Turner, M.D., P. Hiernaux, and E. Schlecht. 2005. The distribution of grazing pressure in relation to vegetation resources in semi-arid West Africa: the role of herding. Ecosystems 8:668-681.

Turner, M.G., Y. Wu, W.H. Romme, and L.L. Wallace. 1993. A landscape simulation model of winter foraging by large ungulates. Ecological Modelling 69:163-184.

Verner, J., M.L. Morrison, and C. J. Ralph, editors. 1986. Wildlife 2000: Modeling habitat relationships of terrestrial vertebrates. University of Wisconsin Press, Madison.

Vesey-Fitzgerald, D.T. 1960. Grazing succession among East African game animals. Journal of Mammalogy 41:161-172.

Walker, B.H., R.H. Emslie, N. Owen-Smith, and R.J. Scholes. 1987. To cull or not to cull: lessons from a Southern African drought. Journal of Applied Ecology 24:381-402.

Walker, B. and N. Abel. 2002. Resilient rangelands - adaptation in complex systems. Pages 293-314 *In* L.H. Gunderson and C.S. Holling, editors. Panarchy: Understanding transformations in human and natural systems. Island Press, Washington D.C.

Ward, D. and D. Salz. 1994. Foraging at different spatial scales: Dorcas gazelles foraging for lilies in the Negev Desert. Ecology 75:48-58.

Weisberg, P., N. T. Hobbs, J. Ellis, and M. Coughenour. 2002. An ecosystem approach to population management of ungulates. Journal of Environmental Management 65:181-197.

Weisberg, P.J., M.B. Coughenour, and H. Bugmann. 2006. Modelling of large herbivore-vegetation interactions in a landscape context. Pages 348-383 *In* K. Danell, R. Bergstrom, P. Duncan, and J. Pastor, editors. Large herbivore ecology, ecosystem dynamics and conservation. Cambridge University Press, Cambridge.

Western, D. 1975. Water availability and its influence on the structure and dynamics of a savanna large mammal community. East African Wildlife Journal 13:265-286.

Western, D. and D. Maitumo. 2004. Woodland loss and restoration in a savanna park: a 20-year experiment. African Journal of Ecology 42:111-121.

Whyte, I.J., R.J. van Aarde, and S.L. Pimm. 2003. Kruger's elephant population: its size and consequences for ecosystem heterogeneity. Pages 332-348 *In* J.T. DuToit, K.H. Rogers, and H. Biggs, editors. The Kruger experience: Ecology and management of savanna heterogeneity. Island Press, Washington D.C.

Widstrand, C.G. 1975. The rationale of nomad economy. Ambio 4:147-153.

Wiegand, R., E. Revilla, and K. Moloney. 2005. Effects of habitat loss and fragmentation on population dynamics. Conservation Biology 19:108-121.

Wilmshurst, J.F., J.M. Fryxell, and P.E. Colucci. 1999a. What constrains daily intake in Thomson's gazelles? Ecology 80:2338-2347.

Wilmshurst, J.F., J.F. Fryxell, B.P. Farm, A.R.E. Sinclair, and C.P. Henschel. 1999b. Spatial distribution of Serengeti wildebeest in relation to resources. Canadian Journal of Zoology 77:1223-1232.

Wilmshurst, J.F., J.F. Fryxell, and C.M. Bergman. 2000. The allometry of patch selection in ruminants. Proceedings of the Royal Society of London, Series B 267:345-349.

Wilsey, B.J. 1996. Variation in use of green flushes following burns among African ungulate species: the importance of body size. African Journal of Ecology 34:32-38.

Wolanski, E., E. Gereta, M. Borner, and S. Mduma. 1999. Water, migration and the Serengeti ecosystem. American Scientist 87:526-533.

Wolanski, E. and I. Gereta. 2001. Water quantity and quality as the factors driving the Serengeti ecosystem, Tanzania. Hydrobiologia 458:169-180.

Wu, J. and O.L. Loucks. 1995. From balance of nature to hierarchical patch dynamics: a paradigm shift in ecology. Quarterly Review of Biology 70:439-466.

Yachi, S. and M. Loreau. 1999. Biodiversity and ecosystem productivity in a fluctuating environment: the insurance hypothesis. Proceedings of the National Academy of Sciences of the USA 96:1463-1468.

Young, T.P., N. Partridge, and A. Macrae. 1995. Long-term glades in acacia bushland and their edge effects in Laikipia, Kenya. Ecological Applications 5:97-108.

Zhaoli, Y., W. Ning, Y. Dorji, and R. Jia. 2005. A review of rangeland privatization and its implications in the Tibetan Plateau, China. Nomadic Peoples 9:31-51.

PART II. CASE STUDIES

Chapter 4

CHANGING PATTERNS OF LAND USE AND TENURE IN THE DALRYMPLE SHIRE, AUSTRALIA

Chris J. Stokes[1], Ryan R. J. McAllister[1], Andrew J. Ash[2], and John E. Gross[3]
[1]*CSIRO Sustainable Ecosystems, Davies Lab, PMB PO Aitkenvale, Q 4814, Australia;* [2]*CSIRO Sustainable Ecosystems, 306 Carmody Rd, St Lucia, Q 4067, Australia;* [3]*National Park Service, 1201 Oakridge Drive, Suite 150, Fort Collins, CO 80525-5589, USA*

1. INTRODUCTION

1.1 Context of Australian agricultural development

Australia is the world's flattest continent, a testament to its ancient, well-weathered geological landforms, and consequently the associated soils are generally of low fertility (Flannery 1994). It is also the world's driest inhabited continent, a situation that is exacerbated by erratic rainfall and high evaporation. In comparison with other countries, agriculture in Australia is characterized by: dependence on low productivity environments that are prone to drought and degradation; the large scale of agricultural activities; concentration on a limited range of products; heavy dependence on overseas markets; and a relatively high standard of living in the agricultural community (Laut 1988). Almost three-quarters of the country is classed as rangelands, mainly arid and semi-arid lands that are not suitable for intensive agriculture. European settlement and agricultural development of the continent, and rangelands in particular, has been marked by bitter experiences of coming to terms with the climatic and edaphic constraints of this environment.

Fragmentation of Australian rangelands has been a relatively recent phenomenon. Although Aboriginal land-use practices shaped Australian

K. A. Galvin et al. (eds.), Fragmentation in Semi-Arid and Arid Landscapes: Consequences for Human and Natural Systems, 93–112.

landscapes for at least 40,000 years, it has been more recent European settlement that has determined the current patterns of fragmented land use. The "First Fleet" arrived in southeast Australia in 1788, bringing not only domestic livestock, but also British land management practices and institutional ideology, and a desire to legitimize British stewardship over a vast land mass, sparsely populated by European standards (Day 2001).

Intensive agriculture dominated early settlement where it was possible, with small, fertile parcels of land being allocated to settlers. When British demand for wool soared in the early 1800s, Australian Merino sheep were well placed to supply quality wool, and pastoralism (ranching) became very profitable. As explorers mapped the interior of the country, pastoralists followed behind, rapidly filling the 'empty spaces' beyond the approved settlement boundaries, displacing existing Aboriginal land-use systems. In 1831, the British government intervened to bring order to land settlement, encouraging close, European-style settlement supported by land 'improvements' to increase productivity. This became a recurring theme in early Australian land policy, and one that gave rise to progressive subdivision of initially vast agricultural properties.

1.2 Geographic patterns of rangeland development

The development of Australian rangelands has not been spatially uniform. A number of factors have influenced the extent to which rangelands have become fragmented in different parts of the continent (McAllister et al. 2006). Demand for land has generally been greatest around population centers where there have been more people, greater access to services, and avenues of arrival for new immigrants. Conversely, settlement has tended to be concentrated on the most productive lands, and these have become highly fragmented by the appropriation of the most productive land for intensive agriculture and the greater capacity of small land units to sustain viable enterprises. The sheep industry has historically experienced several 'boom' periods, accompanied by over-optimistic expectations and property subdivision, whereas periods of strong profitability in the beef industry are a more recent phenomenon. Accordingly, rangelands that have been capable of supporting sheep have been more prone to fragmentation than those that are only suitable for cattle. The date of settlement of rangelands has also been important because of the length of time it has provided for development and fragmentation (and consolidation) to occur. In addition, rangelands that have had a longer history of settlement have generally been exposed to more intense policy pressure for closer settlement, more periods of demand for land from immigrants or soldiers returning from the two World Wars, and more boom periods of exaggerated expectations for pastoralism.

Australia's population is highly urbanized, with 65% living in the capital cities alone, and concentrated in the southeast of the country (ABS 2004). Rangelands in southeastern Australia tend to be highly fragmented because of their proximity to population centers, their long history of development, their greater productivity, and their suitability for sheep production. In contrast, rangelands in the tropical north of Australia are sparsely populated, less suitable for sheep, less productive, have been settled more recently, and are generally less fragmented.

1.3 Types of fragmentation

In Australian rangelands, it is useful to distinguish three types of fragmentation: land tenure units/properties (administration), enterprises (land management), and paddocks (very large, fenced subdivisions within properties determining the scale at which livestock use landscapes). The dominant driver of landscape fragmentation has been government policy, which has dictated the size and arrangement of land tenure units (section 3.2). But superimposed on this is a set of socio-economic factors that affects patterns of land management. Many pastoral enterprises (ranches) consist of more than one property, and there may be several types of business interactions between enterprises. Enterprise structures and interactions between pastoralists that result in joint management of multiple properties increase landscape connectivity and the scale of land use, so that patterns of land management are less fragmented than those of land tenure (see section 3.5 for examples). Such enterprise arrangements are often a response to the constraints of over-fragmentation of land tenure units or an intentional initiative to spread risk associated with climate variability. Conversely, development of infrastructure, particularly fencing and water points, leads to internal fragmentation of enterprises, reducing the size of paddocks and the scale of livestock-landscape interactions (section 3.4). This internal development of enterprises is another response to fragmentation of land tenure units, and is aimed at increasing the productivity of small properties to maintain enterprise viability. Changes in land tenure, land use, and fragmentation are all inextricably linked and have occurred together with each affecting and being affected by the other (section 3).

An initial operating hypothesis for the set of case study sites in this book is that "land fragmentation is taking place at our industrializing study sites (in Africa and Asia) and that this is a dual process of land tenure and land-use change. At our post-industrial sites (in Australia and North America), on the other hand, land consolidation is taking place, and these changes in land use are not accompanied by any major changes in land tenure". We address this hypothesis for the Dalrymple Shire in north-east Queensland (sections 2 and 3) by discussing historical patterns of land use and tenure, and the forces that have driven these to change since European settlement. Two contrasting fragmentation scenarios in other regions of Australian rangelands are then provided for comparison (section 4).

2. SITE CHARACTERIZATION FOR DALRYMPLE SHIRE

2.1 Biophysical site description

Dalrymple Shire covers 6.7 million ha in the northern half of the upper Burdekin River catchment in north-east Queensland. The shire is bounded by the Great Dividing Range to the west and coastal ranges to the east. There is a rainfall gradient, ranging from 500 mm in the southwest to 650 mm in the northeast rangelands (and 2000 mm in the bordering northeast mountain range). Precipitation is highly seasonal with 80% of rain falling in the summer months between December and April. Interannual variation in rainfall is high (CV 30-48%), and El Niño events are associated with drier than average conditions. The shire contains a variety of soil types, from relatively nutrient-rich, self-mulching clays of volcanic origin to nutrient-poor sandy duplexes. These soil types support vegetation communities that differ in species composition, seasonality, and forage production (Rogers et al. 1998). Open eucalypt woodlands dominate the region (Isbell and Murtha 1972). The most extensive grassland community is associated with black speargrass (*Heteropogon contortus*), but these areas were probably dominated by kangaroo grass (*Themeda triandra* var *australis*) prior to European settlement. Over the past 30 years, many of these areas have been invaded by Indian couch (*Bothriochloa pertusa*), an exotic, perennial, stoloniferous grass (Ash et al. 2002). Heterogeneity of resources at the landscape scale results from the complex of soil and vegetation types and other topographic features, such as the border ranges, and the ephemeral and permanent watercourses (Roth et al. 2003).

2.2 Land tenure and use

Eighty-seven percent of Dalrymple Shire is state leasehold land, a proportion that far exceeds the statewide average of 65%. (Leasehold tenure involves leasing land from the State under a set of conditions that prescribe land use, in contrast to freehold title, where land has been purchased from the State and is privately owned.) Aside from a few large freehold pastoral stations, most freehold land is located in urban centers and small rural residential and non-commercial grazing lots surrounding the urban centers. Annual fees for Term Leases in Queensland average less than 1% of a typical extensive cattle enterprise's profits.

Land-use patterns are strongly tied to land tenure, especially because land use on leasehold properties, which dominate Dalrymple Shire, is prescribed by the terms of the lease (section 3.2). The vast majority of the shire is used for extensive cattle breeding and fattening operations within independent, family-run enterprises (Table 4-1, Bortolussi et al. 2005a). Mining (gold, base metals, and dolomite) occupies minimal land area, but is the largest economic sector in the region, generating revenue that was four times that of cattle disposals in 1992/3 (Quirk et al. 1996). Horticultural activities (oranges, grapes, and vegetables) are restricted to a few small enterprises confined mainly to alluvial soils (with revenues valued at 4% of cattle revenues). Conservation land in the shire includes 92,488 ha in national parks and 12,056 ha of environmental reserves (Rogers et al. 1998). Two properties in the northeast of the shire are used by the Australian Defence Force for training activities. Small rural residential and non-commercial grazing properties cover about 30,000 ha.

Table 4-1. Land-use and tenure patterns in the Dalrymple Shire, north-east Queensland.

Land Tenure	Land Use	Typical Areas ('000 ha)	Total Area ('000 ha)	Total Area (% of total)
Leasehold	Extensive cattle grazing	10 to 50	5,804	87.0
	Mining		0.5	
	Other leasehold		32	0.5
Freehold	Extensive cattle grazing	10 to 50	262	4.9
	Sub-commercial grazing	<10	30	0.5
	Other freehold		39	0.6
Other	Military field training		196	2.9
	Conservation		105	1.6
	Water reserve		43	0.6
Total			6,671	100.0

Hinton (1993) found the 271 commercial grazing properties in the shire to be amalgamated into 196 pastoral business entities (1.4 properties per enterprise). More recent data suggest that further amalgamation (2.2 properties per enterprise) has occurred in the past decade (Greiner et al. 2003). Properties range in size from 10,000 to 50,000 ha and carry 2,000 to 5,000 head of cattle (Rogers et al. 1998). The average paddock (fenced grazing area) size is about 3,000 ha, with paddocks tending to be smaller on more productive land. The overall stocking rate for the Burdekin River catchment is about 0.11 cattle/ha (Greiner et al. 2003).

3. DRIVERS OF LAND-USE CHANGE IN DALRYMPLE SHIRE

3.1 Early history of pastoral development

Indigenous land-use and tenure patterns predated European influences by at least 40,000 years. Indigenous Australians lived as semi-nomadic hunters and gatherers with seasonal movements between dry and wet season bases. Activity in much of the catchment was limited to the wet season by the lack of permanent water sources. Aboriginal people made extensive use of fire in their management and use of the land, and the vegetation at the time of European settlement had adapted to these fire regimes. The arrival of European pastoralists in rangelands led to conflict, sometimes violent, over land and water resources (Reynolds 1974). This was followed by a period of coexistence in what has been described as a system of "feudal exploitation" (Anthony 2004), whereby some Aborigines retained ties with ancestral lands by working on pastoral properties. However, the introduction of a minimum wage in the mid-1960s led to widespread lay offs of Aboriginal workers and their forced relocation to settlements disconnected from ancestral lands.

Since European settlement, patterns of land use and tenure in north-east Queensland have been strongly influenced by government policies and by events at local to global scales (Figure 4-1). The first European explorers reached the Dalrymple Shire region in 1845, and the New South Wales government opened the region for settlement in 1859, three weeks before Queensland became a separate colony (Finlay and Lloyd 1983). The push north was led by European pastoralists, miners, and agriculturists in response to the changing economies and geography (Holmes 1963). The New South Wales government had not favored development in northern Queensland, primarily because of concerns over the cost of administering such an isolated region (Bell 2000). Pastoral development following political separation was facilitated by the desire of the Queensland government to generate income from grazing leases, the increased political influence of pastoralists, and the accompanying establishment of new port settlements.

These political changes coincided with a period of rapid expansion and profitability in the wool industry (a 'wool boom'), providing an incentive for

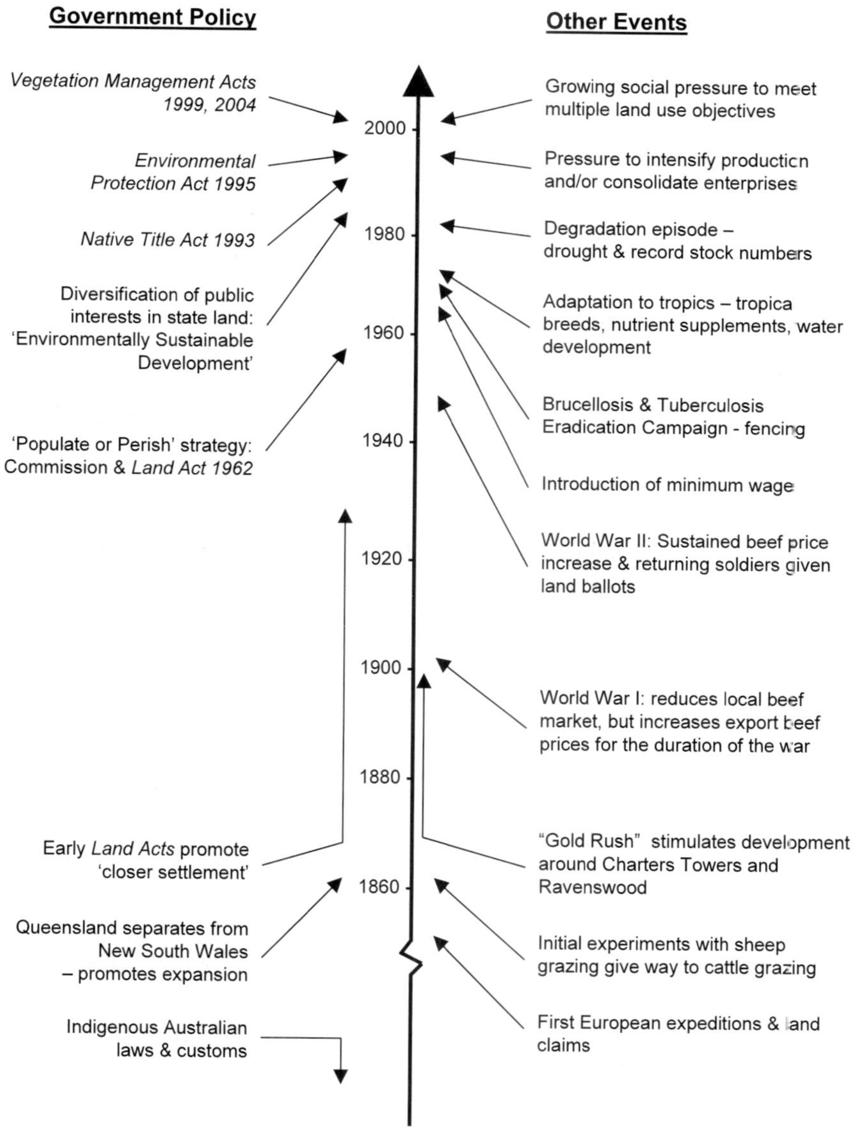

Figure 4-1. Historical events and policies that have influenced land fragmentation, land tenure, and land use in the rangelands of the Dalrymple Shire.

extensive sheep grazing. However, the region was not well-suited to sheep because of disease, parasites, nutritional deficiencies, predators, bushfires, heat stress, and native plants and their seeds that damaged wool and caused health problems for sheep. Thus the primary pastoral land use quickly shifted from sheep to cattle production (Allingham 1977). Initially, transport to markets was a constraint, and many cattle were driven (herded overland) to nearby Townsville and boiled down for tallow and hide, with the meat being discarded (Bell 2000). But, as enterprises further south in the country switched from cattle to more lucrative sheep production, it became profitable to drove (herd overland) cattle vast distances to southern markets. By around 1885 long droving routes were established (Bolton 1963).

In 1865 gold was discovered in the region and a gold rush ensued, gaining considerable momentum with the discovery of gold in Charters Towers in 1871. The mining workforce provided a market for beef in the shire. In parallel, Australia began to develop beef as an export commodity, mainly through the export of tinned beef, but also through export of live cattle from the Northern Territory to the Philippines, and refrigerated beef when the technology became available in the port city of Townsville in 1892 (Bell 2000). The outbreak of World War I led to a decline in the mining work force and a reduced local market for beef but, by this stage, transport infrastructure had been developed and the war boosted the demand for tinned beef.

3.2 Policy and legislation

The high proportion of land leased from the state in Dalrymple Shire means that land legislation, especially that pertaining to lease terms and conditions, has been a major policy instrument for influencing land use. Leasehold tenure was widely used in the settlement of Australian rangelands as an expedient and flexible means for governments to assert their authority over land allocation and use, particularly in limiting unregulated claims to land beyond existing administrative boundaries. Australian land legislation has historically been aimed at controlling the allocation of land, encouraging settlement and land 'improvement', generating revenue, preventing monopolies, promoting social equity, and developing the rights of landholders in relation to the state (Hannam 2000).

As described above, European settlement of the shire coincided with the separation of Queensland from New South Wales and development of a favorable policy environment for pastoral expansion. The *Land Act 1860* promoted settlement, providing leases of 14 years, minimal lease fees, and runs (land tenure units) of 6,600 to 26,700 ha (Mitchell 1997). There was no limit on the number of runs that could be held by a single person, other than that they had to be stocked to a quarter of their carrying capacity within nine months, a measure aimed at preventing land speculation and encouraging development. The *Pastoral Leases Act 1869* extended lease periods to provide greater security for leaseholders. A policy of closer settlement was promoted by the *Crown Lands Act 1884*, which provided for parts of leases to be resumed and reallocated, and later by the *Closer Settlement Act 1906*, which consolidated provisions of several previous Acts (ABS 1910). As a result, many small blocks of land around Charters Towers and Ravenswood were acquired by ex-miners for dairying (Mitchell 1997). Subsequent revisions to the *Land Act* at the turn of the century continued to support a policy of closer settlement. This policy favored the creation of smaller land lots for allocation to family-operators. The policy included grants of land by ballot (socially-equitable, lottery-style allocation) and soldier settler programs to satisfy demand for land by soldiers returning from war (ABS 1925).

During the 1950s, the Queensland "populate or perish" strategy developed from a review of land settlement policy (Land Settlement Advisory Commission 1959). This review recognized the influence of property size on financial security, but sought to balance this against the requirement for increased rural population density for regional community viability. The expectation at the time was that agricultural enterprises would require assistance during establishment, but that enterprises would later become self-sufficient as development and 'improvement' of the land increased its production potential. As an incentive for land development, lease terms were extended, security of tenure was emphasized, and costs of land development were recognized. This applied especially to brigalow (*Acacia harpophylla*) woodlands, which extend into the southeast of Dalrymple Shire, where land 'improvement' involved expensive tree clearing and sowing of pasture. The last major subdivision of properties in the shire occurred during the post-World War II period, when resumptions of land from large leases created several new smaller leases that were allocated by ballot. The influx of new graziers into the region brought with them fresh ideas and a willingness to learn, which facilitated the adoption of new land-use practices in the shire (section 3.3).

With changing societal values over the past two decades, and urban and indigenous communities becoming more vocal about their interests in State land, legislation has been changing to reflect this diversification of views. The *Native Title Act 1993* recognized the claims of indigenous people who can demonstrate an unbroken tie to leasehold land (e.g., encouraging agreements that allow non-exclusive access to land, overlapping with existing tenure arrangements). Recognition of land degradation problems, including soil erosion, declines in water quality, dryland salinity, and changes in pasture species composition (Tothill and Gilles 1992, De Corte et al. 1994, Mortiss 1995), led to the development of more recent environmental legislation. The Land Act 1994 included modifications to reduce the emphasis on closer settlement and promote ecologically sustainable development. This legislation was complemented by the *Environmental Protection Act 1994* and the *Vegetation Management Acts 1999, 2004* (which greatly restrict tree clearing). However, governments and land administration agencies have historically been reluctant to enforce these environmental provisions (Hannam 2000). Despite reductions in the property rights for holders of pastoral leases, and further duties of care being considered in new legislation, anecdotal evidence suggests that the restrictions in lease arrangements are not reflected in lower market values for leasehold compared to freehold properties.

Land policy, with its emphasis on promoting population growth in rural and remote areas, has been the main driver of land fragmentation. Over-optimistic expectations of the productivity of rangelands and the reliability of rainfall have contributed to under-estimating the area of land required to sustain a viable family enterprise. Although most properties remained large enough to support viable enterprises at the time of subdivision, they are nonetheless amongst the smallest in the extensive beef production region of northern Australia (Bortolussi et al. 2005a).

Fragmentation of land tenure units, by itself, does not necessarily constitute fragmentation of land use because human and animal interactions with landscapes do not always conform to these administrative boundaries. Overlaying land tenure patterns are a wider set of socio-economic factors that affect the patterns and scale of land use.

3.3 Enterprise economics

Estimates of the number of cattle required to maintain an economically viable extensive pastoral enterprise in Dalrymple Shire have been steadily increasing during the past decades. It is currently estimated that cattle enterprises need to be able to carry 1,500 to 3,000 adult equivalents (Caltabiano et al. 1999, Roth et al. 1999). This reflects the declining terms of trade for the beef industry; while output prices for pastoral production have remained unchanged, input costs have risen by 1.9%/yr relative to the consumer price index (Centre for International Economics 1997). Over the period 1995-2002, beef enterprises in the Burdekin catchment have had an average annual rate of return of 3.8% (Beare et al. 2003).

In the early pastoral industry, there were many opportunities for improvement in economic efficiency. The northward colonization of Australia from the temperate south meant that land-use practices brought by settlers had not always been appropriate for the development of the tropics. Initial experiments with sheep gave way to cattle operations, but these were based on European breeds of cattle (*Bos taurus*) and interventionist agricultural practices (e.g., tree clearing followed by introduction of exotic grasses and legumes to improve pastures) that had been developed in more intensive production systems in the south of the country. Over time, particularly since the 1970s, the industry adapted its practices to the tropical environment. These changes have included the introduction of hardier Brahman cattle breeds (*Bos indicus*), control of pests and diseases, supplementary feeding, and internal property development (section 3.4). Since the introduction of the minimum wage in the mid-1960s, enterprises have sought to minimize labor costs by dependence on family labor, contractors, and technological substitutes (e.g., helicopter mustering [herding] and better use of fences).

Declining terms of trade are negatively affecting beef enterprises and past fragmentation of land tenure units has placed some constraints on the ability of enterprises to remain viable. Because input costs have already been streamlined, the response of the pastoral industry to these challenges has been to intensify production systems (section 3.4) and/or consolidate properties (section 3.5) (Ash et al. 2003). Recent rises in land prices will likely reinforce these economic pressures and responses. However, social policy considerations for equitable land access and protection of rural industries (e.g., drought assistance measures) have probably sustained less efficient and smaller enterprises and hence worked against the economic drivers for consolidation of sub-economic properties into larger, more efficient, better-managed businesses.

3.4 Property infrastructure development

The availability of permanent water has historically been a major limitation to land use by livestock (Abbott and McAllister 2004), especially for temperate breeds of cattle that do not venture far from water points and riparian areas. Many of the early, large pastoral properties could not be subdivided until new water points could be provided. Limited coverage of water points on properties continues to contribute to uneven patterns of grazing today, and also limits options for internal fencing. Development of new water points and subdivision of paddocks remain top priorities for landholders seeking to intensify production (Bortolussi et al. 2005b). Most internal fencing in the past has been done for animal management reasons (e.g., separation by stock type and to assist with moving and mustering animals). It is only more recently that fencing for vegetation/land management has started to occur. Much of this fencing has concentrated on controlling access to heavily-grazed riparian areas in order to limit land and pasture deterioration and to control weeds. Development of new watering points provides opportunities for strategic use of fencing, such as limiting access by livestock to sensitive areas, encouraging utilization of less desirable pastures, allowing fuel build-up to promote use of fire, spelling (resting pasture), and rotational grazing systems. Although the potential benefits of intensification and increased internal fencing of properties are widely acknowledged/assumed, there could well be less-understood trade-offs that need to be considered (Hobbs et al., Chapter 2). Overcoming environmental constraints allows stocking rates to be increased, but this does not mean that these stocking rates are environmentally sustainable, as previous experiences in the shire have demonstrated (section 3.8). Also, smaller paddock sizes constrain the scale at which livestock interact with resources in the landscape, which may reduce options for animals to select nutritious diets as the quantity and quality of forage resources varies through the seasons (Hobbs et al., Chapter 2).

3.5 Consolidation

Several factors have stimulated consolidation of properties, the most important of which has been a desire to offset rising production costs through efficiencies of scale (section 3.3). Other drivers for amalgamating several properties within an enterprise include opportunities for selecting complementary types of land, specialization of operations between properties, succession planning in large families, speculative property trading, and expansion of successful businesses (Stokes et al. 2004). A recent survey showed that more than half of the landholders in the shire owned more than one property, with an average of 2.2 properties each (Greiner et al. 2003). The average size of individual properties within multi-property enterprises was triple that of enterprises based on a single property, i.e., it appears that the larger properties are the ones being amalgamated.

Official administrative boundaries of land tenure units do not give a complete picture of the scale of land use and fragmentation. First, as described above, most land tenure units are not managed as individual entities but in conjunction with other (often non-adjacent) blocks of land. In addition, there are co-operative arrangements between property owners that offset fragmentation by restoring elements of spatial connectivity (Janssen et al. 2006). These relationships include formal and informal business partnerships, co-operation within kinship networks, and a range of agistment arrangements (leased grazing access to paddocks: from long-term, planned relationships to reactive, drought-coping measures) (McAllister et al. 2005a, 2005b). The development of regional transport infrastructure has facilitated these arrangements by making it easier and less expensive to transfer stock between properties.

3.6 Changing markets

Australia currently exports almost 60% of its beef (ABS 2002) and is the world's largest beef exporter. This leaves Australian beef producers exposed to the risks of changing market access and commodity prices, and subject to fluctuations in international trade relations and exchange rates. Changing markets have had a strong influence on land-use practices. The beef industry was given a boost by the signing of the 15-year meat agreement with the United Kingdom in 1952. As that came to an end, a new market was established supplying low grade manufacturing ('hamburger') beef to the USA. By 1970, 80% of cattle from Dalrymple Shire were exported to this market, with the remainder sold to other enterprises for finishing (Quirk et al. 1996). Brucellosis and tuberculosis had to be eradicated and controlled to meet the requirements of this market, and this led to substantial, widespread fencing improvements to secure property boundaries and to facilitate animal handling. As the USA market declined (to 45% of cattle sales from the shire in 1993), new opportunities arose in live export to Asia and the Middle East (40% of cattle sales) and markets in Japan and Korea (15%) (Quirk et al. 1996). These premium markets have specific demands that have necessitated changes in management, including shifting to younger herd structures and improved management of animal nutrition.

3.7 Diversification

Beef enterprises may also have the option to improve profitability through diversification of income sources as an alternative to intensification or consolidation. Diversification could include establishment of feedlots, farm accommodation, ecotourism, and payments for stewardship of environmental values or ecosystem services (e.g., carbon trading). Pastoral lease conditions currently prescribe land use and restrict some options for diversification. Tourist operations exist on only a few pastoral properties and there has generally been little development of tourism in the shire, outside of the historic gold mining town of Charters Towers. Most enterprises depend on pastoral activities for almost all of their income (Greiner et al. 2003).

There is also increasing pressure for diversification of land use at the regional scale in response to changing societal values and expectations. Many long-term pastoral leases in the shire will be due for renewal within the next decade, and it is likely that lease conditions will change to better integrate property management planning, natural resource management, and indigenous access (QNRM 2001). If it is not possible to meet multiple-use objectives through such overlapping land-use arrangements, then there may be pressure for the government to resume leased pastoral land for allocation to these alternative uses exclusively (which would further fragment pastoral land use).

3.8 Degradation

Degradation of rangeland resources is a consequence of inappropriate land-use practices, often related to constraints of property size on the flexibility of enterprises to respond to unfavorable weather or downturns in markets. In turn, degradation affects land-use practices through diminished productivity and alteration of the land resource base.

A well-documented episode of degradation occurred in the shire in the 1980s (McKeon et al. 2004). During the 1960s and 70s, efforts to improve the profitability of grazing enterprises alleviated some of the environmental constraints on cattle production in the tropics (section 3.3). These factors contributed to a sharp rise in cattle numbers in the mid-1970s when a run of good wet seasons was followed by low cattle prices and depressed sales. Cattle numbers in the shire rose from 300,000 in 1960 to as many as 1,000,000 by 1980 (Mortiss 1995). The coincidence of record cattle numbers with a prolonged drought in the mid-1980s was accompanied by pasture deterioration (e.g., loss of perennial grasses), soil erosion, and invasion by woody weeds. When the drought broke, Indian couch (*Bothriocloa pertusa*), an exotic stoloniferous grass, became established in many of the deteriorated pastures and it has since dominated extensive areas in central Dalrymple Shire. The number of cattle in the shire declined during the 1980s but began to increase during the mid-1990s, and the number of cattle is again approaching a record level.

If the production potential of land diminishes, this will place further pressure on the viability of enterprises, reinforcing pressures to intensify production and consolidate properties. On the positive side, recent episodes of degradation could provide an opportunity for learning, and a motivation for improved land-use practices (Landsberg et al. 1998, Janssen et al. 2000, McKeon et al. 2004, Gross et al. 2006).

4. FRAGMENTATION OF AUSTRALIAN RANGELANDS

The Dalrymple Shire is representative of Australian rangelands in the sense that the shire has undergone an intermediate level of fragmentation in comparison with other areas, and that it broadly demonstrates the suite of factors driving changing patterns of land use. To place this case study in the broader context of rangeland settlement and development across Australia as a whole (section 1.2), we illustrate the range of fragmentation scenarios in the country with brief examples from opposite ends of the spectrum.

4.1 Intensive subdivision: south-west Queensland

The mulga-dominated (*Acacia aneura*) lands of south-west Queensland are located around the Paroo and Warrego Rivers. These rivers form an ephemeral chain of lakes that seldom flow into the large Darling River basin to the south. Early European settlement of Australia was focused around the productive Murray and Darling Rivers to the south, and pastoral expansion followed the Darling River northwards up the catchment along increasingly ephemeral watercourses into the arid rangelands of south-west Queensland. Access to water was a severe constraint to the early pastoral settlement of this region. Initial development was focused on small blocks of land that fronted the rivers, with more speculative purchases of larger blocks that lacked river access. During the 1870s the largest pastoral station in Australia at the time, Tinnenburra, was formed in this region from the consolidation of 864,000 ha of adjacent runs (Blake 1979b). The discovery of artesian water in 1878 and the subsequent development of artesian bores had major implications for the region. It allowed much of the land that lacked surface waters to be developed and settled, and allowed many areas that had formerly only been suitable for cattle to support sheep. Closer settlement was initiated with the passing of the Queensland *Crown Lands Act 1884*, which led to the subdivision of the original large stations. For example, 36 blocks were excised off Tinnenburra by the turn of the century (Blake 1979a). The soldier settlement schemes that followed the two World Wars promoted even greater fragmentation of properties; one 567,000 ha station was fragmented into 56 individual leases (Cameron and Blick 1991). A well-documented episode of land degradation occurred during a drought period from 1964 to 1966 (McKeon et al. 2004). Small property sizes were implicated as a major contributing factor to the stress placed on natural resources and a constraint on the flexibility of enterprises to survive unfavorable periods (Passmore and Brown 1992). Lessons learned from this episode led to the development of the "South-West Strategy" in the 1990s (Hewitt and Murray 1999). This ongoing initiative is aimed at restructuring the pastoral industry, including the assessment of 'safe' livestock carrying capacities (Johnston et al. 1996) and consolidation of properties. A 1979 survey of pastoralists in the region found that about 30% of enterprises had expanded over the period 1970-1978 and that 40% intended to expand further in the future (Holmes 1980). It has been suggested that a viable enterprise would require a minimum flock of 8,000 – 12,000 sheep (Mills 1989) but in 1988, only about 5% of enterprises were large enough to support this number of stock (Passmore and Brown 1992), suggesting that pressure for consolidation will continue.

4.2 The open range: the Victoria River District

The Victoria River District (VRD), in Australia's Northern Territory, was settled by European pastoralists in 1883, making it one of the last pastoral areas in Australia to be occupied (Lewis 2002). Cattle grazing has dominated land use since settlement (Kraatz 2000). The area is a vast 13 million ha mosaic of Mitchell grass (*Astrebla* spp.) plains, rainforest patches, spinifex (*Triodia* spp.)-covered hills, and scrub lands (Lewis 2002). The rainfall pattern is monsoonal and follows a gradient from 1,000 mm in the north to 400 mm in the south (Kraatz 2000). The first eighty years of cattle production in the VRD was characterized by 'open range' production, a low-input system where cattle were managed similarly to a harvested wild herbivore population and left free to roam over large areas with minimal handling. The first cattle were turned loose on the best grazing country while huts, yards, and horse paddocks were constructed in a few convenient locations. Cattle spread out across the countryside in the wet season but converged on the river frontages through the dry season, causing considerable environmental damage (Lewis 2002). The open-range system also meant that cattle were not selectively bred for temperament or handled regularly and therefore stock were difficult to handle. Without controlled mating, calving occurred all year round, including the harsh dry season, which increased calf mortality. Despite its problems, for economic reasons the open-range system persisted relatively intact through to the mid-1950s with minimal development of land (Lewis 2002). The remoteness of the VRD and cost of infrastructure development together contributed to a pattern of land ownership dominated by companies and wealthy absentee owners, who were content to invest in basic infrastructure, pay low lease rents, and pursue low profit, low cost strategies on extensive pastoral properties. Initial property sizes were large and have remained so, currently averaging 400,000 ha. It was not until the mid-1950s that intensification began, with the installation of new bores and extra fencing. As with the Dalrymple Shire, the commencement of the Brucellosis and Tuberculosis Eradication Campaign in 1970 greatly facilitated this fencing effort. The VRD rangelands remain only weakly fragmented, but there is growing interest in intensifying production through internal development of water and fencing infrastructure.

5. CONCLUSION

Returning to the initial hypothesis (stated in section 1.3), Australian rangelands seem to span a continuum between the two hypothetical scenarios of fragmentation (industrializing vs. post-industrial), rather than neatly fitting either alone. The south-west Queensland example seems to fit the "post-industrial" scenario, where a long history of development has led to over-fragmentation and property consolidation is now taking place. In contrast, the Victoria River District example would be closer to the "industrializing" scenario, with large, relatively undeveloped properties that are undergoing changes in land use.

Fragmentation in the Dalrymple Shire is intermediate to that in the VRD and south-west Queensland. Property sizes were probably large enough for viable enterprises at the time of subdivision, but the subsequent decline in enterprise profitability has led to pressures to consolidate properties. Land law in Queensland is still in a state of transition and revisions are being considered that will change land use and tenure, especially on the leasehold land that comprises most of the shire. Strong economic pressures to improve profitability are likely to drive both internal fragmentation of properties through development of water sources and fencing (reducing the scale of livestock-landscape interactions), and regional consolidation of properties (increasing the scale of human-landscape interactions).

ACKNOWLEDGEMENTS

We thank Gordon Landsberg, Tom Mann, Bob Shepard, Lindsay Whiteman, and Jeff Corfield for their historical accounts of the Dalrymple Shire. We were supported by U.S. National Science Foundation Grant #DEB-0119618, the CSIRO Emerging Science Program in Complex Systems Science, and the CSIRO Postdoctoral Fellowship Program.

REFERENCES

Abbott, B. N. and R. R. J. McAllister. 2004. Using GIS and satellite imagery to estimate historical expansion of grazing country in the Dalrymple Shire. Pages 405-406 *In* G. Bastin, D. Walsh, and S. Nicolson, editors. Living in the outback: conference papers. Australian Rangeland Society, Alice Springs, Australia.

ABS (Australian Bureau of Statistics). 1910. Official year book of the Commonwealth of Australia. Australian Bureau of Statistics, Canberra, Australia.

ABS (Australian Bureau of Statistics). 1925. Official year book of the Commonwealth of Australia. Australian Bureau of Statistics, Canberra, Australia.

ABS (Australian Bureau of Statistics). 2002. Year book Australia 2002. Australian Bureau of Statistics, Canberra, Australia.

ABS (Australian Bureau of Statistics). 2004. Year book Australia 2004. Australian Bureau of Statistics, Canberra, Australia.

Allingham, A. 1977. Taming the wilderness: the first decade of pastoral settlement in the Kennedy district. Dissertation. James Cook University of North Queensland, Townsville, Australia.

Anthony, T. 2004. Labour relations on northern cattle stations: feudal exploitation and accommodation. The Drawing Board: An Australian Review of Public Affairs 4:117-136.

Ash, A. J., J. Corfield, and T. Ksiksi. 2002. The Ecograze project - developing guidelines to better manage grazing country. CSIRO, Townsville, Australia.

Ash, A., J. Gross, and M. Stafford Smith. 2003. Scale, heterogeneity and secondary production in tropical grasslands. Pages 569-579 *In* N. Alsop et al., editors. Rangelands in the new millennium. International Rangelands Congress, Durban, South Africa.

Beare, S., R. Bell, A. Blias, P. Gooday, A. Heaney, S. Hooper, D. Langenkamp, G. Love, C. Levantis, C. Mues, E. Qureshi, and C. Riley. 2003. Natural resource management in the Burdekin River catchment: integrated assessment of resource management at the catchment scale - a case study. eReport 03.18. Australian Bureau of Agricultural and Resource Economics, Canberra, Australia.

Bell, P. 2000. A short history of Thuringowa. Thuringowa City Council, Thuringowa, Australia.

Blake, T. W. 1979a. Cunnamulla, a brief history of the Paroo Shire. Paroo Shire Council, Cunnamulla, Queensland, Australia.

Blake, T. W. 1979b. Land settlement across the Queensland border. Volume 12 *In* W. J. Cameron, editor. The history of Bourke. Bourke and District Historical Society, Bourke, New South Wales, Australia.

Bolton, G. C. 1963. A thousand miles away: A history of north Queensland to 1920. Jacaranda Press, Brisbane, Australia.

Bortolussi, G., J. G. McIvor, J. J. Hodgkinson, S. G. Coffey, and C. R. Holmes. 2005a. The northern Australian beef industry, a snapshot: 1. Regional enterprise activity and structure. Australian Journal of Experimental Agriculture 45:1057-1073.

Bortolussi, G., J. G. McIvor, J. J. Hodgkinson, S. G. Coffey, and C. R. Holmes. 2005b. The northern Australian beef industry, a snapshot: 4. Condition and management of natural resources. Australian Journal of Experimental Agriculture 45:1109-1120.

Caltabiano, T., J. R. P. Hardman, and R. Reynolds. 1999. Living area standards. Queensland Department of Natural Resources, Coorparoo, Australia.

Cameron, J. and R. Blick. 1991. Pastoralism in the Queensland mulga lands. Pages 75-116 *In* J. Cameron and J. Elix, editors. Recovering ground: a case study approach to ecologically sustainable rural land management. Australian Conservation Foundation, Melbourne, Australia.

Centre for International Economics. 1997. Sustainable natural resource management in the rangelands. Centre for International Economics, Canberra & Sydney, Australia.

Day, D. 2001. Claiming a continent: a new history of Australia. HarperCollins, Sydney, Australia.

De Corte M. W. M., E. V. Barry, M. J. Bright, M. G. Cannon, and J. C. Scanlan. 1994. Land degradation in the Dalrymple Shire, a preliminary assessment. Methods and results. Project report Q093023, Queensland Department of Primary Industries, Brisbane, Australia.

Finlay, M. C. and P. L. Lloyd. 1983. Dalrymple Shire handbook. Queensland Department of Primary Industries, Brisbane, Australia.

Flannery, T. F. 1994. The future eaters: an ecological history of the Australasian lands and people. Reed Books, Melbourne, Australia.

Greiner, R., N. Stoeckl, C. Stokes, A. Herr, and J. Bachmaier. 2003. Natural resource management in the Burdekin dry tropics: social and economic issues. CSIRO, Townsville, Australia.

Gross, J. E., R. R. J. McAllister, N. Abel, D. M. Stafford Smith, and Y. Maru. 2006. Australian rangelands as complex adaptive systems: a conceptual model and preliminary results. Environmental Modeling and Software 21:1264-1272.

Hannam, I. 2000. Policy and law for rangeland conservation. Pages 165-180 *In* O. Arnalds and S. Archer, editors. Rangeland desertification. Kluwer Academic Press, Dordrecht, The Netherlands.

Hewitt, R. and J. Murray. 1999. South-West Strategy & sustainable rangeland management - it's about attitude. Pages 76-77 *In* D. Eldridge and D. Freudenberger, editors. People and rangelands - building the future. VI International Rangeland Congress, Townsville, Australia.

Hinton, A. W. 1993. Economics of beef production in the Dalrymple Shire. Queensland Department of Primary Industries. Brisbane, Australia.

Holmes, J. M. 1963. Australia's open north. Angus and Robertson, Sydney, Australia.

Holmes, W. E. 1980. Property build-up in a semi-arid grazing area. Dissertation. University of Melbourne, Melbourne, Australia.

Isbell, R. F. and G. G. Murtha. 1972. Vegetation: Burdekin-Townsville region (Queensland): resources series. Department of National Development, Canberra, Australia.

Janssen, M. A., B. H. Walker, J. Langridge, and N. Abel. 2000. An adaptive agent model for analysing co-evolution of management and policies in a complex rangeland system. Ecological Modelling 131:249-268.

Janssen, M. A., Ö. Bodin, J. M. Anderies, T. Enquist, H. Ernstson, R. R. J. McAllister, P. Olson, and P. Ryan. 2005. Toward a network perspective of the resilience of social-ecological systems. Ecology and Society 11:15.

Johnston, P. W., G. M. McKeon, and K. A. Day. 1996. Objective 'safe' grazing capacities for south-west Queensland Australia: development of a model for individual properties. Rangeland Journal 18:244-258.

Kraatz, M. 2000. Managing for healthy country in the VRD. Tropical Savannas CRC, Darwin, Australia.

Land Settlement Advisory Commission. 1959. Report on progressive land settlement in Queensland. State of Queensland, Brisbane, Australia.

Landsberg R. G., A. J. Ash, R. K. Shepherd, and G. M. McKeon. 1998. Learning from history to survive in the future: management evolution on Trafalgar Station, north-east Queensland. Rangeland Journal 20:104-118.

Laut, D. P. 1988. Changing patterns of land use in Australia. Pages 547-556 *In* I. Castles, editor. Year Book Australia 1988. Australian Bureau of Statistics, Canberra, Australia.

Lewis, D. 2002. Slower than the eye can see. Tropical Savannas CRC, Darwin, Australia.

McAllister, R. R. J., I. J. Gordon, and C. J. Stokes. 2005a. KinModel: An agent-based model of rangeland kinship networks. Pages 1624-1630 *In* A. Zerger and R. M. Argent, editors. Proceedings of the International Congress on Modelling and Simulation.

McAllister, R. R. J., I. J. Gordon, and M. A. Janssen. 2005b. Trust and cooperation in natural resource management: The case of agistment in rangelands. Pages 2334-2339 I*n* A. Zerger and R. M. Argent, editors. Proceedings of the International Congress on Modelling and Simulation.

McAllister R. R. J., J. E. Gross, and C. J. Stokes. 2006. Rangeland consolidation patterns in Australia: An agent-based modelling approach. I*n* P. Perez and D. Batten, editors. Complex science for a complex world: Exploring human ecosystems with agents, ANU ePress, Canberra.

McKeon, G., W. Hall, B. Henry, G. Stone, and I. Watson. 2004. Pasture degradation and recovery in Australia's rangelands: learning from history. Queensland Department of Natural Resources, Mines and Energy, Brisbane, Australia.

Mills, J. R. 1989. Management of mulga lands in far south-west Queensland: project report QO89023. Queensland Department of Primary Industries, Brisbane, Australia.

Mitchell, C. 1997. Review of Desert Upland biogeographic region: management issues. Queensland Department of Natural Resources, Brisbane, Australia.

Mortiss, P. D. 1995. The environmental issues of the Upper Burdekin catchment: project report Q905017. Queensland Department of Primary Industries, Brisbane, Australia.

Passmore, J. G. I. and C. G. Brown. 1992. Property size and rangeland degradation in the Queensland mulga rangelands. Rangeland Journal 14:9-25.

QNRM (Queensland Department of Natural Resources and Mines). 2001. Managing state rural leasehold land: a discussion paper. Queensland Department of Natural Resources and Mines, Coorparoo, Australia.

Quirk, M. F., A. J. Ash, and G. McKillop. 1996. Dalrymple Shire, Queensland: case study - present. Pages 71-83 *In* N. Abel and S. Ryan, editors. Sustainable habitation in rangelands. Proceedings of a Fenner Conference on the Environment. CSIRO, Canberra, Australia.

Reynolds, H. 1974. Settlers and Aborigines on the pastoral frontier. Pages 153-162 *In* Anon, editor. Lectures on north Queensland history. James Cook University, Townsville, Australia.

Rogers, L. G., M. G. Cannon, and E. B. Barry. 1998. Land resources of the Dalrymple Shire. Queensland Department of Primary Industries, Brisbane, Australia.

Roth C., J. Aldrick, A. Ash, R. Hook, P. Novelly, D. Orr, M. Quirk, and M. Sallaway. 1999. The north Australia program: NAP occasional Paper No. 4. CSIRO, Townsville, Australia.

Roth, C. H., G. Lawson, and D. Cavanagh. 2003. Overview of key natural resource management issues in the Burdekin catchment, with particular reference to water quality and salinity. Burdekin catchment condition study: Phase 1. CSIRO Land and Water, Townsville, Australia.

Stokes C. J., A. J. Ash, and R. R. J. McAllister. 2004. Fragmentation of Australian rangelands: risks and trade-offs for land management. Pages 39-48 *In* G. Bastin, D. Walsh, and S. Nicolson, editors. Living in the outback: conference papers. Australian Rangeland Society, Alice Springs, Australia.

Tothill J. C. and C. Gillies. 1992. The pasture lands of northern Australia: their condition, productivity and sustainability: occasional paper. Tropical Grasslands Society of Australia, Brisbane, Australia.

Chapter 5

FROM FRAGMENTATION TO REAGGREGATION OF RANGELANDS IN THE NORTHERN GREAT PLAINS, USA

Jill M. Lackett[1] and Kathleen A. Galvin[1,2]
[1]Natural Resource Ecology Laboratory, Colorado State University, Fort Collins, CO 80523-1499, USA; [2]Department of Anthropology, Colorado State University, Fort Collins, CO 80523, USA

1. INTRODUCTION

The Northern Great Plains region has been experiencing a trend towards reaggregation of fragmented land parcels into larger operations since the 1930s. Grasslands were initially fragmented during the settlement of the region in the 1860s, due to settlement policies and the introduction of cropping, and fragmentation continued as roads and fences were built. Today, however, farmers and ranchers find that expansion of operations is one way to stay in business in the face of the challenging environmental and economic conditions of Great Plains agriculture. The control and use of small tracts of land for agriculture and ranching in the Northern Great Plains did not adequately support homesteader families in the region in the 1860s, and judging from the increase in operation size in the region, smaller tracts are often not adequate today.

Much like the Australian case (Stokes et al., Chapter 4), fragmentation of Great Plains rangelands occurs at two levels: those due to land tenure changes and those due to management of individual parcels of land. Although farms and ranches in the Northern Great Plains are expanding as a result of consolidation of land parcels into larger operations, this does not always lead to a reduction of fragmentation in the region because individual

K. A. Galvin et al. (eds.), Fragmentation in Semi-Arid and Arid Landscapes: Consequences for Human and Natural Systems, 113–134.

parcels of land remain small due to fencing and roads. These parcels are sometimes contiguous, but not always. Often an operator will manage different land parcels in the same county, or even in adjacent counties. So while we maintain that reaggregation is occurring in the Great Plains, this does not necessarily mean that fragmentation is still not an important issue for humans, livestock, and wildlife.

We begin this chapter by introducing the region, including ecological, social, and economic characteristics. We next discuss the settlement of the region, including historical land use and land tenure patterns. We then discuss drivers of land-use change in the region, including both fragmentation and consolidation, and we conclude by addressing human and ecological responses to these land use changes.

2. SITE DESCRIPTION

2.1 Biophysical characterization

This case study of fragmentation and reaggregation of rangelands will focus on the Northern Great Plains of the United States generally, while also providing more in-depth data and discussion of two adjacent counties in the region: Adams County, North Dakota and Perkins County, South Dakota (Figure 5-1). These two counties are located in western North and South Dakota on the Missouri Plateau. Adams County covers an area of 2560 km^2, and Perkins County is almost three times as big at 7430 km^2. These two counties were chosen as the focus because of previous work done in these counties by project participants, including the collection of extensive household survey data (Jennings 2000).

The elevation in the Northern Great Plains region ranges from 600-900 m, with high, flat-topped buttes and rolling hills. The soil is characterized by thin topsoils, and is naturally low in soil organic matter. Before European settlement, 95% of North Dakota was covered with grasslands. The rangeland areas of the region are mixed grass prairie ecosystems, composed of midgrasses and shortgrasses characteristic of the Great Plains (Jennings 2000). Species found in the mixed prairie include blue grama, buffalo grass, and western wheat grass. Important grassland herbivore species in the region include elk, buffalo, antelope, and mule deer.

The Northern Great Plains' climate is semi-arid – cold, temperate, and continental with large interannual variation in precipitation (Figure 5-2a), variation that is caused primarily by episodic droughts and winter blizzards.

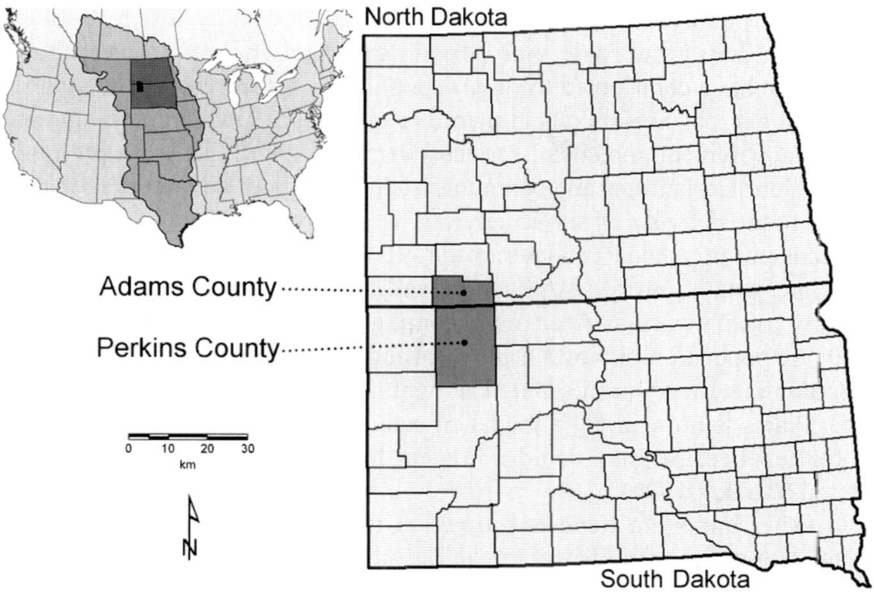

Figure 5-1. The Northern Great Plains region discussed in this chapter includes Adams County, North Dakota and Perkins County, South Dakota. The shaded area depicts the entire Great Plains region.

The average precipitation in Perkins County from 1910-1993 was 310 mm; the average in Adams County was 320 mm. The coefficient of variation of rainfall over the same time period was 22% in Adams County and 24% in Perkins County. Temperatures are also variable, although to a lesser extent than precipitation (Figure 5-2b). The average annual temperature in Perkins County from 1910-1989 was 6.4°C; the average in Adams County was 5.9°C. The semi-arid climate explains much about the state of North Dakota, and indeed about the region as a whole, including the character of the soil and native vegetation, importance of agriculture, both wheat and cattle, increasing size of farms, scattered population, out-migration of young residents, and high costs of public services (Robinson 1966). Also, climatic conditions severely limit the range of alternatives in production in the region, compared to other agricultural areas in the U.S.

2.2 Population trends

Many people who initially settled in the region did not stay long (Gutmann and Pullum 1999) and the population in many areas of the Northern Great Plains has been declining since 1930. This is the case in both Adams County and Perkins County (Figure 5-3). Since 1910, the population in Perkins County has declined 70%, and Adams County lost 52% of its population during the same time period. Structural changes in agricultural production have contributed to rural depopulation by decreasing the number of traditional jobs available, and therefore the tax base, while increasing the cost of government and social services (Jennings 2000). This has profound implications for families and communities in the region, due to the problems related to the delivery of public services, which are compounded by the fact that sparsely populated counties in the Northern Great Plains are often far from interstate highways, large rivers, and larger metropolitan areas. These sparsely populated rural areas are becoming more common. Between 1980-1990, the population of North Dakota shifted from being predominately rural to predominately urban (53.3%). The population fluctuation in the Northern Great Plains is not entirely a matter of economic difficulty, however. Many people left because they could not adapt to the social life that the region offered (Bennett 1990).

Another important trend in the region is the fact that the population is aging. The percentage of people age 65 or over in Adams County in 2000 was 24%, whereas it was 15% in North Dakota as a whole; in Perkins County the respective numbers were 24% and 14% (U.S. Census Bureau 2000). In 1990, 21% of the population of both Adams County and Perkins County was age 65 or over (U.S. Census Bureau 1990). This is an important trend to recognize because of the public services that an aging population requires; also it has implications for the passing on of land to children or other family members, or for the selling or leasing of land, and the potential land use change issues that are associated with the transfer of land.

There are 23 Native American reservations in the Northern Great Plains. These reservations cover 52,000 km^2. Adams County and Perkins County are directly adjacent to the Cheyenne River (5700 km^2) and Standing Rock (3400 km^2) Indian reservations, which are located to the east of the counties. A diversity of Native American tribes share these reservations. They are geographical and socio-economic islands in a region that is already isolated. There is diversity in land tenure on the different reservations. On some reservations over half of the land is owned by a tribal entity, on others, the large majority of land is in individual Indian allotments (The Planning Support Group 1974).

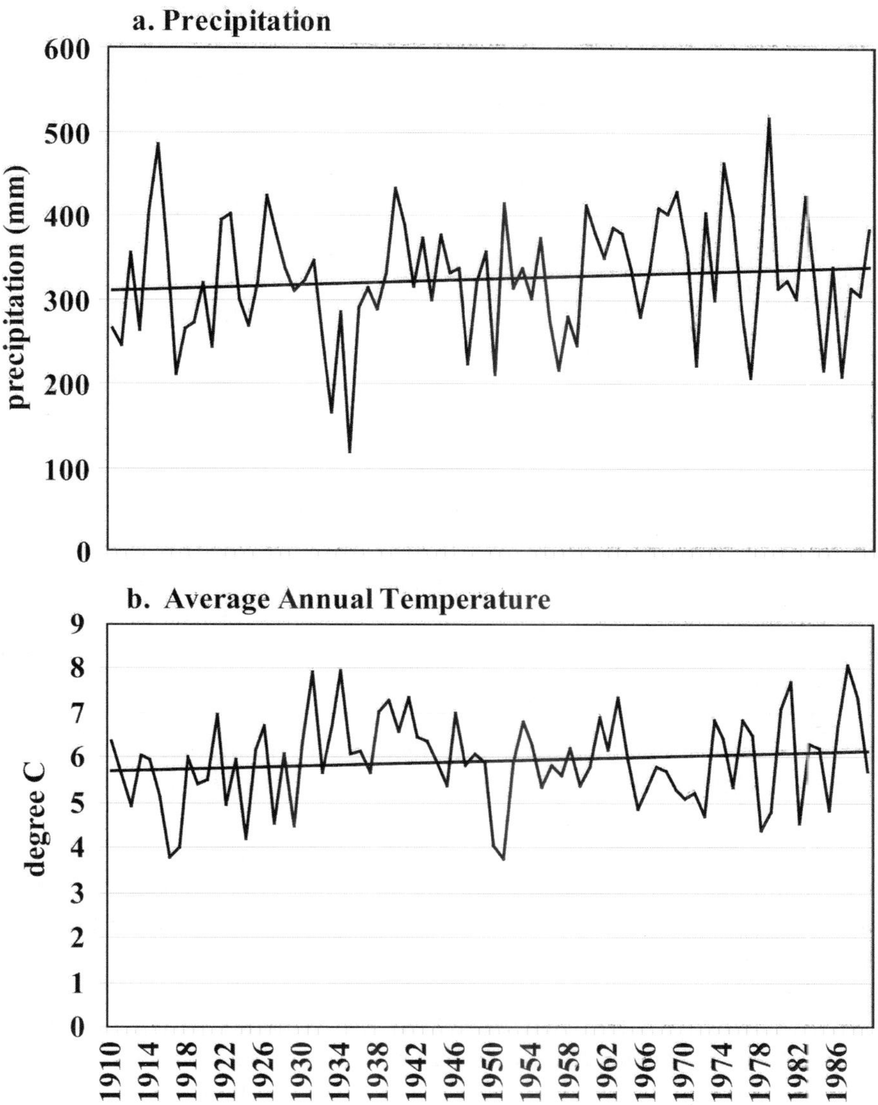

Figure 5-2. Historical precipitation (mm) and temperature (°C), for Adams County, ND. Both variables show an increasing trend from 1910 to the present and substantial interannual variability. The trend is similar for Perkins County, SD.

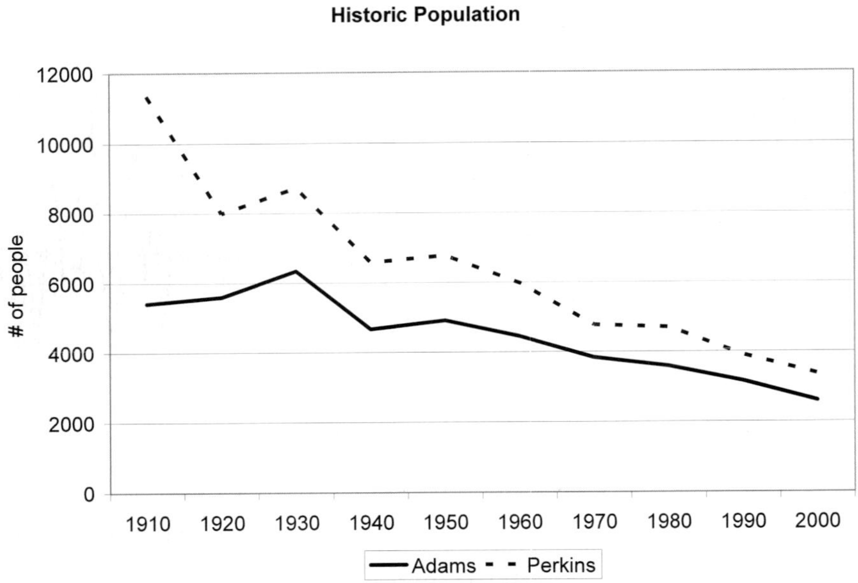

Figure 5-3. The population in Adams County, ND and Perkins County, SD has been decreasing since 1930.

2.3 Economic activities

The main economic activities in Adams and Perkins counties, as well as in much of the Northern Great Plains region, are livestock production, wheat agriculture, petroleum exploration, and mining. The economy in the region has grown and diversified since the late 1950s. There is less reliance on crop and livestock sales today than in decades past, although cropland and pastureland are the most extensive land uses in the region. From 1980-1990, North Dakota's dependence on agriculture for its economic base decreased from 44% to 36% (Coon et al. 1992). In 2002, agriculture's contribution to the gross state output in South Dakota was 23% (Beutler 2003), a decrease from 2001. Concurrent with this trend, direct agricultural employment in North Dakota, including proprietors and wage and salary employment, decreased from 38% to 14% of all state employment from 1980-1990 (Coon et al. 1992). In South Dakota, farm employment decreased from 20% to 7% from 1970-2000 as a percentage of total state employment (Beutler 2003). In North Dakota, energy development, agricultural processing, and manufacturing sectors have expanded, as well as federal government outlays in the state, due to Social Security payments to the aging population and two military bases in the state. In South Dakota, manufacturing has also increased and diversified. In the past the manufacturing of foodstuffs was dominant in the manufacturing sector, but since the 1990s the manufacturing of machinery has increased. Although the federal government spends a considerable amount of money on agriculture in the Northern Great Plains, for example, in 2002, government payments represented 50% of farm net cash income in South Dakota (Beutler 2003), the income in the region is still low compared to other areas of the U.S.

Variability in economic conditions creates challenges for residents in the region. Today, family farms operate at industrial scales and compete in national and international markets. Agricultural sales vary from year to year due to variability in weather, national and international markets, and changes in federal farm programs. This variability, in turn, creates volatility in personal income. There is a need to expand enterprises in these counties, or else be put out of business. The paradox of smaller numbers of more efficient, larger farms, which provide adequate income for families in these rural areas, is that with farm consolidation comes more expensive community services and facilities per capita (Stucky 1961). In 2000, Adams County had 1 resident/km^2 and Perkins County had 0.5 residents/km^2. This is well below the level of 1.5 people/km^2, where delivery of community services, such as schools and medical care, becomes difficult (Popper and Popper 1994).

Operations in Perkins County are over twice as large as operations in Adams County on average (Table 5-1). The average farm in Adams County in 2002 was 621 hectares; in Perkins County, the average farm was 1596 hectares. This is logical because ranches are typically more land extensive than farms, and Adams County has more cropland than pastureland (59% cropland and 41% pastureland), whereas Perkins County is about three-fourths pastureland. Additionally, the increasing aridity and the undulating topography in Perkins County make it less conducive to large scale cropping.

There is a trend in the region towards the reaggregation or consolidation of operations (Figure 5-4). The number of farms in Adams County has been declining since 1935 and in Perkins County the decline has been occurring since 1930. The land in farms in both counties has remained virtually stable after World War II. This exemplifies the fact that operations have been getting larger over the last 70-75 years in the region, leading to the same amount of land being controlled by a smaller number of people.

Operations in the region are extremely intensive, with operators trying to realize as much profit as they can from each piece of land. However, operators who are in the business for the long-term understand the importance of land conservation for continuation of their successful operation. Data indicates that in general, most operators are at their limit in terms of intensification (Jennings 2000). They cannot use their land more intensively with their current time and labor constraints. There is not much room for diversification in the Northern Great Plains region presently and land use is fairly stable as operations are already adapted to their local climatic, topographic, and ecological conditions. Ranching occurs in the driest regions, wheat farming dominates in the wetter areas, and mixed operations are located in between. Policy, climatic, and ecological conditions all contribute to the stability of land use in the region.

There are three conventional categories of tenure in the Great Plains: full-owners, who operate only land that they own; part-owners, who operate land that they own and lease; and tenants, who operate only land that they lease. However, businesses may actually be conducted by an individual (or family), partnership, or corporation. In 2002, in both Adams County, ND and Perkins County, SD, about half of operators were full-owners, with tenants accounting for 10% or less of the operators in each county (Figure 5-5a). However, in both counties, most land area was controlled by part-owners, 71% in Adams County and 54% in Perkins County (Figure 5-5b). Therefore, although full-owners control the most farms and ranches in terms of numbers, part-owners control the most land area in both counties. The majority of operations were family operations, however. In both counties over 80% of operations were run by an individual or family (Adams County – 92%, Perkins County – 85%), and these family operations control over two-thirds of the private land area of each county (Adams County – 90%, Perkins County – 68%).

Although most land is privately owned in Adams County, North Dakota and Perkins County, South Dakota, there are also extensive federal and state lands throughout the Northern Great Plains region. For example, the Grand River National Grassland administered by the U.S. Forest Service, covers approximately one-third of Perkins County, and about one-fifth of the operators in the county own grazing leases for these grasslands. These leases are very important to ranchers in the region, and many operators could not survive without grazing their cattle on federal land.

Table 5-1. Farms and land use (USDA: 2002 Census of Agriculture).

	Adams County, ND	Perkins County, SD
Farms (number)	394	452
Land in farms (hectares)	245000	721000
Avg. size of farm (hectares)	621	1596
Total cropland (# of farms)	359	396
Total cropland (hectares)	145000 (59%)	199000 (28%)
Pastureland (# of farms)	239	378
Pastureland (hectares)	100000 (41%)	522000 (72%)

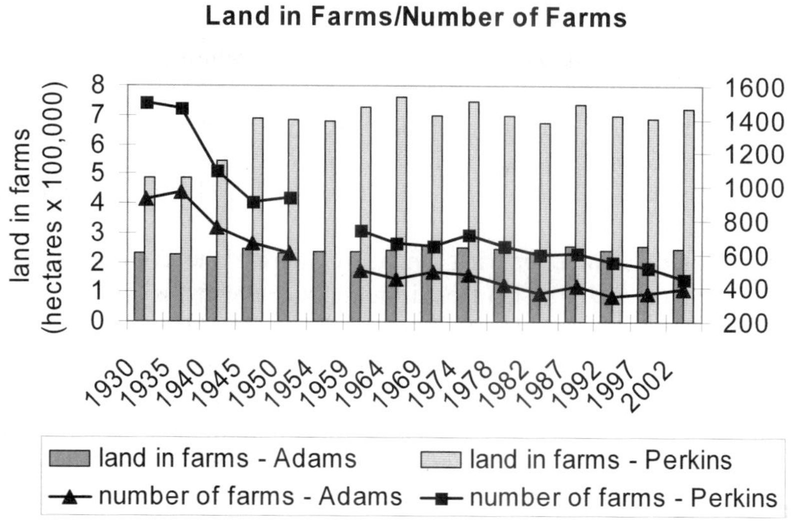

Figure 5-4. Total land area devoted to agriculture in the Northern Great Plains has been relatively steady since the mid-1940s; however, the number of operations has been steadily declining.

3. SETTLEMENT: LAND USE HISTORY/LAND TENURE DEVELOPMENT

3.1 Native Americans

The earliest settlers in the Northern Great Plains region were the ancestors of Native Americans who arrived from Asia between 11 and 15 thousand years ago. Starting around 2,000 years ago, Native American tribes moved into or through the region from the east and south. As early as 1700, white trade goods arrived in the region (Robinson 1966), and the period between 1700-1850 saw drastic changes in Indian territories. Horses and guns were introduced into the region. Horses became important in Plains Indian culture after they were acquired from Spanish outposts in New Mexico around 1650. They then spread north from tribe to tribe, reaching the Northern Plains by the end of the 1700s. The Tetons, Crows, and Cheyennes had large herds of horses, and therefore, they were more nomadic than other tribes (Robinson 1966). North American bison provided the bulk of resources for Plains Indian tribes. Seasonal patterns of bison numbers and movements also determined patterns of tribe mobility, aggregation, and dispersion (Reher 1977). The size of bison herds in the region was determined by the availability of food in the winter. Therefore, the tribes supplemented their diet when necessary by gathering and storing wild plants, trading for wild or cultivated plants, or by hunting antelope and deer (Reher 1977, Bamforth 1988). After 1850, government payments, conflicts with white settlers, and the destruction of the bison herds all severely altered Plains Indian life (Bamforth 1988).

There is some debate over what kind of pressure Plains Native American groups exerted on the grassland environment in which they lived (Truett 1996). Some scholars assert that Native Americans were voluntary conservationists (Martin 1978), whereas others view them as opportunists, driven by the simple need to survive (Hawkes 1992). Martin and Szuter (1999) assert that both Native Americans and Europeans had a significant human impact on the environment. Based on analyses of Lewis and Clark's journals, activities of Native Americans had some regulatory effect on the range and numbers of large animals present, before any overhunting by Europeans. For example, in the land separating nations that were at war, large numbers of large animals were found (Martin and Szuter 1999). These animals congergated in these buffer zones that human groups tended to avoid unless traveling on a war raid.

3.2 European settlers

The Gold Rush in California in 1849 led to white travelers passing through the region. European-Americans arrived to settle in the region beginning around 1850, although in some areas white farmers and ranchers did not arrive in significant numbers until after World War I. During the settlement of the Northern Great Plains by white settlers, grazing preceded grain production and general farming (homesteaders). Land, livestock, and water were all critical to the economic endeavors of white settlers on the Plains. The level of scarcity of these factors, together with the variable costs of establishing and enforcing ownership, determined the system of property rights in the Great Plains (Anderson and Hill 2001).

In the early 1880s, the short grass region of the Little Missouri River in western North and South Dakota was recognized as ideal cattle country. This attracted large cattle companies from Texas and the Southwest. These were wealthy ranching entrepreneurs, and ranches flourished during this time. However, the winter of 1886-1887 was harsh and many grasslands were overstocked. This made competition for grass fierce, and therefore, holding a title to land was necessary. Settlers became more sensitive to overstocking and developed methods to sustain production under climate variability, and they also tried to capitalize on land by claiming rights to land they had used in the past (Starrs 1998). This contributed to a more geographically stable, fenced, and rooted ranching system (Starrs 1998), but it also led to land fragmentation.

In the 1890s, dirt farmer immigrants arrived in the region. This was a period of extensive railroad construction, great immigration, expansion in

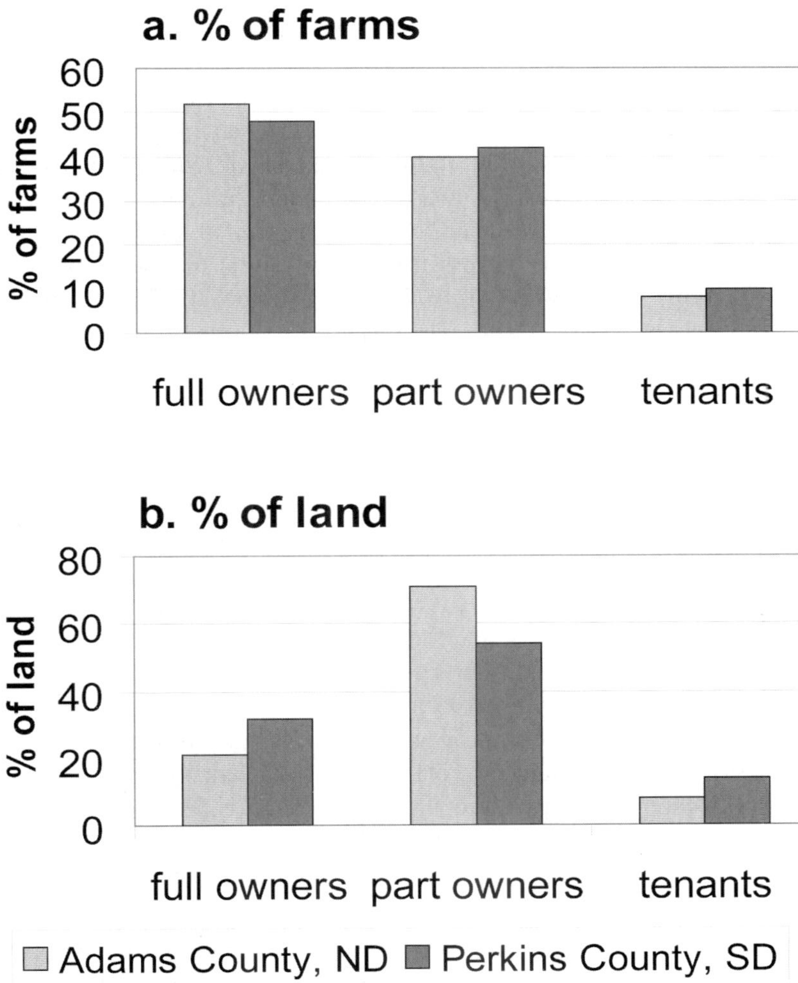

Figure 5-5. Tenure arrangements for farms and land operated in Adams County, ND and Perkins County, SD for 2002. Full owners operate the largest number of farms, but part owners control the most land in both counties.

manufacturing, and growth in cities (Robinson 1966). This boom led to extensive wheat farms in eastern North Dakota. These homesteaders, who came from the eastern U.S., Norway, Canada, Germany, England, Ireland, Sweden, and Russia for free land, did not arrive to an uncharted frontier, however. Many laws, railroads, and businesses were already established in the region (Jennings 2000).

3.3 Land tenure

When cattle ranching began in the Great Plains, cattle grazed on open and unclaimed pasture. By the end of the 1800s, the unrestricted use of free, public lands ceased. The first formal enclosed ranches were in the Southern Great Plains, and this was where the Great Plains ranching tradition began. Introduction of barbed wire in 1874 ushered in the end of the extensive cattle kingdom on public lands, and the beginning of the modern system of ranching in the Great Plains. The western system of extensive livestock ranching, on thousands of borrowed or leased acres, instead of on smaller private pastures, was a break from the federal policies of land acquisition and use practices (Starrs 1998). This system developed in response to the physical, cultural, and economic conditions in the arid West. Although the trend during this period was towards exclusive ownership of land, there was no way to stop livestock from crossing range boundaries, until the invention of barbed wire in the 1870s. From 1860 to 1900, changing land costs and values led to more exclusivity as more individuals and groups devoted resources to definition and enforcement activity. This included a move towards exclusivity and control of grazing on public lands, by trying to lease unclaimed public land from the government. Much land was also granted to the transcontinental railroads, and much of this was eventually transferred to private landowners (Anderson and Leal 1998).

Alternatives to Eastern property rights laws had to be developed in the Great Plains, including voluntary local agreements and extralegal institutions. In the 1860-1870s, squatter sovereignty was sufficient to control a tract of land, but as population pressure increased, range rights were recognized although settlers still did not own the land. Early property rights laws did provide for punishments for those who drove stock from their "accustomed range." As the value of grazing lands increased, cattlemen organized in groups and used the coercive authority of the government to protect private property. From this grew stockgrower associations, which restricted entry onto the range by controlling access to limited water supplies. There was a gradual move towards private property by restricting entry onto land once held in common (Anderson and Leal 1998).

Access to water is essential in the Great Plains. Initial settlements were traced to river stream bottoms. Water rights were initially controlled by riparian doctrine, where all owners had coequal rights to flows. As settlement pressure increased, and as water was starting to be used for irrigation and mining, more effort was put into redefining water property rights. These property rights were not static, however. They evolved through social arrangements, laws, and customs which governed asset ownership and allocation. As the value of the asset increased, the incentive to establish property rights also increased (Anderson and Leal 1998).

3.4 Policy

There were a number of federal policies that shaped the settlement of the Great Plains. The first was the *Homestead Act of 1862*. This act provided settlers with a 160 acre (65 hectare) parcel of land, which was 'free' to them if they built a house, made improvements, farmed, and lived on the land for five years. Therefore, the U.S. system of land laws made it difficult to acquire large tracks of land, and land ownership was encouraged on small parcels (Starrs 1998). The act was passed during a period of relatively favorable rainfall. Farms failed when drought set in, because in the semi-arid climate of the Northern Great Plains many were too small to produce enough to support a family. It was soon recognized by some that large tracks of land were needed for extensive cattle ranching in the arid West, and that the *Homestead Act* was not the best way to settle the Northern Great Plains. As early as 1878, John Wesley Powell advocated a revision of 160 acre (65 hectare) homesteads in the region, along with modification of the rectangular system of land survey in the region where water was concentrated in a few parcels instead of being widely distributed on many parcels. He contended that traditional grid surveys, while effective in the more humid eastern U.S., were ineffective in the West for irrigated agriculture and ranching (Powell 1962). Powell was a forward thinker, but had only minimal success influencing land use and land tenure policies in the region. It was also recognized that the policies and laws under which the arid West was settled were meant to be orderly, effective, and replicable, but they did not allow for flexibility to accommodate different geographical and climatic conditions, such as those found in the Western U.S. (Starrs 1998).

Many policies were established through the years in order to stabilize the population in the region. Stabilizing of the population in the early years after settlement became important after droughts and the end of large speculative wheat operations led to a declining population in the region. One stabilizing policy was the *Reclamation Act of 1902*, which developed irrigation in the West. In 1934, another stabilizing force was the passage of the *Taylor*

Grazing Act, which allowed for 575,000 km^2 of unclaimed land to come under Federal control. This act also required ranchers to have a privately-owned home ranch in order to have access to public lands. These home ranches were only a small part of the grazing land needed for most operators (Starrs 1998). The Great Plains Program was started in 1956. This program allowed operators credit payments in order to make changes on their farms to enhance the conservation of soil and water resources. In 1992, the U.S. Fish and Wildlife Service began the Great Plains Initiative, where U.S. states and Canadian provinces work together in order to manage wildlife species and habitat on a regionwide, or Plainswide basis, instead of focusing on smaller areas or single species (Popper and Popper 1994). Currently, the Great Plains region is highly subsidized by the government through crop payments and set-aside programs. Even with this support, the region is still losing population and many farms are foreclosed each year under a system that may be less and less likely to be able to sustain traditional farming and ranching households.

Policies put in place to settle the land and to transform the grasslands to croplands had large impacts on the biodiversity of the region. According to recent land cover estimates, 70% of grasslands have been lost in the Great Plains. Tallgrass prairie areas have declined 87%; mixed prairie has declined 71%, while shortgrass prairie has declined 48% (Samson et al. 2004).

4. DRIVERS OF LAND FRAGMENTATION

There are many references to the fragmentation of land in the Northern Great Plains, and to the detrimental effects that fragmentation often brought. John Wesley Powell in 1878 saw that land was "…being chopped to ruinous bits by the advancing front of the rectangular surveys and the tradition-bound, hopeful, ignorant, and doomed homesteaders" (Stegner 1962:x). Likewise, sources of fragmentation, such as fences, are also discussed in the literature. Samson et al. (2004:11) note that "fences are the problem in, not the solution to, conservation of historically grazed ecosystems."

Many factors that will be discussed below contributed to the fragmentation of rangelands in the Northern Great Plains. Important drivers of fragmentation are the historical legacy of settlement and policy in the region, population and society, and climate and the natural resource base. These drivers will be discussed, followed by a brief discussion of drivers of consolidation, including government programs and technology.

4.1 Settlement

Ultimately, the pattern of land use and landscape fragmentation in the region extends from the legacy of the settlement of the region. The system of land settlement, which allowed for each nuclear family to claim 160 acres (65 hectares) of land, encouraged extended families to split into nuclear families to claim land in order to maximize the size of land holdings for the entire extended family. This was a break from tradition for many families new to the region who historically lived and worked land together as extended families. These homestead plots were plowed for wheat cultivation; therefore, agricultural fields are also an important source of fragmentation. Fencing, another source of fragmentation, was used to obtain exclusive rights over land or to exclude others from desirable land, such as land with water sources or good forage. Today the system of rotational grazing used on many ranches keeps fences up, because pastures need to be delineated in order for cattle to be moved around seasonally. Likewise, fences around cropped areas are also prevalent in order to keep wildlife and domestic livestock out.

As settlement progressed, and as parts of the grasslands were transformed into urban settlements, roads have become a major source of fragmentation. Roads have resulted in 70% of the land parcels ranging between 100 and 1000 km^2 in size. Without roads, 90% of the land parcels in the Great Plains region would be greater than 10,000 km^2 in size (White et al. 2000).

4.2 Population/social drivers

Farms in North Dakota are not typically inherited by children as intact units from parents due to many factors, including socio-economic ones, such as family size and wealth (Tauxe 1992). Therefore, land holdings become fragmented as land owners retire and eventually pass away and divide their land among their children as an inheritance. The population is currently decreasing in both Adams County and Perkins County, while the land in farms and ranches has remained relatively stable (see Figure 5-3 and 5-4). Therefore, although dividing up operations as inheritances is causing some land fragmentation, there is still a trend towards aggregation of fragmented parcels of land in order to create larger operations. Larger operations are necessary today, as they were over a century ago following the failures associated with the *Homestead Act*, to sustain a viable family operation in this region (Jennings 2000). The landscape is still fragmented; the land is just managed in larger blocks.

It is the goal of many operators in the region to stay in business or pass the operation on to their children or other family members (Jennings 2000). This may lead to further fragmentation of land if operations are split among children when their parents retire. However, it is becoming increasingly likely that children will leave the area and not take over farms or ranches from their parents. In this case, the land may be sold or leased to another operator. This scenario may have a neutral effect on land fragmentation.

4.3 Climate/natural resource base drivers

The natural resource base of the region, which is a semi-arid region of short and mid-grasses with scarce water resources, also contributes to land fragmentation. In the settlement process, parcels of land with access to water, both wells and surface water, were privatized and excised in order to gain exclusive control over these resources. Geographic features, such as rivers, are also natural sources of fragmentation in the region (White et al. 2000).

The climate in the region, including the frequent droughts and harsh winters, also can lead to land fragmentation. It is desirable for an operator in the region to own or control parcels of land in spatially heterogeneous areas, in order to survive undesirable weather events. Spatial heterogeneity of parcels included in an operation is beneficial in an area of high climate variability. It is possible for droughts or storms to only affect localized areas, so an operator can spread his/her risk by being spatially diversified.

5. DRIVERS OF LAND CONSOLIDATION

5.1 Government programs

Crop payments, set-aside programs, and conservation programs that are part of the *U.S. Farm Bill* all contribute to certain patterns of land use for owners participating in the programs. These government programs may encourage the trend towards consolidation. For example, the Conservation Reserve Program (CRP), which pays operators to replant cropland in native grasses, may help reaggregate certain parcels of land, reduce soil erosion, and provide habitat for wildlife. Additionally, federal programs benefit large operators more than small-scale ones, as payments are based on the number of acres enrolled in the particular program (Tauxe 1992). This may also lead to further consolidation of land parcels.

5.2 Technology drivers

Technology is an important driver of consolidation. Mechanization, including larger machinery, allows for a family operator to manage a larger operation with little or no hired labor or family help. This is only possible to a point, however, because there are only so many hours in a day for a single person or family to operate machinery and successfully sustain a farm or ranch operation.

6. RESULTS OF FRAGMENTATION

6.1 Human responses

There are both short and long-term human and ecological responses to, and outcomes of, fragmentation in the Northern Great Plains. The need for increased inputs and consolidation due to fragmentation leads to long-term outcomes, such as changes in enterprise viability and human welfare.

Many agricultural operations in the Northern Great Plains rely heavily on inputs, such as machinery, fertilizer, policy, and technology in order to stay in business and make a profit. These inputs are more common in intensive cropping operations than in ranching operations. Operators in the Northern Great Plains also compete in both national and international markets. This market access often acts as a double-edged sword. Both productivity and debt are increased simultaneously through the expansion of land holdings and bigger machinery. The resulting overproduction results in low crop prices, and therefore, difficultly paying debt. The economics of running a successful operation in the Northern Great Plains is dictated by prices, prices both for products and for inputs. A fragmented landscape requires more inputs than an unfragmented one in order to realize the same amount of profit from the land (Ellis and Peel 1995). For example, water sources may need to be developed if access to water is limited by fragmentation. Inputs are costly for operators and can be a burden for farmers and ranchers operating on the margin of survival.

Consolidation of operations is presently common for survival of many farms and ranches, along with income earned from off-farm sources by the operator or his/her spouse. These adaptations required to deal with fragmentation have implications for the long-term viability of the operation and for human welfare. For example, data have shown that operators in the region feel stressed with the lifestyle they must live in order to successfully manage these larger consolidated operations. The management of a larger number of fragmented, often non-contiguous parcels also results in fragmentation of time and labor for the operator. This stress can also lead to dissatisfaction with life. Half of all operators surveyed in Adams and Perkins counties reported that they were only slightly satisfied, or they were dissatisfied with their life (Jennings 2000).

6.2 Ecosystem responses

The initial settlement of the region and establishment of agriculture led to the plowing up of most of the native grasslands in the region. Short-term impacts of fragmentation on wildlife and domestic livestock include altered movements, access to resources, nutrition, and survival/reproduction rates, leading to long-term impacts on populations. Vegetation changes also lead to long-term impacts on the biological diversity in the region.

Drivers of ecosystem functioning in the Great Plains have historically included drought, grazing, and fire. Changes in the Great Plains since white settlement include a reduction in the number and distribution of native herbivores and replacement with domestic livestock, and fire suppression regimes. Diversity of native grasslands have been reduced by past management strategies, including fragmentation. Non-native species have often thrived at the expense of native species (Samson et al. 2004).

Klement et al. (2001) document three changes in Northern Great Plains vegetation over the last eighty years. These changes are largely due to human activities in the region, such as road-building and agriculture, with less of an impact from fencing. First, there is an increased density and cover of woody plants. Second, there have been changes in plant community structures and species composition due to human modifications, including tillage, haying, and road building. Third, non-native species have invaded the region due to roadside and agronomic plantings. These impacts due to roads have been documented in other areas of the world, such as in semi-arid areas in Utah (Gelbard and Belnap 2003) and in South American pampas grasslands (Ghersa et al. 2002). These shifts in plant species composition are linked to changes in the structure and function of the ecosystem. The World Wildlife Fund (2004) cites habitat fragmentation as one of the major threats to biodiversity in the Northern Great Plains.

Wildlife may be impacted differentially by fragmentation due to roads and fencing. Grazing intensity also differentially affects wildlife species (James 2003) and vegetation (Richardson-Kageler 2004). Organism size - large, small, or microscopic - is important to consider when assessing impacts (Saunders et al. 1991), as well as the type of mobility the organism has (Keller et al. 2004).

Although there is not abundant literature on the impact of fragmentation on wildlife species in the Northern Great Plains, impacts on some species of grassland birds has been documented. Many grassland bird species in the region are not thriving under current management schemes that strive for equal grazing pressure (World Wildlife Fund 2004, Samson et al. 2004). Grasslands in their native state consist of a patchwork of varying amounts and qualities of forage, with the amount and quality of the forage depending on rainfall (Bamforth 1988). The system of rotational grazing, where livestock are moved to different pastures at different times of the year, is an adaptation ranchers use to mimic the resource tracking that an animal would naturally perform in an unfragmented system to access temporally and spatially variable resources. This grazing system can lead to a homogeneous landscape, instead of a patchy one, as the rancher will often graze each parcel at the same intensity and move the animals at regular intervals before overgrazing of any one parcel is allowed to occur. This system does not meet the habitat needs of many bird species in the region that thrive in a patchy landscape due to needing open spaces or densely vegetated areas for breeding or nesting purposes.

Other impacts on wildlife due to fragmentation include the tendency for carrying capacity to decrease on smaller land parcels (Boone and Hobbs 2004) and often for species that require large home ranges, such as elk, to be completely absent from the area (World Resources Institute 2004). Fragmentation also can result in reduced populations that are often genetically isolated, fewer native species, and, if reaggregation of fragmented parcels does occur, there is a decreased incidence of recolonization by wildlife (White et al. 2000).

7. CONCLUSIONS

The Northern Great Plains region has been experiencing a trend towards consolidation of land parcels into larger agricultural operations. Land was originally fragmented, mainly by fencing and roads, during the settlement process; however, today viable operators need to have access to more land to sustain traditional farming or ranching operations. The size of operations has been increasing since the 1930s, with individual operators controlling larger numbers of land parcels that are either purchased or leased from neighbors as they migrate to urban areas and get out of the business of agriculture. These land parcels may or may not be contiguous, so land blocks are still fragmented in the region; they are just managed together as larger operations. Therefore, this consolidation does not necessarily solve the problems created by, or influence the short-term and long-term outcomes of, fragmentation of grasslands in the region.

As land was privatized in the Northern Great Plains, and exclusive use was solidified, the movements of people, domestic livestock, and wildlife were restricted and their access to resources was limited. This left both humans and animals with fewer options to capitalize on temporal and spatial variability in water and vegetation. Inputs of policy and capital can offset some of these effects of fragmentation, but there are both economic and social costs to the operator. For example, many operations in the region are on the margin of economic survival, and operators in the region are increasingly stressed with their lifestyle. Wildlife, on the other hand, generally must contend with the fragmentation without buffering inputs.

Policies of settlement in the Great Plains that fragmented land were adopted because of the ease of management and security of investment that they afforded. However, the downside is severe and includes altered ecosystem function and viability of grazing systems by reduced access to resources due to restricted movements of species, including humans, who rely on the rangelands.

With the continuation of the historical trend towards wetter and warmer conditions in the region (see Figure 5-2), as well as the projected increase in these trends in the future (Ojima et al. 2002), there is the potential for increased fragmentation of land parcels in the future across the Northern Great Plains as farming may become viable in areas that are not plowed presently. This will be an important potential trend to consider in the future.

ACKNOWLEDGMENTS

We would like to acknowledge Tori Jennings for her collection of household data in Adams County, ND and Perkins County, SD. We also thank Steve Ogle and Bill Parton for reviewing this chapter. This research is supported by a grant from the U.S. National Science Foundation (DEB-0119618).

REFERENCES

Anderson, T. L. and D. R. Hill. 2001. Free market environmentalism. Palgrave: New York.

Anderson, T. L. and P. J. Leal. 1998. From free grass to fences: Transforming the commons of the American West. Pages 119-134 *In* Baden, J.A. and D.S. Noonan, editors. Managing the commons. 2nd ed. Indiana University Press, Bloomington.

Bamforth, D. B. 1988. Ecology and human organization on the Great Plains. Plenum Press, New York.

Bennett, J. W. 1990. Human adaptations to the North American Great Plains and similar environments. Pages 41-80 *In* P. A. Olson, editor. The struggle for the land: Indigenous insight and industrial empire in the semiarid world. University of Nebraska Press, Lincoln, NE.

Beutler, M. K. 2003. Impact of South Dakota agriculture 2002. South Dakota State University, Brookings.

Boone, R. B. and N. T. Hobbs. 2004. Lines around fragments: effects of fencing on large herbivores. African Journal of Range and Forage Science 21:79-90.

Coon, R. C., F. L. Leistritz, and T. A. Majchrowicz. 1992. The role of agriculture in the North Dakota economy. Agricultural Economics Statistical Series Report 50, North Dakota State University, Fargo, ND.

Ellis, J. E. and M. Peel. 1995. Economies of Spatial Scale in Dryland Ecosystems. *In* Arid Zone Ecology Forum, Kimberly, South Africa.

Gelbard, J. L. and J. Belnap. 2003. Roads as conduits for exotic plant invasions in a semiarid landscape. Conservation Biology 17:420-432.

Ghersa, C. M., E. de la Fuente, S. Suarez, and R. J. C. Leon. 2002. Woody species invasion in the Rolling Pampa grasslands, Argentina. Agriculture, Ecosystems and Environment 88:271-278.

Gutmann, M. P. and S. M. Pullum. 1999. From local to national political cultures: social capital and civic organization in the Great Plains. Journal of Interdisciplinary History 24:725-762.

Hawkes, N. 1992. Myth of the noble savage. World 57:36-38.

James, C. D. 2003. Response of vertebrates to fenceline contrasts in grazing intensity in semi-arid woodlands of eastern Australia. Austral Ecology 28:137-151.

Jennings, T. L. 2000. Living with uncertainty: adaptive strategies for sustainable livelihoods in the Northern Great Plains. M.A. Colorado State University, Fort Collins, CO.

Keller, I., W. Nentwig, and C. R. Largiader. 2004. Recent habitat fragmentation due to roads can lead to significant genetic differentiation in an abundant flightless ground beetle. Molecular Ecology 13:2983-2994.

Klement, K. D., R. K. Heitschmidt, and C. E. Kay. 2001. Eighty years of vegetation and landscape changes in the Northern Great Plains. Conservation Research Report 45, U.S. Department of Agriculture, Agricultural Research Service.

Martin, C. 1978. Keepers of the game: Indian-animal relationships and the fur trade. University of California Press, Berkeley.

Martin, P. S. and C. R. Szuter. 1999. War zones and game sinks in Lewis and Clark's West. Conservation Biology 13:36-45.

Ojima, D. S., J. M. Lackett, and the Central Great Plains Steering Committee and Assessment Team. 2002. Preparing for a changing climate: The potential consequences of climate variability and change - Central Great Plains. Report for the US Global Change Research Program, Colorado State University.

Popper, F. J. and D. E. Popper. 1994. Great Plains: Checkered past, hopeful future. Forum for Applied Research 9:89-100.

Powell, J. W. 1962. Report on the lands of the arid region of the United States. The Belknap Press of Harvard University Press, Cambridge.

Reher, C. A. 1977. Adaptive process on the shortgrass plains. Pages 13-40 *In* L. R. Binford, editor. For theory building in archaeology: Essays on faunal remains, aquatic resources, spatial analysis, and systematic modeling. Academic Press, New York.

Richardson-Kageler, S. J. 2004. Effects of large herbivore browsing on the functional groups of woody plants in a southern African savanna. Biodiversity and Conservation 13:2145-2163.

Robinson, E. B. 1966. History of North Dakota. University of Nebraska Press, Lincoln.

Samson, F. B., F.L. Knopf, and W.R. Ostlie. 2004. Great Plains ecosystems: Past, present, and future. Wildlife Society Bulletin 32:6-15.

Saunders, D. A., R. J. Hobbs, and C. R. Margules. 1991. Biological consequences of ecosystem fragmentation: a review. Conservation Biology 5:18-32.

Starrs, P. F. 1998. Let the cowboy ride. The Johns Hopkins University Press, Baltimore.

Stegner, W. 1962. Editor's Introduction. *In* Powell, J.W. Report on the lands of the arid region of the United States. The Belknap Press of Harvard University Press, Cambridge.

Stucky, H. R. 1961. Characteristics and trends in the Great Plains. Pages 1-21 *In* Land tenure in the Great Plains. North Dakota State University, Fargo, ND.

Tauxe, C. 1992. Family cohesion vs. capitalist hegemony: cultural accommodation on the North Dakota farm. Dialectical Anthropology 17:291-317.

The Planning Support Group, Bureau of Indian Affairs. 1974. Indians in the Northern Great Plains: Anticipated socio-economic impacts of coal development. U.S. Department of Interior, Billings, MT.

Truett, J. 1996. Bison and elk in the American Southwest: in search of the pristine. Environmental Management 20:195-206.

University of Texas Population Research Center Great Plains Population and Environment Database: Version 1.0 1998. Austin: Texas Population Research Center, University of Texas at Austin.

U.S. Census Bureau: State and County QuickFacts. Data derived from Population Estimates, 2000 Census of Population and Housing, 1990 Census of Population and Housing, Small Area Income and Poverty Estimates, County Business Patterns, 1997 Economic Census, Minority- and Women-Owned Business, Building Permits, Consolidated Federal Funds Report, 1997 Census of Governments.

U.S. Department of Agriculture. 2002 Census of Agriculture. National Agricultural Statistics Service.

White, R. P., S. Murray, and M. Rohweder. 2000. Pilot analysis of global ecosystems: grassland ecosystems. World Resources Institute, Washington D.C.

World Resources Institute. 2004. Grassland Fragmentation by Roads in the Great Plains. http://www.earthtrends.wri.org/maps_spatial/maps_detail_static.cfm?map_select=253&theme=9.

World Wildlife Fund. 2004. Northern Great Plains: Threats to Biodiversity. http://www.worldwildlife.org/wildplaces/negp/threats.cfm.

Chapter 6

LAND USE, FRAGMENTATION, AND IMPACTS ON WILDLIFE IN JACKSON VALLEY, WYOMING, USA

Jill M. Lackett[1] and N. Thompson Hobbs[2]

[1]Natural Resource Ecology Laboratory, Colorado State University, Fort Collins, CO 80523, USA; [2]Department of Forest Rangeland Watershed Stewardship, Colorado State University, Fort Collins, CO 80523, USA

1. INTRODUCTION

The Jackson Valley in northwest Wyoming, USA, contains 4,000 km^2 of public and private land used for agriculture, grazing, forestry, recreation, conservation, and housing (Figure 6-1). The region offers an unusually complete history of changes in land uses and their implications for access to heterogeneity for humans, domestic livestock, and wildlife. Unlike most of the other case studies in this volume that focus on impacts of fragmentation on people and livestock in classically pastoralist systems, this chapter will include a discussion of impacts of fragmentation on wildlife in an area that is increasingly being developed.

In this chapter, we will discuss three historical periods: the period before settlement by Europeans, the rise of agriculture, and the emergence of recreation-based economies. For each of these historical periods we will discuss land tenure and use, economic forces, and sources of fragmentation. We will then discuss the consequences of fragmentation for wildlife in the region. We close by describing how the Jackson case (and the more general situation in agricultural areas of western North America) fits the model of progression of land use and land tenure described in this book (see Behnke, Chapter 13), a model that predicts consolidation of privately-owned land

K. A. Galvin et al. (eds.), Fragmentation in Semi-Arid and Arid Landscapes: Consequences for Human and Natural Systems, 135–150.

following its fragmentation. We also show how the Jackson case departs from that model in a fascinating way.

The Jackson Valley, which is part of the 77,000 km^2 Greater Yellowstone Ecosystem, includes the Snake River alluvial plain and surrounding highlands ranging between 1,800 and 4,200 m in elevation. Annual temperature averages about 3°C and annual precipitation averages 390 mm. Predominant vegetation includes grassy meadows and marshes in the valley bottom, forests along the rivers and on some of the upslopes, and sagebrush and rock outcroppings along the foothills of the Teton Range in the west and the Gros Ventre Range in the east. Historically, fire had a major impact on ecosystem structure and function in the Jackson Valley before 20th century suppression efforts (Gruell 1974). The policy of fire suppression has led to a more homogenous landscape, which can support fewer wildlife species than were present historically. There is now a move towards incorporating fire back into the ecosystem through fire management, not automatic suppression, and prescribed burns (Clark 1999).

2. BEFORE EUROPEAN SETTLEMENT

2.1 Land tenure/use

The Jackson Valley has a rich, enduring history of varied land use by people, even though there were climatic and topographic constraints on unrestricted access into and out of the Jackson Valley before settlement by Europeans (Wright 1984). Artifacts from hunter-gathers suggest the valley has been used by native peoples in the late spring, summer, and early autumn for as long as 5000 years. More recent evidence shows use of the valley from 10,000-13,500 years ago (Connor 1998, Love et al. 2003). During this era, Native American tribes, including Shoshone, Bannock, Blackfeet, Crow, and Gros Ventre, hunted communally in the valley during the growing season, but apparently left the area during winter, which tended to be severe at this elevation and latitude. Likewise, much of the valley's game may also have been migratory, wintering in more snow-free areas to the south (Smith et al. 2004). The availability of key resources, both inside and out of Jackson Hole, dictated the patterns of movement of human groups (Wright 1984).

2.2 Economy

The development of commerce in furs during the early 1800s created a confluence of trade routes for European trappers around what now is the

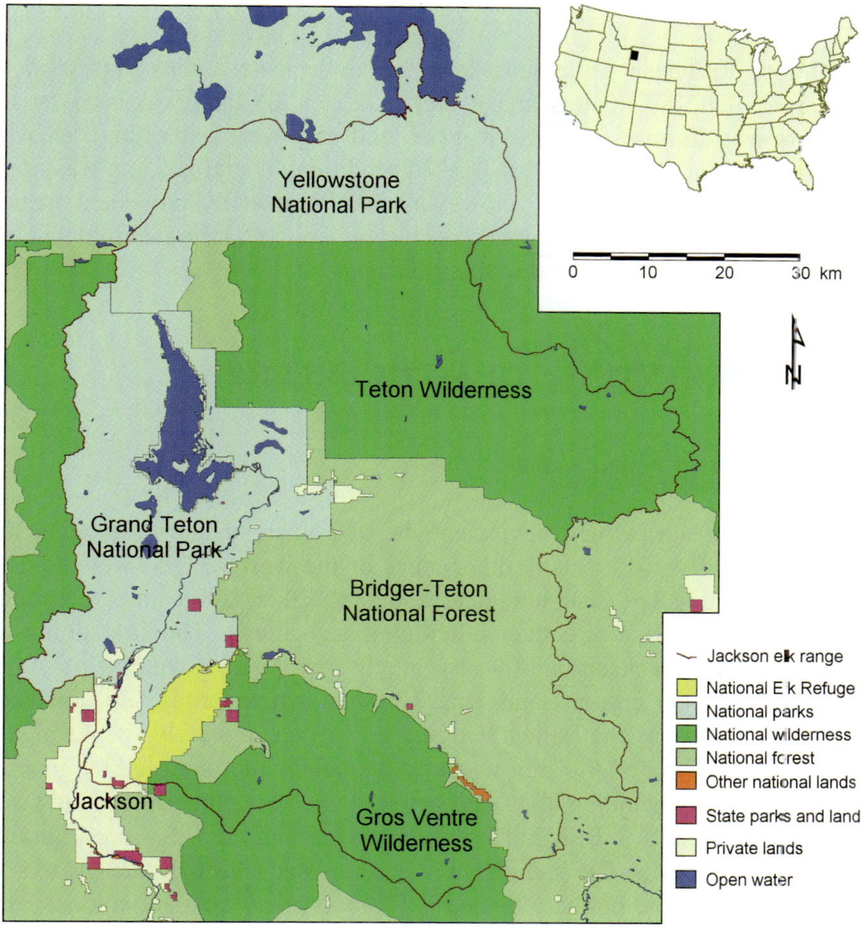

Figure 6-1. Land tenure patterns in the Jackson Valley in 2003 (U.S. Department of the Interior 2004). The area of the Jackson elk herd range is 6100 km^2.

town of Jackson, Wyoming. During the 1830s, the beaver population was declining from over-harvest. The consequent reduction in supply of furs and the decline in market demand resulting from shifts in European fashion caused dramatic declines in the fur trade (Saylor 1970). Little is known about the use of the valley between the mid-1830s, when the trappers departed, and 1860, when continued exploration began in preparation for agricultural settlements.

2.3 Sources of fragmentation

Before European settlement, land in the Jackson Valley was used as communal land by hunters and gatherers. Later the land was used by fur trappers who did not formally own the land they were exploiting, but who established routes that were 'owned' by tradition. Therefore, this was not a time of land fragmentation. However, snow seasonally prevented human groups and wildlife from accessing resources, and both migrated into and out of the valley as resources and weather dictated.

3. BEGINNINGS OF AGRICULTURE

3.1 Land tenure/use

Permanent agricultural settlement did not occur in the valley until the 1880s and 1890s, after many other areas of the western U.S. were settled, because of the valley's harsh winters, geographical isolation, and topography less suited to agriculture than other western valleys. The first domestic cattle arrived in Jackson in 1884. Eventually, the original patterns of settlement became the current land tenure system where productive lowlands on alluvial soils in the valley bottoms are owned privately, while higher elevations with poor soils are owned by the public and managed by government agencies, including the National Park Service, the Bureau of Land Management, the U.S. Fish and Wildlife Service, and the U.S. Forest Service (Table 6-1). During the summer, privately-owned livestock are grazed on allotments on some publicly-owned U.S. Forest Service and National Park Service lands.

Although publicly-owned lands in the Jackson Valley are used for grazing, they are not used communally, at least not in the strict sense of the word communal. The *Taylor Grazing Act of 1924* mandated that public land be allotted for grazing to individual ranchers who own land proximal to the allotment. Grazing allotments are akin to private ownership because they provide vested entitlements for land use that is associated with private land ownership. Private land owners pay fees to state and federal agencies for using these allotments, but fees are far below market values for use of private grazing land (Bryner 1998). Ranchers in the region often have grazing rights in the National Forest and rights to move livestock across some National Park Service lands (Saylor 1970). These patterns of land tenure and use, where private and public lands are used to graze privately-owned livestock, are typical of rangelands in the western United States.

3.2 Economy

During the period of agricultural settlement, cattle formed the mainstay of Jackson Valley's economy. Ranchers grew hay in the valley bottom during the summer growing season to feed cattle and horses from late fall through early spring. Dude ranching, guiding hunting trips, and outfitting for wealthy Europeans were also important sources of income for early settlers, although these activities were often supplemental to the cattle industry (James 1936). Some of these visitors eventually built houses or bought ranches in the valley. The character of Jackson Hole was strongly influenced by the influx of these "outsiders" (Betts 1978). The region's economy fluctuated over time with booms and busts in the energy extraction and timber industries, which also began during this period (Jobes 1991).

3.3 Sources of fragmentation

Fragmentation began during this period, as lands were converted to private ownership under the *Homestead Act* and the *Desert Land Act* and were subsequently fenced and farmed (Smith et al. 2004). Road construction accommodated the emerging recreation, timber, and energy extraction Industries. Because agricultural settlement, domestic livestock grazing, and urban settlements occurred on areas of the landscape that were traditionally used as wintering grounds by native ungulates, particularly elk, human uses of the land historically created conflict between wildlife and people. This was particularly true during severe winters where heavy snow limited access of wildlife to native forage (Wilbrecht and Robbins 1979).

These conflicts occurred on two scales – both within and beyond the Jackson Valley. Historically, thousands of elk that spent summer and fall in the foothills and mountains surrounding the Jackson Valley, including southern Yellowstone National Park, migrated to wintering areas in southwest Wyoming. By 1895, market hunting, barbed wire fencing, and retaliation by ranchers over conflicts with livestock operations decimated large numbers of these migrant elk that traveled 150-450 km between seasonal ranges. Elk cut off from winter forage began raiding haystacks or were provided handouts by local ranchers, habituating them to hay in the Jackson Valley (Smith et al. 2004). To mitigate this conflict between elk and ranchers, the National Elk Refuge was established in 1912 when the federal government purchased native rangelands to provide winter habitat for elk. By 1916, 1,100 ha of public and private lands were subsumed into the refuge.

Table 6-1. Patterns of land use and tenure in the Jackson Valley.

Land Use	Land Tenure	Pasture Type	Season of Use	Soil	Size (km^2)
Wildlife protection, recreation, grazed	Public	National Elk Refuge (wildlife only)	Fall, winter, spring	Mountain valley – good alluvial soils	100
Recreation, conservation, grazed	Public	National Park (wildlife and domestic, by permit)	Mostly summer, fall, spring	Mountain slopes and uplands – poorer, sandy soils	1952
Recreation, watershed protection, logging, wildlife habitat, grazed	Public	National Forest (wildlife and domestic, by permit)	Domestic: mostly summer, fall, spring Wildlife: all seasons	Mountain slopes and uplands – poorer, sandy soils	1672
Recreation, watershed protection, wildlife habitat, grazed	Public	National Wilderness (wildlife and domestic, by permit)	Domestic: mostly summer, fall, spring Wildlife: all seasons	Mountain slopes and uplands – poorer, sandy soils	2085
Recreation, watershed protection, wildlife habitat, grazed	Public	State parks and other public lands (wildlife and domestic, by permit)	Wildlife: fall, winter, spring Domestic: all seasons	Most in mountain valley – good alluvial soils	17
Hay production (grazed); housing (not grazed)	Private	Ranches – grazed by domestic and wildlife; residential or ranchette	Mostly winter, sometimes all seasons; some grazing of horses in all seasons on ranchettes	Mountain valley – good alluvial soils	128

The refuge proved insufficient in size to prevent elk depredations on private hay crops and periodic high mortality during severe winters. Private and public investment in land acquisitions significantly increased the size of the refuge by 1940. A boundary fence was completed in 1938 to restrict down-valley movement of elk toward the privately-owned ranches to the south. Since its establishment, elk have been supplementally fed alfalfa hay on the refuge every winter, save nine (Smith 2001). After winter feeding by the federal government became an annual practice, an increasing proportion of the Jackson elk herd learned to migrate to the refuge, and the herd memory of distant winter ranges to the south were lost. This resulted in reduced migrations, and a poor distribution across potential range with the elk forced to winter on a fraction of their former range, compared to the 1800s (Smith et al. 2004).

Today, the National Elk Refuge totals 100 km^2 and between 60 and 70% of the Jackson elk herd winters there (USDI 2005). The refuge accounts for only a small portion of the Jackson herd's traditional winter range. A maximum of 7,500 animals is the target for winter elk population to be supported on the National Elk Refuge, although numbers have varied from 5,000 to 11,000 in recent years (USDI 2005). Numbers exceeding 7,500 are due to mild winters and low elk harvests from the fall hunt.

The conflict between elk and agriculture could have been resolved, as happened repeatedly under similar circumstances in the pioneer West, by eliminating the Jackson elk herd (Smith et al. 2004). By 1900, elk and other ungulates were virtually driven to extinction throughout North America. However, eliminating the herd was not a viable solution to the conflict because the native ungulates were themselves becoming an important communally-used commodity that supported the rapidly growing recreation sector of the economy. The economic benefits accrued primarily from fee-based hunting, where the right to hunt publicly-owned wildlife was sold by the government, mainly to wealthy Easterners and Europeans. Substantial secondary economic benefits to local economies from healthy and visible wildlife populations are well described for the western U.S. and for Jackson in particular (Merrifield and Gerking 1982, Power 1991, Culver 2003). The number of animals maintained on the National Elk Refuge now exceeds the capability of developed habitat to support wintering animals in the absence of development (Hobbs et al. 2002), illustrating the strong economic forces shaping policy and emphasizing the external supplements to the system.

4. EMERGENCE OF RECREATION-BASED ECONOMIES

4.1 Land tenure/use

As the Jackson Valley became more popular as a recreation destination, ranches were subdivided, and housing tracts began to dominate the landscape on private land. These privately-owned parcels are now used for residential and commercial development, in order to support the tourist industry.

The Jackson Valley eventually became the center of human settlement in Teton County, Wyoming. Agricultural and census data from Teton County show dramatic changes in human demographics and economies, changes accompanied by alterations in patterns of land use. Although the population of Teton County grew steadily following settlement of the region in the late 1800s, dramatic increases in the human population did not occur until a century later in the 1980s and 1990s when many ranches were subdivided (Figure 6-2). The population in Teton County in 2000 was 18,250, a 39% increase from 1990, and a 73% increase from 1980 (U.S. Census Bureau 2000). Likewise, the number of housing units has more than doubled since 1980, from about 5000, to almost 11,000 in 2000. This increase in population produces a high functional density of people because only 3% of the land in Teton County is privately-owned; the remaining 97% is federally or state owned. This rapid expansion of the human population has increased fragmentation of the valley, heightened pressure on natural resources, and led to an emerging concern for the sustainability of the region (Clark 1999).

4.2 Economy

The economy of the Jackson Valley today is largely based on recreation and tourism, due to its proximity to Yellowstone and Grand Teton National Parks and other federal lands. The importance of agriculture and ranching is much diminished. In 1930, 47% of employed people in Teton County worked on farms or ranches. In 1990, this percentage had dropped to 4% (University of Texas 1998). The sale of goods and services to visitors to Grand Teton National Park is the single largest portion of the economy of Teton County (Merrifield and Gerking 1982), although unearned income is increasing in importance as will be discussed subsequently. Cattle ranching persists, but it contributes only a small portion of the economic activity.

Valley residents rely on the economic benefits derived from natural resource-based tourism and recreation for their economic welfare (Power 1991).

Although the Jackson economy is expanding, jobs available are typically low-paying service employment. Often sales revenues are funneled outside of the region as businesses are increasingly branches of out-of-state stores (Jobes 1991). The service industry workforce is largely transient or seasonal, moving on when they can no longer sustain a living in the area. However, due to the influx of resort-style hotels and golf courses that cater to wealthy visitors and residents, Teton County has the highest per capita income of any county in the U.S. (Grand Teton Park and Jackson Hole Visitor's Guide 2005).

Unearned income is also flowing into the region, especially income from investments and pensions of retirees (Power 1991). Unearned income has increased so significantly over the last several decades that it is now about three-quarters as large as wage and salary income in the area (Power 1991). Therefore, the economic impacts of retirees in the area must not be overlooked.

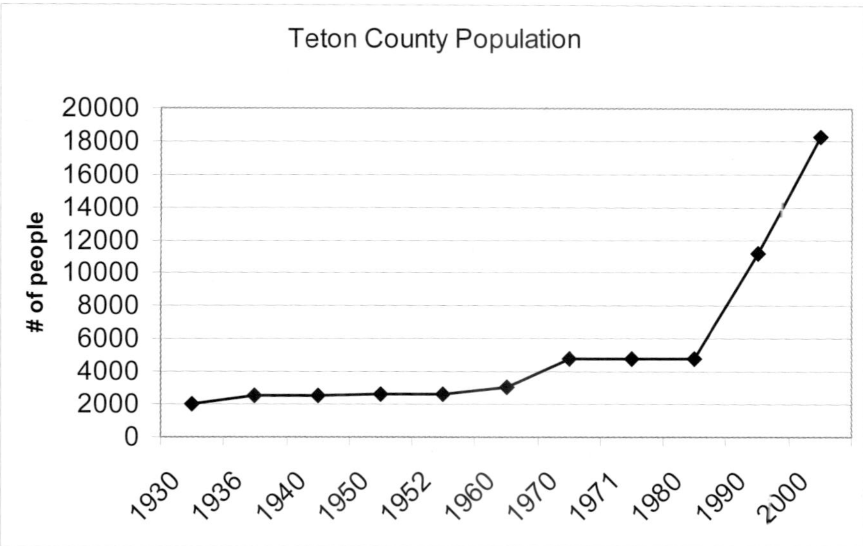

Figure 6-2. Population in Teton County, WY has been increasing since 1930, with the steepest increase since 1980.

4.3 Sources of fragmentation

Sales of farms and ranches to support housing and infrastructure associated with tourism have diminished the importance of ranching and agriculture. After initial growth in agricultural holdings since settlement of the region by Europeans, the average size of agricultural operations in Teton County decreased since the 1940s. In 1940, the average operation was 2,023 hc. From 1970 to the present, the average agricultural holding is about 10% as large, ranging from 200-280 hc (University of Texas 1998). Therefore, fragmentation of agricultural parcels into farms and ranches has increased through time. Likewise, the number of agricultural operations in Teton County has declined since 1930 (Figure 6-3) due to subdivision for housing tracts. In 2002, there were only 110 farms or ranches in the county (USDA 2002).

Jackson Valley offers a telling case of fragmentation produced by a land tenure system. This fragmentation, caused in part by fencing and in part by the attraction of supplemental feeding, severs historic patterns of migration by native ungulates between summer and winter concentration areas. The motivation for these fragmenting effects is to separate use of land by wild and domestic ungulates, and in so doing, resolve conflict between publicly-owned wildlife and privately-owned livestock. The consequences of this disruption of migrations by wild ungulates are well documented and will be discussed in the following section.

5. CONSEQUENCES OF FRAGMENTATION

Consequences of the severe fragmentation of the Jackson Valley include conflicts between ranchers and ungulates (Wilbrecht and Robbins 1979), the need for external inputs to support the system, high disease incidence in the Jackson elk herd due to crowding on the National Elk Refuge in the winter (Smith 1991, 2001), and the degradation of some plant communities from extensive browsing where crowding occurs (Dieni et al. 2000, Smith et al. 2004).

The conflict between ranchers and ungulates was discussed previously. As more humans settled in the region and as the landscape became more fragmented with fences and roads, wildlife were cut off from their traditional winter ranges — both within and beyond the Jackson Valley. Therefore, elk often raided haystacks on ranches in severe winters and competed with livestock for forage. Supplemental feeding was instituted to resolve the conflict.

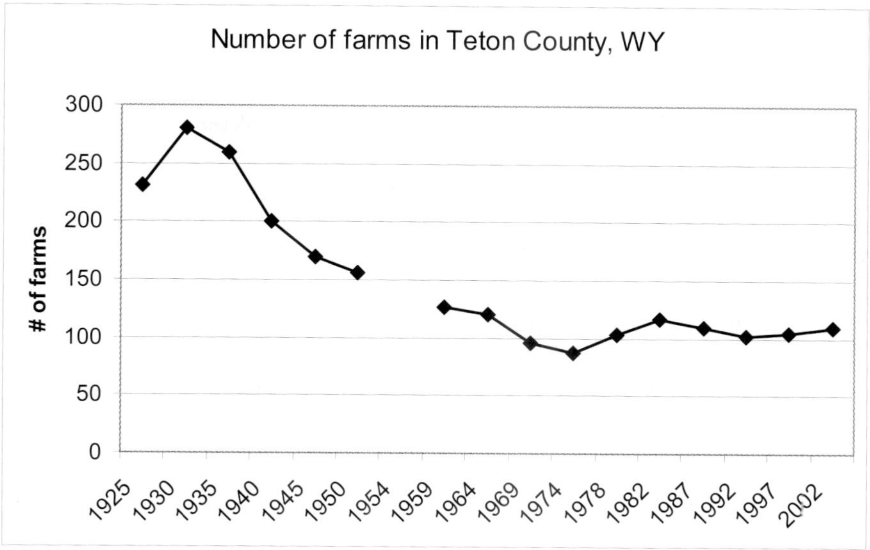

Figure 6-3. The trend in Teton County, WY since 1930 is towards a decreasing number of agricultural operations.

The external inputs of labor and capital required to sustain this system are well documented (Wilbrecht and Robbins 1979, Clark 2001, Smith 2001). For example, the National Elk Refuge is intensively managed to produce as much forage as possible for the elk herd. Inputs include irrigation, seeding of grasses, and prescribed burning (Smith et al. 2004). However, for approximately 2½ months during an average winter, supplemental feed is required, in the form of pelletized alfalfa. During their six month stay on the National Elk Refuge, about 35% of food requirements are met by the supplemental feed (Smith 1991). Reasons cited for the supplemental feeding of elk include: satisfying the public's demand for a large elk herd, compensating for the loss of winter range, limitations on winter forage due to snowpack, minimizing winter mortality of elk, and the poor distribution of elk on the landscape due to habitat fragmentation (Wilbrecht and Robbins 1979, Smith 1991). About 27,000 kg of hay is fed per day to a herd of 7,500 elk (U.S. Fish and Wildlife Service 2005). The cost of the feed is shared by the Wyoming Fish and Game Department and the U.S. Fish and Wildlife Service. The winter feeding program supports an artificially high number of elk, and a controversial controlled hunt is required in late fall to cull some animals on the refuge and on adjacent public lands, including in Grand Teton National Park (Smith and Robbins 1994).

High densities of elk on the National Elk Refuge contribute to high infection rates of brucellosis. About 28% of elk on the refuge test positive for brucellosis antibodies, whereas in adjacent areas where there is no supplemental feeding only about 1% of elk test positive for the antibodies (Halverson 2000, Smith 2001). This is an additional source of conflict between ranchers and the elk herd, as there is fear that the elk could infect domestic cattle herds. Likewise, there is a fear that a bovine tuberculosis outbreak on the refuge could infect other elk and wild bison and spread throughout the region (Smith 1991, 2001). This also has important implications for human populations as this disease can also be transmitted to people.

An additional impact of high densities of elk on the National Elk Refuge is the degradation of plant communities, especially willows, which results from extensive browsing. There is considerable evidence that elk herbivory has a negative impact on willow communities. These impacts are manifested through negative effects on nitrogen dynamics (Schoenecker et al. 2004) and through decreased size, stature, growth, biomass, and seed production of willows (Singer et al. 1994, 1998, Baker et al. 2005). These impacts are exacerbated when elk herds reach high densities.

There is also evidence that current levels of grazing and browsing by elk during the winter are exerting harmful effects on other plant committees, particularly aspen and riparian shrubs (Kay 1995, 1997, Wagner et al. 1995, Smith et al. 2004). These harmful effects on plant communities are believed to impact a range of animal species, such as beaver and birds, which use plant communities impacted by elk grazing and browsing. There is a feedback between willow declines and beaver declines, with depressed willow populations leading to declines in beaver, and declines in beaver resulting in further willow reductions (Kay 1997, Singer et al. 1998, Baker et al. 2005). These impacts together also result in a loss of biodiversity (Matson 2000, Smith et al. 2004).

6. CONCLUSIONS

The Jackson Valley case is important because of the severity of fragmenting effects, but it also stands as an excellent example of near one-to-one correspondence between land tenure and land use. The use of land on privately and publicly-owned parcels is tightly constrained by policies and laws that govern the types of acceptable uses on each land tenure type (see Table 6-1). For example, land use is tightly prescribed on government-owned land in the empowering acts that created management agencies and in subsequent legislative guidance. Zoning laws regulate use of private land in distinct categories, predominantly agriculture, commercial, and residential use.

The progression of land tenure systems in the Jackson Valley models the trajectory of land fragmentation discussed in this book (see Behnke, Chapter 13), particularly the observation that the most highly valued and productive lands are those that are likely to be most privately-owned and fenced. Historically, lands were used communally for hunting wild game. As settlement began, private land parcel sizes initially grew and were fenced. This progression has been observed in other parts of the western U.S., especially in agricultural areas of the Great Plains. To this point, our case follows the trajectory of the land fragmentation model perfectly. However, in the "post-industrial" conditions that prevail in the Jackson Valley, we are now seeing a reduction in parcel size as large agricultural holdings are subdivided for residential development and to service the increasing tourist population, where the aesthetic and recreational amenities offered by proximity to public land make real estate development far more profitable than ranching (Figure 6-4). This progression reemphasizes communal use of public land for conservation, recreation, and natural beauty and a concomitant shift from an agricultural to an amenity-based economy. This refragmentation of land is occurring in other areas of the western U.S. where land once used for agriculture is being converted to urban housing where natural beauty and relatively uncrowded conditions prevail. Of course, this will change as more people discover these scenic areas. One real challenge in this region today is balancing the preservation of open spaces and scenic beauty with continuing expansion as more people find the area to be a desirable place to live (Betts 1978). The irony is that the region, once protected from urban development because of its geographic isolation, harsh winters, and severe terrain, is now attractive to tourists and residents because of these characteristics (Jobes 1991).

In conclusion, the history of the Jackson Valley offers well-documented insight into the effects of fragmentation for three reasons. First, there is a very tight connection between land tenure and land use. Second, the two sources of fragmentation discussed in the chapter — policy-related choices, such as fences, roads, and feeding, and snow depth in severe winters limiting access to resources — amplify one another. And, third, the Jackson Valley story provides a superb example of the external policy and capital inputs that are needed to offset the effects of fragmentation and of ecosystem stresses on both plant and animal communities, due to fragmentation.

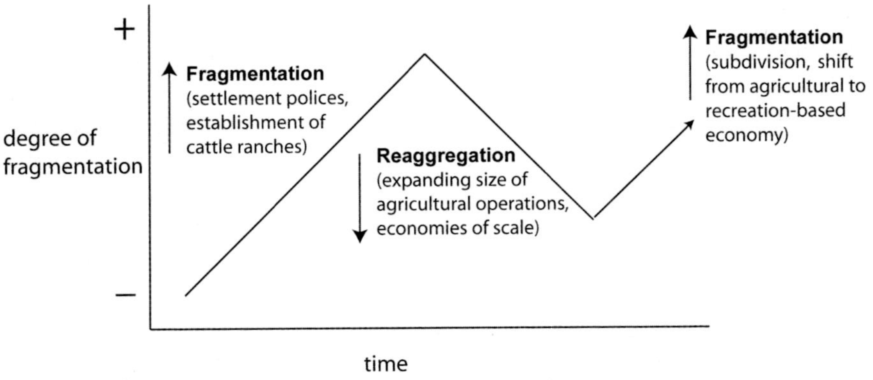

Figure 6-4. Trajectory of land fragmentation and reaggregation in Jackson Valley. The area is again being fragmented as subdivision is occurring.

ACKNOWLEDGMENTS

This work is supported by a grant from the U.S. National Science Foundation (DEB-0119618). We would like to thank R. Easterbrook and R. Boone for help generating the map and the table. Also, the helpful comments of two external reviewers are appreciated.

REFERENCES

Baker, B. W., H. C. Ducharme, D. C. S. Mitchell, T. R. Stanley, and H. R. Peinetti. 2005. Interaction of beaver and elk herbivory reduces standing crop of willow. Ecological Applications 15:110-118.

Betts, R. B. 1978. Along the ramparts of the Tetons: The saga of Jackson Hole, Wyoming. Colorado Associated University Press, Boulder.

Bryner, G. 1998. U.S. land and natural resources policy: A public issues handbook. Greenwood Press, Westport, CT.

Clark, T. W. 1999. The natural world of Jackson Hole: An ecological primer. Grand Teton Natural History Association, Moose, WY.

Clark, T. W. 2001. Wildlife resources: The elk of Jackson Hole, Wyoming. Pages 91-108 *In* J. Burger, E. Ostrom, R. B. Norgaard, D. Policansky, and B. D. Goldstein, editors. Protecting the commons: A framework for resource management in the Americas. Island Press, Washington D.C.

Connor, M. A. 1998. Final report on the Jackson Lake archeological project, Grand Teton National Park, Wyoming. USDA, National Park Service, Midwest Archeological Center, Lincoln, Nebraska.

Culver, L. 2003. From "last of the old West" to first of the new West. Pages 163-180 *In* L. Nicholas, E. M. Bapis, and T. J. Harvey, editors. Imaging the big open: Nature, identity, and play in the new West. The University of Utah Press, Salt Lake City.

Dieni, J. S., B. L. Smith, R. L. Rogers, and S. H. Anderson. 2000. Effects of ungulate browsing on aspen regeneration in Northwestern Wyoming. Intermountain Journal of Sciences 6:49-55.

Grand Teton Park and Jackson Hole Visitor's Guide. 2005. Other History. http://www.jacksonholewy.net/area_info/jh_other_history.php.

Gruell, G. E. and L. L. Loope. 1974. Relationships among aspen, fire, and ungulate browsing in Jackson Hole, Wyoming. US Department of Agriculture, Forest Service, Intermountain Region.

Halverson, A. 2000. The National Elk Refuge and the Jackson Hole elk herd: Management appraisal and recommendations. Pages 23-52 *In* T. W. Clark, D. Casey, and A. Halverson, editors. Developing sustainable management policy for the National Elk Refuge, Wyoming. Bulletin No. 104. Yale School of Forestry and Environmental Studies, New Haven, CT.

Hobbs, N. T., F. J. Singer and G. Wockner. 2002. Assessing management alternatives for ungulates in the Greater Teton ecosystem using simulation modeling. Natural Resource Ecology Lab, Colorado State University.

James, P. E. 1936. Regional planning in the Jackson Hole country. Geographical Review 26:439-453.

Jobes, P. C. 1991. The Greater Yellowstone social system. Conservation Biology 5:387-394.

Kay, C. E. 1995. Browsing by native ungulates: effects on shrub and seed production in the Greater Yellowstone Ecosystem. Pages 310-320 *In* B. A. Roundy, E. D. McArthur, J. S. Haley, and D. K. Mann, compilers. Proceedings: wildland shrub and arid land restoration symposium; 1993 October 19-21; Las Vegas, NV. Gen. Tech. Rep. INT-GTR-315. Ogden, UT: U.S. Department of Agriculture, Forest Service, Intermountain Research Station.

Kay, C. E. 1997. Viewpoint: Ungulate herbivory, willows, and political ecology in Yellowstone. Journal of Range Management 50:139-145.

Love, J. D., J. C. Reed, and K. L. Pierce. 2003. Creation of the Teton landscape. Grand Teton Natural History Association, Moose, Wyoming.

Matson, N. 2000. Biodiversity and its management on the National Elk Refuge, Wyoming. Pages 101-138 In T. W. Clark, D. Casey, and A. Halverson, editors. Developing sustainable management policy for the National Elk Refuge, Wyoming. Bulletin No. 104. Yale School of Forestry and Environmental Studies, New Haven, CT.

Merrifield, J. and S. Gerking. 1982. Analysis of the long-term impacts and benefits of Grand Teton National Park on the economy of Teton County, Wyoming. University of Wyoming – National Park Service Research Center, Laramie, WY.

Power, T. M. 1991. Ecosystem preservation and the economy in the Greater Yellowstone Area. Conservation Biology 5:395-404.

Saylor, D. J. 1970. Jackson Hole, Wyoming: In the shadow of the Tetons. University of Oklahoma Press, Norman.

Schoenecker, K. A., F. J. Singer, L. C. Zeigenfuss, D. Binkley, and R. S. C. Menezes. 2004. Effects of elk herbivory on vegetation and nitrogen processes. Journal of Wildlife Management 68:837-849.

Singer, F. J., L. C. Mark, and R. G. Cates. 1994. Ungulate herbivory of willows on Yellowstone's northern winter range. Journal of Range Management 47:435-443.

Singer, F. J., L. C. Zeigenfuss, R. G. Cates, and D. T. Barnett. 1998. Elk, multiple factors, and persistence of willows in national parks. Wildlife Society Bulletin 26:419-428.

Smith, B. L. 1991. Jackson: The big herds. Bugle 8:48-58.

Smith, B. L. 2001. Winter feeding of elk in western North America. Journal of Wildlife Management 65:173–190.

Smith, B.L. and R.L. Robbins. 1994. Migrations and management of the Jackson elk herd. National Biological Survey Resource Publication No. 199. USDI, Washington, D.C.

Smith, B. L., E. K. Cole, and D. S. Dobkin. 2004. Imperfect pasture: A century of change at the National Elk Refuge in Jackson Hole, Wyoming. Grand Teton Natural History Association, Moose, Wyoming.

University of Texas Population Research Center Great Plains Population and Environment Database: Version 1.0. 1998. Austin: Texas Population Research Center, University of Texas at Austin.

U.S. Census Bureau: State and County QuickFacts. Data derived from Population Estimates, 2000 Census of Population and Housing, 1990 Census of Population and Housing, Small Area Income and Poverty Estimates, County Business Patterns, 1997 Economic Census, Minority- and Women-Owned Business, Building Permits, Consolidated Federal Funds Report, 1997 Census of Governments.

USDA, National Agricultural Statistics Service. 2002 Census of Agriculture. County Summary Highlights.

U.S. Department of the Interior, National Park Service, Grand Teton National Park, Science and Resource Management, GIS Office. 2004. Land tracts at Grand Teton National Park and the John D. Rockefeller, Jr. Memorial Parkway, WY. http://science.nature.nps.gov/nrdata/.

USDI. 2005. Draft bison and elk management plan and environmental impact statement. National Elk Refuge and Grand Teton National Park. US Fish and Wildlife Service, Denver, Colorado.

U.S. Fish and Wildlife Service. 2005. National Elk Refuge highlights, history, feeding, and management. http://nationalelkrefuge.fws.gov/NERHistMgt.html.

Wagner, F. H., R. Foresta, R. B. Gill, D. R. McCullough, M. R. Pelton, W. F. Porter, and H. Salwasser. 1995. Wildlife policies in the U.S. national parks. Island Press, Washington D.C.

Wilbrecht, J. and R. Robbins. 1979. History of the National Elk Refuge. Pages 248-255 *In* M. S. Boyce and L. D. Hayden-Wing, editors. North American elk: Ecology, behavior, and management. University of Wyoming, Laramie.

Wright, G. A. 1984. People of the high country: Jackson Hole before the settlers. Peter Lang, New York.

Chapter 7

IDEOLOGY, LAND TENURE AND LIVESTOCK MOBILITY IN KAZAKHSTAN

Iliya I. Alimaev[1] and Roy H. Behnke, Jr.[2]

[1]Department of Pasture and Fodder, Scientific Centre for Animal Production and Veterinary Research, 51 Jandosov St., Almaty Kazakhstan; [2]Macaulay Institute, Craigiebuckler, Aberdeen AB15 8QH U.K.

1. INTRODUCTION

This chapter examines the importance of land tenure in causing changes in the scale of livestock movement. 'Land tenure' refers here to the legal principles, written or oral laws, or (more broadly) culturally accepted rights and privileges with respect to property in natural resources. Land tenure is an institutionalized system of ideas. 'Land use' refers to observable patterns of land holdings, access to and exploitation of resources. Livestock mobility is an important component of land use in pastoral systems. This analysis focuses on livestock mobility because it is the best available indicator of the extent of rangeland fragmentation across several centuries in Kazakhstan, a country roughly the size of western Europe.

In Kazakhstan, as elsewhere in the semi-arid zone, agricultural intensification and nomadic settlement have been caused by a variety of factors including changes in agricultural technology, population pressure, and commercial considerations (Niamir-Fuller 1999). The decline of mobile Kazak pastoralism has also been punctuated by periodic revolutions in the property systems that external political authorities imposed on rural communities. In Kazakhstan as in few other settings, the state and its policies have forced changes on pastoral systems of land use.

Four idealized property systems provide the theoretical justification for different types of land tenure regimes. Based on European Enlightenment

K. A. Galvin et al. (eds.), Fragmentation in Semi-Arid and Arid Landscapes: Consequences for Human and Natural Systems, 151–178.

philosophy and systematized by modern scholarship, these idealized systems are distinguished by their characteristic property-owning units and by the distinctive mechanisms that control rates of resource exploitation for each property type (Bromley 1989, Table 7-1).

The theorizing summarized in Table 7-1 has been invoked to understand property rights and, more polemically, to create these systems by influencing policy. Each of the property types depicted here has been advocated by one or more of the 'grand' theories of political economy – capitalism, communism, or Euro-American notions of the 'primitive'. The popularity of the different property types has waxed and waned with the fortunes of these competing ideologies.

Pastoral Kazakhstan participated unwittingly in this European philosophical debate. In the 500 years that the Kazakhs have existed as a distinct people, their rangeland tenure systems have incorporated features from all the major ideal types of property regimes – common, private, state property and open access. Pastoral mobility also declined over the long term, often in association with major shifts between types of tenure systems. But the correlation between a kind of tenure system and a particular level of livestock mobility has not been simple. Periods of migratory expansion or contraction have also coincided with periods of stability in the overall tenure system, most recently under the Soviets (Table 7-2).

Until around 1800 most Kazakhs possessed a clan-based political organization, common property resource management, and an extensive system of migratory livestock production. Changes to these traditional forms of politics, tenure and mobility came in three major waves. Russian colonial penetration began the process. From 1800 until 1929, the decline of large-scale pastoralism followed a standard colonial pattern: an expansive European power used diplomacy and force to occupy what it perceived to be 'free', 'underused' or 'excess' land that native pastoralists were incapable of defending. Pastures suitable for cultivation were lost initially to imperial military settlements and eventually to colonial settlers. As alien land tenure laws and administrative

Table 7-1. Alternative tenure regimes.

Tenure type	Owners	Regulatory mechanism
State property	State (Hardin 1968)	Administrative control
Common property	Corporate groups (Runge 1981)	Collective restraint – 'stinting'
Private property	Individuals (Gordon 1954)	Internalization of resource rents
Open access	No one (Ciriacy-Wantrup and Bishop 1975)	Low levels of resource utilization

restrictions were imposed, native pastoralists defended their shrinking resource base by adopting agricultural practices and adapting their land tenure system to European expectations based on settled farming. The result was a hybrid land tenure regime combining elements of both Kazakh and Russian legal traditions, common and private property. Early Soviet rule softened the impact of this system on native pastoralists, but did not reverse imperial land policy.

In a second wave of change, the process of gradual accommodation stopped in 1929, about a decade after Soviet rule was established. In 1929 Stalin imposed collectivization and involuntary settlement on Kazakh pastoralists. Nowhere in the Soviet Union, in no Soviet satellite state, and in no part of Communist China was collectivization begun earlier or carried out more thoroughly. If state socialism was the great political and economic experiment of the 20th century, then Kazakhstan was one of the most radical expressions of what state ownership of natural resources could mean for indigenous pastoralists.

The experiment with state ownership ended abruptly in 1991 in a third wave of change that began with the collapse of the USSR followed by the creation of an independent Kazakhstan and market reforms. In the turmoil that ensued, Kazakhstan lost about two-thirds of its national flock, the state

Table 7-2. Property rights systems and periods of expansion or contraction in pastoral mobility.

Events	Date	Changes in migratory scale	Tenure regime
Traditional clan-based pastoralism	Pre-1800	Large-scale mobility	Common property
Contact with Russian military and settlers	1800-1917	Contraction	Common and private property
Civil War and Bolshevik New Economic Policy	1917-1929	Contraction	Common and private property
Collectivization	1930-1940	Severe contraction	State property
Re-emergence of seasonal pasture use	1941-1964	Expansion	State property
State farm intensification	1965-1990	Contraction	State property
Collapse of Soviet Union and decollectivization	1991-1999	Severe contraction	Open access and private property
Market economy established	2000-	Partial expansion	Open access and private property

farms disintegrated, and limited forms of private land ownership took their place. Today, pastoral Kazakhstan once again has a hybrid tenure system that combines elements of unregulated open access and private property modeled, as before, on the Russian peasant farm.

2. HOW DOES LAND TENURE MATTER?

Ideas about what constitutes legitimate property obviously affect patterns of land holding and settlement, but the relationship is indirect. Elizabeth Colson (1966) provided an elegant illustration of this point in a study of land use among African agro-pastoralists who were relocated to make way for the flooding caused by the construction of a large dam. Colson looked at patterns of land holding before and after relocation and found them to be quite different, though there had been no change in customary law. What had changed were the circumstances in which the law was applied – in this instance, from a riverine to an upland ecology, and from a situation in which most arable land was already under cultivation to one in which farmers were claiming and opening new fields at a previously unoccupied site. Colson concluded that legal uniformity was consistent with considerable variation in settlement and landholding patterns, and that there was no direct, necessary or obvious connection between legal rules and observable patterns of land use:

> The same legal rules may …be conjoined with quite different patterns of land holding depending upon the circumstances within which they are applied. A knowledge of the land law does not permit an observer to predict the likelihood of finding any particular pattern of land holding unless he can also predict a great many other influences which may be operating upon the community (Colson 1966:1).

Because of the gulf that separates legal principles from behavioral patterns, we should not expect to find a simple correlation between a particular kind of land tenure and a given level of fragmentation or integration in the use of a semi-arid landscape. But if – in Colson's words – "many other influences" intervene, then how important are different 'ideal' types of tenure systems, and how might these systems operate to constrain or promote fragmentation?

The rangelands of Kazakhstan provide an appropriate setting in which to examine these questions. Kazakh history has been dominated by long periods of incremental change interrupted by periodic revolutions in property rights systems. Struggles over land and pastoral mobility have been a persistent feature of every historical period. If different types of rangeland tenure do indeed influence migratory land use, then we should be able to observe these effects in Kazakhstan.

The following sections discuss in chronological sequence each period of migratory contraction or expansion. The discussion examines the policy initiatives that have governed the livestock sector, illustrates how policies were implemented, and assesses the reasons for the occasional reversals of official thinking on migratory stock keeping. Whenever possible we describe how pastoralists and collective farm workers responded to a process of change, driven in large measure from above, by government policies.

3. KAZAKHSTAN BEFORE RUSSIAN PACIFICATION

The Kazakhs emerged from an amalgamation of tribal groups in the late 1400s and within a century established control over a territory roughly comparable to modern Kazakhstan. At this time the Kazakhs were highly mobile pastoralists who kept mixed herds of sheep, goats, camels and horses, migrating annually on circuits of a couple hundred to a couple thousand kilometers (Olcott 1995). Pastures were held as a common clan patrimony:

> If times were stable, each clan and its member *auls* [encampments] had its traditional territory in which it migrated between summer and winter pastures. However, land was not 'owned'; it was only loosely identified with the clan for as long as its members pastured there. This loose territorial affiliation was a feature of Eurasian nomadic land use that can be traced back to…the Mongol empire of Chingis Khan (Martin 2001:21).

In this system of resource control, both property owning groups and the territories they controlled were flexible. Although a group's home base could usually be identified, this core area was surrounded by a periphery used by other pastoral groups, where ownership was contested and boundaries were uncertain (Khodarkovsky 2002). Group membership could be equally vague. Tribal and clan groups were organized around notions of descent, but kinship links were largely fictive and subject to opportunistic reformulation. Groups coalesced for concerted political and military action only in certain seasons or situations, and the groups themselves were made up of both core members and a shifting constellation of clients, allies and fictive kin (Bacon 1958, Krader 1963, Khazanov 1984).

The natural environment favored these arrangements. The Kazakhs occupied a mountain-steppe-desert environment in which pastoral resources were widely dispersed and erratically available. The dispersal of natural resources mitigated against the creation of a centralized political system. The unpredictable distribution and productivity of resources, due to drought, epizootic disease, and episodes of severe winter weather, favored the survival of groups that were prepared to appropriate whatever they needed to survive. It also required groups to maintain claims to large areas and to ceaselessly dispute access with other competing groups. Strong incentives for political action were therefore coupled with weakly institutionalized political authority. With no paramount political power to referee resource allocation, groups were organized to pursue their vital interests in a competitive environment made up of similarly constituted groups. These were societies organized for war (Khodarkovsky 2002). The instability inherent in this system was further exacerbated by external perturbations generated by neighboring settled states – the growth and periodic collapse of oasis-states to the south of the Kazakh steppes and cycles of Chinese expansion and contraction (Lattimore 1951, Khazanov 1992). Themselves rendered unstable by nomadic predation, the fluctuating fortunes of these states, and pastoral competition for control of the trade routes upon which they depended, sent repeated waves of dislocation rippling across the steppes (Grousset 1970).

Land relations in these uncentralized common property systems were sustained by intense political competition rather than administrative edict, a widespread organizational pattern across the semi-arid zone of pastoral Africa and Asia (Behnke 1994).

4. THE IMPERIAL PERIOD

After 1700, the Kazakhs increasingly came into regular contact with the Russian Empire as it expanded east into Siberia and south into Central Asia. This contact was eventually to transform their migratory way of life.

The abandonment of long-distance nomadism was a gradual process. Beginning in the early 18th century, grazing land was expropriated to create Russian lines of fortification and to provide farms for the Cossak military personnel who manned these installations. These forts, which frequently stood astride migratory routes, were used to control movement, which was initially viewed as a military threat and later as an administrative annoyance (Khodarkovsky 2002:215-216). From the 1820s to the 1860s, territorial units were superimposed on clan groups; provinces, districts, and villages were given defined boundaries and permission was required from the authorities to move outside these borders. The pastoralists therefore lost land on several fronts – some land was directly alienated to Russian defensive settlements and military settlers, and additional land was rendered inaccessible due to movement controls (Martin 2001).

After 1860, the appropriation of land by incoming Russian peasants gradually replaced land expropriation by force. Imperial land policies provided the legal cover for this process. In 1867-8 Russian legislation declared all land used by Kazakh nomads to be the property of the state. In 1891 further legislation specified that each Kazakh household was entitled only to the amount of land it needed, "such that everything above the level of need [the meaning of which was often debated and adjusted] could be deemed 'excess', and therefore available to the growing non-nomadic steppe population" (Martin 2001:72; see also Olcott 1995:87). Nearly three million Europeans, most of them peasants, settled in Kazakh territory in the decade prior to World War I (Olcott 1995:81). By 1916, forty percent of the population of the four northern provinces of Kazakhstan was Slavic, and the wave of settlement was moving progressively southward.

Imperial land tenure regulations did not in themselves cause this massive demographic shift. Much of the settlement was spontaneous, extra-legal and chaotic, with colonial administrators struggling to regularize it after the fact (Martin 2001:68-70, Kendirbai 2002). But Imperial laws legitimated the transformation of pastures into farms and probably accelerated the process of pastoral land loss. As one Russian administrator bluntly put it "not to actively promote the sedentarisation of the …[Kazakhs] means to neglect the needs of the Russian people" (Kendirbai 2002:55). But it is difficult to conceive of how different laws could ultimately have led to any other outcome, given the overwhelming disparities of power between the Russian Empire and the Kazakh clans. By the middle of the 18th century, the development of gunpowder weaponry meant that the steppe nomads were no longer a threat to an industrializing European state (Saunders 1971:168, 191). By the late 19th century, developments inside Russia – the abolition of serfdom, a growing rural population and land hunger – had created powerful incentives for peasant immigration into sparsely populated border areas. Eventually the development of railroads helped to efficiently deliver the colonists and to link the border lands into the expanding Imperial economy.

In addition to displacing nomads with peasants, Imperial land laws also fostered settlement by the nomads themselves. Haymaking for winter feed replaced movement to seasonal pastures, herd sizes declined with the loss of access to pastures, and former pastoralists farmed to make up the shortfall. It was estimated that by 1919 half of all Kazakhs migrated only from May to September and 90% grew at least some grain (Olcott 1995:98). The composition of Kazakh herds also changed, especially in northern Kazakhstan. "As Kazakh grazing land shrank, the relative importance of small animals for subsistence declined and cattle became the primary animal in the herd, bred for profit. Whereas in the mid-eighteenth century sheep and goats had accounted for over 90 percent of the total herd, by the revolution they made up only about half the herd" (Olcott 1995:98).

The impact of Russian land law and policy was great because pastoralists became active agents in its promulgation (Martin 1996:2). Kazakh customary law had always recognized more enduring rights to winter home pastures than to seasonal summer pastures, which were occupied on a first-come, first-serve basis (Martin 2001:116, 137, 202 note 3). Customary law did not equate assured access with permanent ownership, but this changed as Kazakhs exercised their right to appeal to Russian law. Especially in the north, some entrepreneurial Kazakhs engrossed their holdings, sowed grain as cattle feed, and switched to commercial cattle production for Russian markets (Olcott 1981:19). Changes also occurred in areas of heavy settlement, high grazing pressure, or potential land expropriation. At these sensitive locations, pastoral households subdivided and demarcated their winter pastures as hay fields, adopted agricultural practices that conferred security of tenure, prolonged their periods of residence on their home properties, and constructed permanent structures in order to establish exclusive ownership claims against both Russian settlers and other pastoralists (Martin 1996:4, Martin 2001:74-83). At the regional level, Kazakh community leaders used Russian administrative boundaries and movement controls to avoid their obligations in the clan-based system of pasture access. The regions that tended to opt out were situated in favored environments with an unusual concentration of resources, a dense population, and the capacity to sustain flocks year round (Martin 2001:121). With their resources already under pressure, these potentially self-sufficient communities had much to lose and little to gain from continued participation in any regional system of resource exchange.

With respect to a rangeland enclosure movement in East Africa, Manners noted that "Although the circumstances under which the change in land tenure discussed here took place are unique…the form and consequences of the change are not so distinctive" (Manners 1964:266). Much the same can be said for Kazakhstan. Up until collectivization, the evolution of land tenure systems in Kazakhstan ran in a well-worn colonial groove:

- Exclusive forms of land tenure made inroads into collective rangeland ownership by large kinship or tribal groups.
- Unusually productive resources were excised from the migratory system and appropriated for more intensive forms of land use by colonists, by entrepreneurial Kazakhs who saw new opportunities for making money, or by pastoralists seeking to defend their land rights in the face of growing land pressure.

- Coerced change combined with legal accommodation as pastoralists adopted for their own purposes the legal vocabulary of their rulers.

Cross continental regularities of this magnitude suggest that the fragmentation of the steppe landscape was in some sense inevitable, a question that will be examined by Behnke (Chapter 13).

5. REBELLION, CIVIL WAR, AND THE NEW ECONOMIC POLICY: 1916-1929

The Bolshevik October Revolution of 1917 and the incorporation of Kazakhstan into the Soviet Union transformed Kazakh pastoral systems. But this transformation was not immediate. Wide scale economic disruption and the immediate problems of establishing order preempted attempts at radical reform by the early Bolshevik administration (Olcott 1995:160). Initially the new regime bolstered its authority by accommodating Kazakh interests and incorporating the traditional leadership into the new system. The period went through two contrasting phases:

- First, 1916-21 was a time of herd loss and famine caused by rebellion (1916), civil war (1917-20), and a severe winter in 1920-21. By late 1922, the size of cattle herds was only one-third of that of 1916 and there had been a population decline of over a million people due to out-migration and death. Destocking and the inaccessibility of pastures led to forced settlement as a result of impoverishment (Olcott 1995:159), since many households no longer had enough livestock to make migration feasible or necessary.
- From 1922 to 1929, the livestock economy recovered but this recovery was not associated with increased levels of migratory movement. Official policy played a part in this result. Relief was available to Kazakh pastoralists from the public land reserves, which were gradually turned over to common grazing land, and from tax relief on wool and meat until 1922. Private ownership was also permitted in agriculture through Lenin's New Economic Policy, announced in 1921. But nomadism was not encouraged. The *Resolution of Land Construction Among the Nomadic, Semi-Nomadic, and Transformed to Sedentary Population of the Kazakh SSR*, issued in Moscow in 1924, called for regional governments to establish how much land was needed to support a household, and to provide this land to every household that agreed to settle. In addition, these households were to be given free construction material, loans for the purchase of farm implements and work animals, advice on farming, and relief from state and local taxes for five years.

By the time of the 1926 census, just under a quarter of the Kazakh population was engaged solely in agriculture; 38% depended only on livestock and about a third on mixed cropping and livestock production. Only 10% of the Kazakh population was classified as pastoral in the sense of migrating year-round, though two-thirds were semi-nomadic, migrating with their animals in summer (Conquest 1986:191). In this period Kazakh participation in farming was also increasing – in 1925, about 25% of all fields were sowed by Kazakhs, up from 15% in 1920. The same trends emerge from Table 7-3, which documents the falling proportion of the population engaged in some migratory movement between 1916 and 1928.

Lenin's policies had redressed some of the injustices of Imperial land policy, but they retained the long-standing Russian commitment to nomadic settlement, which increased in this period.

Table 7-3. Nomadism before and after the Revolution.

Province	1916 – percentage of population migratory	1928 – percentage of population migratory
Aktyubinsk	40	14
Semipalatinsk	73	37
Turgai	63	-
Uralsk	64	12
Akmolinsk	-	17
Semirech'e	72	Less than 40
Syr Darya	61	Less than 40

Source: Olcott 1995 appendix 4.

6. COLLECTIVIZATION AND *OTKACHOVKA*: 1930-1941

The accommodations reached with pastoral communities in the early period of Soviet rule were overturned with collectivization. The authorities created large collective farms by expropriating private land and moveable agricultural assets such as livestock. Rural elites were executed, deported, or fled and nomadism was suppressed. The impact of collectivization was immediate and appalling. Ninety-two percent of the national sheep flock was lost between 1929 and 1933, and over half the households in Kazakhstan disappeared during this period, falling in number from 1,350,000 in January 1927 to 626,950 in July 1933. Some fled Kazakhstan. Others shifted between collective farms in an effort to locate more grain or livestock, so that a quarter of those who had settled in 1930-32 were on the move again by the end of 1932, though without livestock (Conquest 1986: 195, 196).

The resultwas return migration, *otkachovka,* as starving people fled the rangelands and fell back on cities and farming areas in an attempt to stay alive. Whole rangeland areas were emptied of people and animals, and would remain empty for over a decade.

Collectivization was a rapid process; 95% of the rural population was collectivized by 1933, up from 7% in 1929. Each of the collective farms created in the early 1930s contained anywhere from 10 to 20 nomadic encampments with 10 to 15 families in each, settled several kilometers apart on a territory of about 20,000 hectares (Conquest 1986:193). The conditions on the collectives were rudimentary. There was little agricultural equipment, few houses or stock shelters, and little in the way of improved water supplies for desert or semi-desert settlements. Some collectives had neither seed nor livestock. Livestock were scarce because the Kazakhs had slaughtered them rather than hand them over to the authorities. In some areas half of the herd was lost in the first weeks of collectivization, and national estimates put total losses in the first year of collectivization at about 35% (Conquest 1986:194). A renewed wave of collectivization led to the winter slaughter of herds in 1931; hidden meat remained frozen and provided food until the spring thaws in 1932, when there was famine.

For those collective farms that were able to maintain a herd, little fodder was available and "driving the herds to the pasture was forbidden" (Conquest 1986:193). On one large state farm, inadequate shelter was provided for cattle and only 13,000 out of 117,000 head survived their first winter (Conquest 1986:194). Sedentarisation therefore played a direct role in collectivized herd losses due to starvation, in addition to prompting preemptive mass slaughter. As a proportion of the population, more people died during collectivization in Kazakhstan than in any other part of the Soviet Union, which must be attributable, at least in part, to the combined impact of collectivization with nomadic settlement.

The propaganda treatise *From Nomadic Life to Socialism* by I.A. Zveriakov explains why the authorities felt compelled to combine settlement and collectivization. Written in 1931 in the middle of the collectivization campaign, this book was intended as a theoretical guide for those implementing collectivization and settlement in Kazakhstan. Its rhetoric was uncompromising. It assumed that there existed a single scientific theory that could solve the "difficult question of the Soviet and world economies" (Zveriakov 1932:46). Those who disagreed were not simply mistaken, but were malicious or depraved. Various saboteurs played a role in the analysis, from "bourgeois restorer and right opportunist" to the "vermin circle.. great-power chauvinists and bourgeois opportunists" (Zveriakov 1932). The errors of these opponents of sedentarisation included the belief that it was necessary to build on existing livestock husbandry systems, that these systems would gradually and inevitably change, that current agricultural technology permitted only grazing in the steppe areas, and that sedentarisation would actually delay progress. Running through these errors, it was argued, was a fundamental misconception:

> The harm of [gradualist] theory is that the question [of sedentarisation] is transferred from the plane of the class struggle into the plane of relations between abstract man and abstract nature. It is necessary to remember that the intention of the vermin is to distract working people from class struggle by transforming class struggle into the struggle between man and nature (Zveriakov 1932:41).

The mechanisms meant to sustain settlement were mundane – veterinary inputs, hay making, tractors, rural electrification – but the goal was heroic: "sedentarisation…releases Kazakh husbandry from dependence on natural conditions…" (Zveriakov 1932:48). To cite environmental limitations or observable facts as reasons for not adopting an unrealistic agricultural policy was, within this ideological system, to be anti-revolutionary, defeatist and ignorant of the primary role of class struggle in determining reality.

The political and technical dimensions of collectivization were, therefore, closely linked: "Solution of the animal husbandry problem is absolutely impossible without ruthless class struggle" (Zveriakov 1932:20). Kazakh pastoralism was equated with feudalism; it was therefore exploitative by definition and any positive overtones of primitive or tribal communism were denied (Zveriakov 1932):

> Sedentarisation is a class question; if it solves the problem of the elimination of semi-feudal and tribal relations in the [Kazakh encampments], if it releases poor and middle classes from exploitation, this will only be accomplished as a result of class struggle (Zveriakov 1932:51).

But the definition of the offending class was treacherously elastic, and simply being poor was no assurance of being on the safe side of the class divide. Impoverished *bai's*, the former pastoral elite, became an even greater threat, it was asserted, when their wealth was lost or expropriated and their class affiliations were less visible. Poor livestock owners who did not cooperate with collectivization were deemed to be the *bai's* assistants and, hence, class enemies. In sum, no evidence from the natural or social worlds could be adduced to cast doubt on the sedentarisation process. The fact that without irrigation settled agriculture was impossible in much of Kazakhstan was irrelevant, as was the suffering or death of the purported beneficiaries of sedentarisation.

7. THE RE-EMERGENCE OF MIGRATORY PASTURE USE (*ISPOLZOVANIE OTGONNOGO JIVOTNOVODSTVA*): 1941-1965

After some final adjustments in the balance between private and collective ownership, the collectivization drive was concluded in 1938 with 98% of the rural population in the collective sector. Barely three years later, in a remarkable about-face, support for migratory livestock husbandry was adopted as USSR policy in 1941. (See Annex 1 for an extract from the official decree.)

Sedentarisation had been a way for the authorities to establish political and economic control over rural Kazakhstan. By 1941, this control was undisputed and it was safe for the authorities to abandon their old policies in order to explore more effective technical options for raising livestock. With the rehabilitation of the migratory option, the extensive livestock sector gradually assumed the institutional structure it was to maintain until the command economy was dismantled by the market reforms of the mid-1990s: centrally directed seasonal livestock movement organized by large state farms.

The re-emergence of migratory livestock husbandry in the 1940s was encouraged by a combination of factors:

- *The technical and economic limitations of settled livestock husbandry.* Following collectivization, livestock could, depending on circumstances, spend from 100-180 days or more per year stall-fed (Zalsman 1948). Extended periods of winter stall feeding did not at first create problems since there were so few animals, but the limitations of this management system became more apparent as livestock numbers increased. The costs of fodder preparation, transportation, and storage were high. Irrespective of costs, some collective farms did not control enough haylands to produce sufficient fodder to sustain their herds over the winter period. In 1940 there was on average within collective farms (*kolkhozs*) 0.5 ha of pasture per sheep in Almaty Province, in Southern Province 1 ha, and in Jambul Province on average 1.5 ha per head, when average pasture requirements were about 4 ha per sheep. At the same time vast areas of seasonal pastures outside the farm boundaries were unused. And the more land that was plowed in the vicinity of a settlement, the less that remained for pastures; only 20% of the pastures allocated to farms in Jambul Province, for example, remained as pastures by 1940 (Matveev 1950).
- *The indigenous technical knowledge of Kazakh shepherds.* Migratory herding systems were reinvented, spread, and became permanent during World War II, a process of spontaneous experimentation and diffusion

that was thought likely to continue well into the 1950s, according to a senior scientific observer of the process (Zalsman 1948). For socialist Inner Asia, Sneath attributes this policy reversal to the survival of traditional pastoral values among the collectivized labor force, an explanation that also applies to Kazakhstan:

> The legacy of [the pre-collective period]... to the collectives was to accord highest value to the 'yield-focused' or 'specialist' pastoral strategy. This involved high mobility, and was seen as a 'professional' and diligent herding technique, associated with high positions of trust and prestige. The 'model herder' of the collectives inherited these values, and strove to get yields above the norm set in the plan. The collectives supported the long moves basically because they produced an excellent return of livestock products (Sneath 1999:268).

In Kazakhstan commentators spoke of "studying and using the centuries-old experience of former nomads" and following the example of "leading collective farms" (Zalsman 1948, Borodyn 1950), or simply cite old Kazakh shepherds as their authorities (Balmont 1950). The scientists of this time leave no doubt that their early attempts at reoccupation of deserted pastures were guided by the indigenous knowledge of local shepherds.

- *The combination of technical arguments with an acceptable political rationale.* Migratory herding was suppressed during collectivization on the grounds that it was both technically inefficient and politically unacceptable. By the 1940s, the practical demands of livestock management in a war economy had counteracted simplistic ideological misgivings. It remained to be shown, however, that migratory stock keeping could be reorganized according to socialist principles. This argument was won by directing attention to the overriding deficiency of traditional nomadism – its exposure to catastrophic winter livestock losses due to *djut*, icing events that prevented animals from grazing beneath the snow cover. Under socialism, it was argued, collective farms could support specialized fodder production brigades that would provide emergency winter fodder for shepherds, a division of labor that had been beyond the capacity of all but the richest traditional pastoral families. Mechanization, improved social services, and cultural amenities were made possible by collective economies of scale and further reinforced the advantages of a new industrial and distinctly socialist form of migratory herding (Borodyn 1950).
- *The successful application of science to the problems of migratory pastoralism.* Soviet livestock and pasture sciences refined and extended the local knowledge of Kazakh shepherds. Research concentrated on

practical problems – how to manage stock in winter conditions, breed selection for different production systems and locations, the engineering of water supplies and, above all else, pasture inventories to identify underused grazing resources, their period of optimal utilization, and appropriate stocking rates.

National maps of flock migration suggest that by the early 1950s most pre-collectivization patterns of migration had been reestablished. Contemporary observers classified the resulting movement patterns into four types differentiated by the season of movement and the tenure arrangements under which farms obtained access to distant pastures (Table 7-4).

Table 7-4. Soviet systems of migratory movement and rangeland tenure.

Movement type	Land tenure on distant pastures	Distance from home base in km	Region	Notes
Stable occupation of distant pastures, not returning to farm base	State land	400-450	Central and northern KZ; Karaganda, Akmola and Semipolatinsk Provinces	Mainly horses, some sheep and cattle
Distant winter pastures	Collective farm land	250-300	Central KZ; Akmola and Torgai Provinces	Home area fully plowed
Distant summer pastures	State Land	200-250	Southwestern and south KZ; Guriev Province	Home area deficient in summer pastures
Four-season movement system, all on distant pastures	Collective farm land	250-350	Almaty, Jambul, Chimkent, Taldi Kurgan Provinces	Farms own separate sections in each major pasture zone

Source: Asanov et al. 1992.

8. *'VIHADNOE POGOLOVIE'* AND INTENSIFICATION: 1965-1990

Another policy shift, this time to less nomadic systems of sheep husbandry, was signaled by the enactment Occupation of desert and semi-desert *pastures for developing sheep production in the Kazakh SSR*, issued in 1964 by the USSR and Kazakhstan Council of Ministers and by the Central Committee of the Communist Party of Kazakhstan. (See Annex 2 for a summary of this act). The act mandated the creation within five years of over 150 specialized sheep breeding *sovkhozes* or state farms, each with 50-60,000 head, on state land in the semi-desert regions of the republic. A second enactment in 1979 by the Central Committee and Council of Ministers of Kazakhstan renewed the intensification program by setting more targets and allocating additional money. According to the *1979 Act*, between 1980 and 1990 the number of sheep in Kazakhstan was supposed to increase by 42% to 50 million head. Both acts provided funding for a program of intensification that included the building of new settlements, irrigation works, road building, and the extension of electricity to rural areas.

The *1964* and *1979 Acts* did not intentionally reverse previous policies, but rather dealt with the emerging implications of those policies. From the 1930s to the early 1960s the national small ruminant flock grew steadily, as it recovered from collectivization by utilizing reopened grazing areas. By 1961 small stock numbers exceeded pre-collectivization levels, but the rate of flock growth had fallen off. Continued 'lateral' expansion was becoming increasingly difficult since the best pastures had already been reclaimed.

But the Soviet system did not readily admit to environmental limits, and stable stock numbers were not an option for most state farm directors. The Soviet system (as Zveriakov, quoting Engels, had argued with respect to sedentarisation) would lift the nomadic encampment "from the kingdom of necessity to the kingdom of freedom" (Zveriakov 1932:93). In mundane terms, the denial of natural limits lay behind the unspoken implications of a standard Soviet reporting category for farm managers - *vihadnoe pogolovie* 'quantity [of livestock] at year end'. It was assumed that ambitious farm managers would see to it that next year's flock was larger than that of last year. Once the rangelands were full, the only option was to exploit already used pastures more intensively.

In practice, intensification meant resettlement and irrigation (Kazgiprozem 1983:36). Under both the 1964 and 1979 enactments, large villages and state farms were created in semi-desert rangeland areas that had not previously been permanently inhabited. Both sheep and the humans that cared for them were now to live in large numbers on seasonal pastures, and the fodder deficiencies of these marginal areas were to be offset by increased irrigated fodder production. Available statistical data suggest that such a shift did occur. In 1970, about half of the national livestock feed supply came from natural pastures; by 1980, over 60% was artificial.[1] The second intensification initiative aimed to continue this process; there was a planned increase in total irrigated area of 48%, in improved pastures of 69%, and in irrigated pastures of 180% (Kazgiprozem 1983). Cultivated fodder was to substitute for natural forage supplies that were becoming incapable of feeding the growing ruminant population. Massive increases in both planted area and forage output occurred between the mid-1950s and 1980.

Seasonal pasture use did not stop, but it came under increasing pressure as all available grazing niches were occupied. Increasingly, only fragments of an integrated livestock route might be used, or the portion of a farm's flock that migrated to seasonal mountain pastures might now return to pastures that had been grazed by flocks that remained behind. Spring-autumn pastures were occasionally used in three seasons or year-round, the intervals between use were shortened, or movement largely stopped in spring during the period of rapid pasture growth. Limited access to seasonal pastures was especially a problem for the new semi-desert farms that had been created under the intensification program decades after the most valuable rangelands had been allocated.

Accompanying intensification was a steady growth in farm size. In 1953, there were in Kazakhstan 291 *sovkhoz* or 'state farms' in which workers were government employees, and 2,966 *kolkhoz* or 'collective farms' which were nominally owned by the workers. By 1982, this ratio was reversed, with 2,112 *sovkhoz* and 394 *kolkhoz*. Since the average state farm was twice as large as a collective farm (about 80,000 versus 40,000 ha), the territorial size of farm units across Kazakhstan had very nearly doubled in two decades. The transformation of collective into state farms and their enlargement started before but continued throughout the period of official intensification initiated in 1964. By 1980, specialized sheep farms kept flocks of about 27,000 head and held 98% of the national flock. Farm enlargement was accompanied by improved infrastructure in the form of more sheep shelters and larger areas of pasture accessible to new water points, all built at an increased rate in the 1970s (Central Statistics Directorate Kazakh SSR 1984, Olcott 1995:239). Improvements in shelter, water, and pasture availability meant that livestock were able to spend more time within the boundaries of farms that were, in any case, now much larger. This meant, in turn, that it was possible for these farms to raise specialized sheep breeds adapted to one ecological zone rather than traditional nomadic breeds with the ability to range widely across zones (Kazgiprozem 1983:39-41).

The consequences of increased grazing pressure and reduced mobility were not immediately apparent. Up to 1984 the national flock size increased, particularly in desert and semi-desert zones. As total stock numbers grew and finally peaked at around 35 million head in the early 1980s, so did the total national production of meat, wool, and karakul lamb pelts. But rapidly increasing production costs were not compensated by what were now relatively modest improvements in livestock output. Wool yields per sheep may have improved in the 1960s but were stable thereafter, and the average live weight of sheep sold to the state actually declined by 10 kg, from 46 to 36 kg per head between 1960 and 1983 (Central Statistics 1984:25). After the early 1980s, any substantial benefits from increased investment had been realized, and the national sheep flock began, gradually at first, to decline in size.

The intensification program was also running out of steam. In the first intensification drive between 1964 and 1969, 155 new farms were planned and 158 were built; in the second wave of intensification, 78 new specialized fodder or sheep breeding *sovkhozes* were slated for construction between 1980 and 1990. Over half, 37, of these 'new' farms were in fact conversions, of collective into state farms, or of existing sheep farms into specialized sheep farms; in the end only 29 of the 78 planned farms were either built or converted (Kazgiprozem 1983).

The Soviet experiment was coming to a close. Stalin's settlement and collectivization policies had been a thorough if inhumane test of the capacity of social engineering to prevent 'inevitable' capitalistic development, and the experiment had yielded mixed results. Some aspects of collectivization endured for half a century, such as the creation of large farms and economies of scale in the use of machinery and other support services for shepherds. Other elements, such as restrictions on livestock movement, were officially abandoned within a decade of their imposition.

When more settled forms of livestock production re-emerged in the 1970s, intensification was driven not by ideological fervor but by the practical need to accommodate ever more animals on limited pastures. There was, nonetheless, an ideological component to the dilemma facing managers at this time. The authorities demanded ever higher levels of aggregate output; somebody, it would seem, still believed in Zveriakov's assertion that "sedentarisation…releases Kazakh husbandry from dependence on natural conditions" (1932:48).

Initial successes in fulfilling this ambition were, in the end, to render the entire system more unstable. When flock sizes were small, cheap, and relatively simple interventions – additional water points, well placed barns or winter forage supplies – sufficed to prevent winter losses in poor weather. Farm managers adopted these improvements because they were profitable, but these successes ultimately resulted in perverse consequences. Because they kept more sheep alive, more elaborate and expensive precautions had to be taken to sustain performance or prevent losses with every passing year.

Weather-driven fluctuations in animal numbers and performance are a persistent feature of extensive livestock production systems in semi-arid environments (Behnke 1992, Kerven et al. 2003). In the half-century between collectivization and market reform, Soviet Kazakhstan presented a different picture – multiple decades of constant flock expansion followed by stability.

Intensification had suppressed the effects of climatic variability in a command economy that did not encourage a close calculation of immediate profits and losses. But because it led to the maintenance of larger numbers of animals, intensification restricted the operation of migratory production systems. It also made individual pastoral households dependent on a lavish state support system. Finally, it eroded the profitability of state sheep farms, leaving these farms indebted, overstaffed, and exposed to sudden collapse in the 1990s when state subsidies were removed.

Stalin's experiment had indeed settled the nomads, but not in the way or on the time scale that its proponents had foreseen. As we will show in the next section, it was the collapse of the Stalinist alternative and the reintroduction of a market economy that, in the end, did the job.

9. AFTER 1991: DECOLLECTIVIZATION AND COLLAPSE OF LIVESTOCK MOBILITY IN POST-SOVIET KAZAKHSTAN

Government support mechanisms ceased after state farms were denationalized in the mid-1990s (ADB 1996, Robinson 2000). Many state farms were already in debt prior to decollectivization and their capital and infrastructure were disbursed in several ways – sold to pay debts, given to preferred farm employees as part of their farm share, stolen or abandoned. A few state farms were reborn as cooperatives, and managed to retain some control over assets that could still be used to move livestock seasonally, but most centralized services for moving livestock simply collapsed (Robinson and Milner-Gulland 2002, Behnke 2003, Kerven et al. 2003).

Concurrent with the loss of state support to collective farms, an economic crisis in Kazakhstan accompanied the shift from a centrally planned to a market economy. One of the casualties of this crisis was the national small stock population, which crashed by two-thirds in a couple of years. Over a quarter of the national flock disappeared in 1994 alone.

In the mid-1990s the state also passed legislation that made it possible for individuals or groups to lease agricultural land, initially for a period of 99 years, and later amended to 49 years, with the state remaining the owner. By 2000, the leasing of land for cultivation was common whereas leasing of rangeland was not. This pattern reflected the relative value of the two types of land. Land registration procedures were cumbersome, corrupt and expensive, and exposed the lessee to land taxation. Only relatively valuable land was worth the trouble and expense of registration. Arable land was intrinsically more productive than rangeland, and the low value of rangeland had been exacerbated by the drop in livestock numbers. Until stock numbers stopped falling in 2000, rangeland resources were abundant relative to stocking pressure and there was little competition among pastoralists to control land by registering it (Behnke 2003).

The reform tenure laws that were designed to hand land back to private individuals and groups had therefore not worked as intended in rangeland areas. A few large flock owners had registered small areas of land containing dwellings for shepherds, animal barns, water points, and hay fields. These leases controlled valuable resources while exposing the lessee to a minimum of land taxation. Privately registered rangeland was invariably insufficient to support the lessee's flock which grazed open rangeland that had reverted to the state. Typically the land rights of small flock owners were even more tenuous. Some owned shepherds' dwellings or barns but not the land these structures stood on; many operated small flocks from the back yard of their (privately-owned) house in the village. In all these cases the flocks grazed on land and used water sources that were unallocated and had reverted to the state, though these resources could be under the informal control of particular individuals or groups.

National statistics give some indication of the extent of rangeland abandonment. Prior to reform in 1991, about 80% of all land in Kazakhstan was registered to agricultural users; by 2001 the proportion of registered agricultural land had fallen to 39%. At the same time, state reserve land, the bulk of which was unoccupied rangeland or seasonal pastures, had risen from 7% in the Soviet period to 44% of all rural land by 2001 (Sabirova 2001:32-33). The state therefore remained in control of rangeland despite market reforms. But because it anticipated transferring this property to lessees, the state had developed no institutions for managing it.

National estimates of the extent of flock mobility are not available for the Independence period, but between 2000 and 2004 an interdisciplinary study of livestock grazing patterns was conducted in six former collective farms in two provinces to the south (Almaty Province) and center (Jambul Province) of the country.

Figure 7-1 illustrates the propensity of flocks owned by these villagers to graze in the immediate vicinity of their home village. Villages 1-3 lay along a transect to the west of Almaty city stretching north from the Alatau mountains into the semi-desert adjacent to the Ili River. In the pre-Soviet and Soviet periods, these villages had been linked together into a regional system of seasonal migration from high to low elevation, a north-south distance of about 200 km. Along this transect at least 40% of all the grazing by village-owned flocks took place within 5 km of the central village in 2003. Villages 4-6 lay along the Chu River, with the Moinkum Desert to their south and the Betpackdalla plains to their north. In pre-Soviet and Soviet times, these villages had also been part of a regional north-south migratory system that linked the steppes north of the Betpackdalla with the Moinkum Desert, a distance of about 500 km. By 2003, the extent of grazing around the Chu villages was highly variable depending on the pastures around the village. Villages 4 and 5 were situated along parts of the Chu floodplain that afforded poor and seasonally unstable forage availability. Most of the grazing around these villages came from transient flocks passing through the area. In contrast, natural forage around Village 6 was plentiful, diverse and of reasonably good quality, and over half of all the animals owned by the residents of Village 6 spent all of their time around the village.

According to Figure 7-1, the propensity to migrate seasonally is highly variable. It is also linked to the availability of pasture resources around large settlements (Kerven et al. 2003). Kazak pastoralists move away from large settlements, but only when they are forced to do so by the scarcity of forage at these sites. Following decollectivization, the maximum amplitude of these movements has also been reduced, down from a north-south distance of about 500 km to under 200 km in the Chu area, and from around 200 to 30-50 km along the transect west of Almaty.

Table 7-5 looks at the kinds of flocks engaged in mobility versus settlement in the six villages previously discussed. The extent of mobility in these villages depends in part on how mobility is measured. Examined from the perspective of animal numbers, these are fairly mobile husbandry systems, with nearly three-quarters of all sheep and goats based outside the main villages and nearly half of all sheep and goats (46%) seasonally mobile as well. The picture looks somewhat different if one tabulates results in

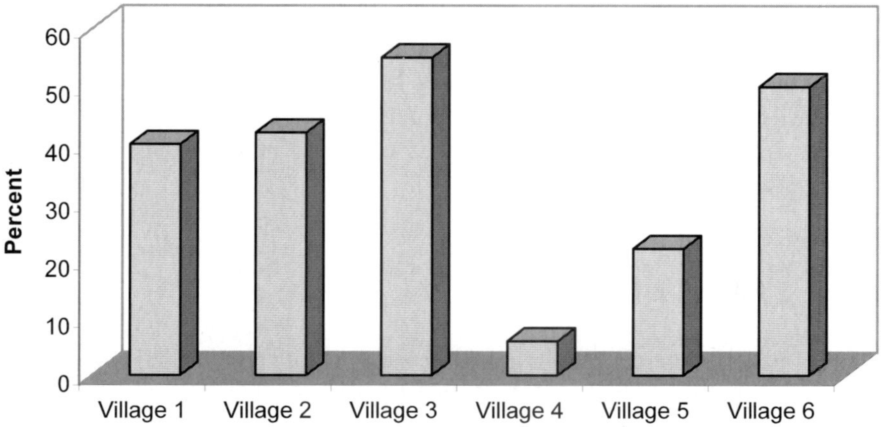

Figure 7-1. Percentage of all grazing that occurs around six villages.

Table 7-5. Mobility in six former collective farms in 2002-03.

Movement category	Number of sheep and goats	% of sheep and goats	Number of flocks	% of flocks	Mean flock size – head of sheep and goats
Based in village and do not move	7666	19	174	38	44
Based in village and move	2896	7	138	31	21
Based outside village and do not move	11110	28	74	17	150
Based outside village and move	18597	46	62	14	300
Total	40269	100	448	101	90

terms of the husbandry practices of flock owners, 65% of whom are based in the central villages and keep their animals on village pastures at least part of the year. This apparent discrepancy is explained by the correlation between flock size, mobility, and residence. Flocks that are based outside the village are on average about six times larger than those based in the village. This finding suggests the possible re-emergence of less fragmented patterns of land use. If shepherds adopt geographically scattered home bases and more mobile migratory patterns as their flocks grow in size, then they will voluntarily spread their livestock into remote seasonal pastures as flocks recover from the population crash that occurred in the 1990s following privatization.

10. CONCLUSIONS

We began this chapter by asking how ideas about land ownership affected patterns of land use. Colson provided an initial response to this question: the relationship was indirect and mediated by "many other influences" (Colson 1966:1). Colson's case study was elegant because it was simple. Concepts of ownership were generally agreed upon by all land users, there was a clear before and after, and small numbers of people were involved.

The history of mobility in Kazakhstan was very different. How land should be owned was contested, and the temporal and spatial scale of events was vast. There are, nonetheless, aspects of Kazak history that potentially clarify the relationship between ideology, land tenure and mobility. One of these is the six-decade period between 1930 and 1990, when Kazakhstan was governed by a stable ideology and its pastoral tenure system evolved gradually. Did state socialism as practiced in Kazakhstan have a consistent impact on levels of herd mobility, or did mobility in this period vary significantly in response to Colson's 'many other influences'? The ideological shifts that opened and closed the Soviet experiment also allow us to ask the question in another way: To what extent did abrupt changes in ideology and tenure, collectivization and decollectivization, precipitate equally rapid changes in mobility patterns?

In the sixty years of Soviet collectivization, communist ideology was compatible with remarkably different levels of pastoral mobility:

- settlement (1930-40)
- the re-emergence of nomadism (1941-64)
- the erosion of mobility (1965-90).

These changes were sanctioned by a Soviet ideological commitment to collectivization and 'modern' industrial techniques of agricultural production: first sedentarisation, then mechanized fodder production, and finally irrigated agriculture. Throughout the six decades of collectivization, neither the ideology nor the level of ideological commitment changed fundamentally, but the means chosen to implement ideological objectives did evolve, with unpredictable consequences for mobility patterns. The ideology of state socialism was, in short, subject to multiple interpretations. It rationalized agricultural policies but was not specific enough to dictate their contents, which could either encourage or impede mobility.

Collectivization was initiated in the 1930s and terminated in the 1990s by abrupt shifts between state and private ownership of natural resources. These upheavals were remarkably similar, despite their antithetical intentions. Each was followed by an economic depression that depopulated the countryside, destroyed rural infrastructure, and caused a massive crash in livestock numbers. The most apparent difference between the two episodes was quantitative rather than qualitative, the Stalinist revolution being more brutal and more destructive than the capitalist revival of the 1990s. But the immediate implications of both revolutions for pastoral mobility were identical. Impoverished pastoralists settled, at least temporarily, and began the slow process of rebuilding their herds and reoccupying remote pastures. The disruptive impact of radical reform would seem, at least in the short term, to be more important than the content of the reforms.

The history of pastoral mobility in Kazakhstan therefore provides little evidence of a direct relationship between levels of mobility and different idealized tenure regimes. State property was consistent with both the suppression and maintenance of mobility, and the shift between communism and capitalism had equally deleterious impacts irrespective of the direction of change.

The survival of mobility in Kazakstan owes less to any particular regime of resource ownership than it does to the tenacity of pastoral communities. With respect to pastoral access to natural resources, there has in Kazakstan been a fair degree of *de facto* continuity over the centuries. Irrespective of the ideology behind the land tenure system prevailing at a particular time, pastoralists have managed, with a few notable and disastrous exceptions, to hold on to the geographically extensive areas needed to sustain mobility.

Prior to Russian contact, both collective and individual land rights were recognized by customary law, with individual rights encapsulated within and sustained by collective political responsibility. After pacification, both Russians and Kazakhs began excising from the collective patrimony those resources that were susceptible to privatization or commercially attractive, basing their expropriations on the protection offered by laws and government officials, rather than membership in territorial groups. Increasing individualization of land rights was therefore occurring in northern Kazakhstan under the stimulus of Russian settlement at the end of the Imperial period, and was associated with fragmented patterns of resource ownership and reduced scales of migratory movement. But the bulk of the country's semi-arid rangelands in central and southern Kazakhstan were never privately owned and managed. As Soviet control took root at the local level in the 1920s, pasture ownership passed from the clans to clan-based soviets; these were replaced by *kolkhozes* in the 1930s, which were replaced by *sovkhozes* between 1950 and 1980. *De facto* collective pasture use remains in place to this day. Despite land laws to encourage the private leasing of agricultural land, privately-owned flocks obtain the bulk of their feed from open access to pastures legally controlled by the state. It would seem that any of the four major land tenure types – communal, state, private or open access – can be interpreted to provide the geographically extensive units of resource control that are compatible with nomadism.

ACKNOWLEDGEMENTS

Research was supported by U.S. National Science Foundation Grant DEB-0119681 and by European Commission grant ICA2-CT-2000-10015. Nurlan Malmakov translated Zveriakov's *From Nomadic Life – To Socialism*; Aidos Smailov translated into English all other sources originally in Russian.

ANNEX 1

Extract from: The enactment by the Peoples' Commissioners of the USSR and the Central Committee of the Whole Union of the Bolshevik Communist Party: 'Measures for preserving young animals and increasing the number of livestock in kolkhozes and sovkhozes'

Section 4: Organizing distant pasture livestock production

45 Attaching importance to utilization of natural rangelands that are in several regions, in order to increase the number of livestock and obtain cheaper livestock products, the following institutions – the USSR People's Commissioners of Land Use (now Ministry of Agriculture), People's Commissioners of Sovkhozes of USSR (Ministry of Sovkhozes), Soviet People's Commissioners (Cabinet of Ministers) of Kazakh, Kyrgyz, Uzbek, Tadjik and Azirbaijan SSRs and Regional Executive Commissioners of Altya, Krasnoyarsk and Ordjonikedzevsk Adminsitrative Units – are obliged to:

a. To devise a plan within two months for developing measures to organize distant pasture management of livestock on summer and winter rangelands. To give permission to kolkhozes of Chkalovsik, Chelyabinsk, Omsk and Novosibirst Oblasts and Altai Administrative Unit to use as pastures the land of State Land Fund in the regions of Kazakh SSR in accordance with Soviet People's Commissioners (Cabinet of Ministers) of Kazakh SSR.
b. To distribute and allocate rangeland (before July 1 1942) to kolkhozes that are organizing distant pasture livestock breeding for periods of not less than 10 years.
c. To establish livestock movement tracks for accessing distant pastures, and to organize stopping points along these routes which are provided with water and necessary amounts of fodder.
d. In 1942 to provide livestock on distant pastures with water points by reconstructing old watering points, building new wells and water reservoirs, and also to build the necessary livestock barns and houses of kolkhoz shepherds looking after the animals.
e. To provide insurance fodder on winter pastures at a rate of 7-8 centners [1 centner = 100 kg] per head of cattle and 10-12 centners per head of horses and 1 centner per head of sheep.

46 To recommend to kolkhoz administration to pay 50% above the normal salary for kolkhoz workers caring for livestock.

47 To establish additional salary of 50% for veterinary technicians, doctors and medical attendants above the ordinary wage.

48 To give permission to the Agricultural Bank of the USSR to provide in 1942 the kolkhozes that are in the regions of distant pasture livestock breeding with credits for a 5 year period in the amount of 5 million rubles for building wells, barns and houses.

49 To free kolkhozes from obligatory supply of agricultural goods from all temporarily allotted territory located on State Fund lands and used for distant pastures.

50 For the Peoples' Commissioners of Finance, USSR, to stipulate the expenses in 1942 from the Republican budget to finance measures for construction of seasonal pastures.

Published in Pravda, 13 March 1942

ANNEX 2

Summary of *'Occupation of desert and semi-desert pastures for developing sheep breeding in Kazakh SSR'*, an enactment by the Central Committee of the Communist Party of Kazakhstan and the Counsel of Ministers of the Kazakh SSR, issued in 1964.

The objective of the enactment was to promote the rapid development of specialized sheep breeding in the desert and semi-desert regions of Kazakhstan. This was to be achieved by increasing sheep and goat numbers up to 50 million head through the creation of 155 specialized sheep breeding state farms with 50-60,000 head each on land taken from the state land fund. Specifically the act require agricultural scientists and administrators to:

- Zone and re-allocate land among the different agricultural subsectors and production systems in Kazakhstan.
- Identify land suitable only for pasture and improve the efficiency of its use.
- Increase productivity from irrigation through the control of soil erosion, occupation of saline land, improvements to natural rangelands and the creation of artificial pastures and hay lands.
- Determine the correct volume of production of the main agricultural commodities.
- Develop specific proposals for the construction of new sheep breeding and fodder producing state farms.
- Design a rational system for the location of human populations in the sheep breeding regions.
- Design a road, rural electrification and communication network for the sheep breeding regions.
- Specify the costs of these measures and set priorities for their implementation.

ENDNOTES

[1] Sheep depended on cultivated forages for much less than 60% of their diet. Most cultivated feed was consumed by cattle but data currently available do not permit us to separate fodder use by sheep from consumption figures for all types of livestock.

REFERENCES

ADB (Asian Development Bank). 1996. Working Paper No. 1 on Land Reform. Strengthening the implementation of agriculture sector reforms. Danagro Adviser a\s and Landell Mills Ltd, Almaty.

Asanov, K. A., B. P. Shakh, I. I. Alimaev, and S. N. Pryanishnikov. 1992. Pastures of Kasakhstan. Gylym-Science , Almaty.

Bacon, E. E. 1958. Obok: A study of social structure in Eurasia. Wenner-Gren Foundation for Anthropological Research Inc, New York.

Balmont, V. A., V. I. Matveev, A. I. Polyakove, and N. I. Suvorov. 1950. Funds of winter pasture for occupation by migratory livestock husbandry: Southern part of pre-Ili Moinkun (Sary-Tau-Kum). *In* Works on the organization of distant pasture livestock breeding. Kazakh State Publishing House, Alma-Ata.

Behnke, R. H. 1992. New directions in African range management policy. Pastoral Development Network Paper 32c. ODI, London.

Behnke, R. H. 1994. Natural resource management in pastoral Africa. Development Policy Review 12:5-27.

Behnke, R. H. 2003. Reconfiguring property rights in livestock production systems of western Almaty Oblast, Kazakstan. Pages 75-107 *In* C. K. Kerven, editor. Prospects for pastoralism in Kazakstan and Turkmenistan: From state farms to private flocks. Routledge and Kegan Paul, London.

Borodyn, I. E. 1950. The distant pasture livestock breeding system in Betpak-Dalla Desert conditons. *In* L. M. Zalsman and B. L. Blomkvest, editors. The experience of distant pasture livestock management in Kolkhozes. State Publishing House for Agricultural Literature, Moscow.

Bromley, D. W. 1989. Economic interests and institutions: The conceptual foundations of public policy. Basil Blackwell Inc, New York and Oxford.

Central Statistics Directorate Kazakh SSR. 1984. Thirty years of cultivation: A short statistical compilation. Central Statistics Directorate Kazakh SSR, Alma-Ata.

Ciriacy-Wantrup, S. V. and R. C. Bishop. 1975. Common property as a concept in natural resource policy. Natural Resources Journal 15:713-27.

Colson, E. 1966. Land law and land holdings among the Valley Tonga of Zambia. Southwestern Journal of Anthropology 22:1-8.

Conquest, R. 1986. The harvest of sorrow: Soviet collectivization and the Terror-Famine. Hutchinson, London.

Gordon, H. S. 1954. The economic theory of a common-property resource: the fishery. Journal of Political Economy 62:124-42.

Grousset, R. 1970. The empire of the steppes: A history of Central Asia. Rutgers University Press, New Brunswick, NJ and London.

Hardin, G. 1968. The tragedy of the commons. Science 162:1243-48.

Kazgiprozem (Kazakh State Institute of Land Tenure and Use). 1983. General scheme for the occupation and use of desert, semi-desert and mountain haylands and pastures of the Kazakh SSR, by complexes. Ministry of Agriculture Kazakh SSR, Alma-Ata.

Kenderbai, G. 2002. Land and people: The Russian colonization of the Kazakh steppe. Klaus Schwarz Verlag, Berlin.

Kerven, C., I. I. Alimaev, R. Behnke, G. Davidson, L. Franchois, N. Malmakov, E. Mathijy, A. Smailov, S. Temirbekov, and I. Wright. 2003. Retraction and expansion of flock mobility in Central Asia: costs and consequences. VII International Rangelands Congress, Durban South Africa, July-August 2003.

Khazanov, A. M. 1984. Nomads and the outside world. Cambridge University Press, Cambridge.

Khazanov, A. M. 1992. Nomads and oases in Central Asia. Pages 69-89 *In* J. A. Hall and J. C. Jarvie, editors. Transition to modernity: essays on power, wealth, and belief. CUP, Cambridge.

Khodarkovsky, M. 2002. Russia's steppe frontier: The making of a colonial empire, 1500-1800. Indiana University Press, Bloomington and Indianapolis.

Krader, L. 1963. Social organization of the Mongol-Turkic pastoral nomads. Mouton & Co, The Hague.

Lattimore, O. 1951. Inner Asian Frontiers of China. Capitol Publishing Co, Inc. and the American Geographical Society, New York.

Manners, R. A. 1964. Colonialism and native land tenure: a case study in ordained accommodation. *In* R. A. Manners, editor. Process and pattern in culture: Essays in honor of Julian H. Steward. Aldine, Chicago.

Martin, V. 1996. Nomads, borders and the resolution of land disputes in the Middle Horde Kazakh Steppe. Association for the Advancement of Central Asian Research Bulletin 9:2-5.

Martin, V. 2001. Law and custom in the steppe: The Kazakhs of the Middle Horde and Russian colonialism in the nineteenth century. Curzon, Richmond, Surry.

Matveev, V. I. and A. I. Polyakov. 1950. Distant pasture livestock breeding in the sands of Moinkum and in Chu-Ili Mountains of Jambul Oblast: Direction and specialization of livestock over several years. *In* Works on the organization of distant pasture livestock breeding. Kazakh State Publishing House, Alma-Ata.

Niamir-Fuller, M. 1999. Managing mobility in African rangelands: The legitimization of transhumance. Intermediate Technology Publications Ltd, London.

Olcott, M. B. 1981. The settlement of the Kazakh nomads. Nomadic Peoples 8:12-23.

Olcott, M. B. 1995. The Kazakhs. Hoover Institution Press, Stanford, California.

Peoples' Commissioners and Central Committee of the Whole Union of the Bolshevik Communist Party. 1942. Measures for preserving young animals and increasing the number of livestock in Kolkhozes and Sovkhozes. Pravda 13 March 1942.

Robinson, S. 2000. Pastoralism and land degradation in Kazakhstan. Ph.D. thesis, Department of Biological Sciences, Warwick University.

Robinson, S. and E. J. Milner-Gulland. 2003. Contraction in livestock mobility resulting from state farm re-organization. Pages 128-145 In C. K. Kerven, editor. Prospects for pastoralism in Kazakhstan and Turkmenistan: From state farms to private flocks. Routledge and Kegan Paul, London.

Runge, C.F. 1981. Common property externalities: isolation, assurance and resource depletion in a traditional grazing context. American Journal of Agricultural Economics 63:595-606.

Sabirova, A. I. 2001. Structural reformation of the land fund of the Republic of Kazakhstan in the process of adaptation to market relations. Pages 31-51 *In* A. I. Sabirova, V. V. Gregorik, and T. M. Arshidinov, editors. Land issues in the Republic of Kazakhstan. Almaty.

Saunders, J. J. 1971. The history of the Mongol conquest. Routledge & Kegan Paul, London.

Sneath, D. 1999. Spatial mobility and Inner Asian pastoralism. Pages 218-277 *In* C. Humphrey and D. Sneath. The end of nomadism? Society, state and the environment in Inner Asia. Duke University Press, Durham.

Zalsman, L. M. 1948. Next tasks of distant pasture livestock breeding. *In* L. M. Zalsman, and B. L. Blomkvest, editors. The experience of distant pasture livestock management in Kolkhozes. State Publishing House for Agricultural Literature, Moscow.

Zveriakov, I. A. 1932. From nomadic life – to Socialism. Kraevoe Izdatelstvo Ogiza v Kazakhstane, Alma-Ata.

Chapter 8

POLICY CHANGES IN MONGOLIA: IMPLICATIONS FOR LAND USE AND LANDSCAPES

Dennis Ojima[1] and Togtohyn Chuluun[2]
[1]Natural Resource Ecology Laboratory, Colorado State University, Fort Collins, CO 80523-1499, USA; [2]Environmental Remote Sensing and Geographic Information System Laboratory, National University of Mongolia, Ulaanbaatar 210646, Mongolia

1. INTRODUCTION

The Mongolian rangelands encompass a diversity of ecosystems, ranging from forest-steppe in the north, to the Gobi desert in the south, with the steppe ecosystem dispersed in between. The Altai Mountains in the southwest, and the Khangai and Khentii Mountains in the north-central part of the country add to the diversity of landscapes, habitats, and resource availability.

The Mongolian nomadic pastoral cultures occur as an emergent feature of the variability of resources and ecosystem dynamics of these temperate arid and semi-arid systems (Chuluun 2000, Fernández-Giménez 2006). These pastoral systems have adapted to resource variability in space and time by utilizing movement to access resources across these landscapes. Formal state and customary (informal) institutions have influenced livestock ownership, management of livestock, and land-use patterns of pastoralists for at least the last five hundred years (Sneath 2003). There has been flexibility with this dual formal and informal regulation of pastoralism that is disappearing today. Political changes in the 20th century have altered the size and boundaries of administrative units which in turn have altered access to resources and restricted seasonal movements of livestock. This fragmentation of grazing lands has led to overuse of resources in the smaller

K. A. Galvin et al. (eds.), Fragmentation in Semi-Arid and Arid Landscapes: Consequences for Human and Natural Systems, 179–193.

administrative units. With privatization of livestock and talk of land private-zation (GISL 1997, Fernández-Giménez 2006), it is increasingly becoming difficult to move livestock across the landscape to access water and forage. These processes are discussed within an historical context followed by suggestions for change.

2. MOVEMENT PATTERNS

Traditional Mongolian pastoralist movement patterns were oscillatory or transhumant in regions where the climate and rangeland production dynamics have been relatively predictable and could accommodate the need to move only between summer and winter camps. The forest-steppe areas are typical of this oscillatory pastoral movement. Local level and state institutions together controlled, variously through time, livestock and movement patterns.

More frequent movements with more than one movement during the summer season occured in the mountain steppe and in the wetter regions of the steppe. In regions with relatively higher climate variability, pastoral movements tended to be more chaotic and followed more opportunistic strategies to secure forage. These movements were associated with drier parts of the steppe and desert areas, such as the Gobi desert and desert steppe region, where non-equilibrium ecosystem dynamics are observed (Ellis and Chuluun 1993, Fernández-Giménez 1999, Chuluun 2000, Bedunah and Schmidt 2004). The herders from these regions moved to places where better rangeland conditions existed, especially during the summer season.

Hierarchical pastoral networks or cooperative groups based on common location of grazing or family relationships ensured that people could gain access to forage and water for their livestock. A hierarchy of informal social networks and institutions including the *hot ail* (a network of households sharing resources within a particular region), *neg golynhon* (people from one river area), and *neg nutgiinhan* (people from one living place) existed in the traditional Mongolian nomadic pastoral system (Bazargur et al. 1993). These networks based their livestock management and spatial sharing of key resources through consideration of common seasonal camping areas, water points, or meadow areas. This system of social organization served as a regulatory function for land-use management and as a mechanism to provide safeguards against natural hazards (Bazargur et al. 1993).

During the 1600s through the 1800s, grazing systems were modified by the Manchu administration of the Mongolian territory (Bawden 1968).

The territory of Mongolia was partitioned, reaching 86 county-level administrative units by the beginning of the nineteenth century (Bawden 1968, Information Mongolia 1990). In this period, what Sneath (2003) calls the neo-feudal period, the land and livestock were owned by nobles and Buddhist monasteries. Pastoralist households herded single-species livestock herds owned by the elite and were granted a share of the surplus livestock. A combination of formal and customary regulations was used to ensure movement across the landscape. During this period prior to establishment of Soviet collectives, cross-territorial use of rangeland resources were negotiable between tribal and community leaders to reduce vulnerability during times when forage was reduced due to harsh weather coniditions.

2.1 Pre-collective movement patterns

The trend to smaller administrative units was advanced during the early Soviet period in the first half of the 20th century (Bawden 1968) (Table 8-1). During this time, reduction in the size of county-level administrative territories took place with establishment of 324 *sums* (county-level administrative units) in 1931. Currently Mongolia is divided into 331 *sums* and 1,671 *bags* (administrative units similar to municipalities). The *sum* and *bag* administrators maintained control over the movement of livestock and many other livestock management decisions (Bawden 1968, Fernández-Giménez 2006). Before the formation of the collectives (called *negdels)* in the late 1950s, movement of the Mongolian herders incorporated traditional pastoral management concepts associated with using forage and water resources within a broader landscape context and using community-based movement decisions (Bazargur et al. 1993, Fernandez-Gimenez 1999, Chuluun and Ojima 2002).

2.2 Movement patterns under socialism

During the *negdel* period between 1960 and 1990 (see Table 8-1), livestock were owned in common and controlled by collectives. Pastoralists moved collective animals, along with a smaller number of personally-owned animals (Sneath 2003). *Negdel* herders were specialized according to the class of livestock herded characterized by livestock species, sex, and age class. The *negdels* took control over land-use regulations within *bag* and *sum* boundaries. The result of these smaller administrative units was that nominal livestock and household movements tended to be shorter and restricted to *sum* boundaries. However, the provincial and national government also established a strategy for short-term, long-distance moves (*otor*) to safe-gauard against drought and rangeland overuse, as a strategy to fatten the stock in summer and fall, and to avoid drought-affected areas and *zud* (severe winter conditions of various kinds) conditions (Humphrey 1978).

Provisioning of machinery and transport equipment heralded in the time when herders no longer needed to move their livestock long distances with their own labor. Machinery was also important in the advancement of hay making. Herders eventually moved less frequently; though when necessary and ordered to make long-distance moves (i.e., *otor*), mechanized transport vehicles (e.g., trucks and tractors) were utilized. Shelter construction, well-building, and haymaking were also strategies to reduce the vulnerability of these pastoral systems to climatic extremes.

Table 8-1. Changes in land-use patterns, land-use regulation and land tenure from pre-*negdel* to post-*negdel* period. *Negdel* refers to pastoral cooperative (Bazargur et al. 1993, Fernandez-Gimenez 1999, Chuluun 2000).

Attributes	Pre-*negdel* period (Until late 1950s)	*Negdel* period (1960 - 1990)	Post-*negdel* period (Since 1990)
Land-use patterns	Nomadic movements with ecological conditions	Less frequent and more distant movements, but often with conservation of cultural landscapes; *Otor* enforced; Many shelters and wells built	Further reduced distance and frequency of moves; Less *otor*; Year-round use of riparian and reserve pastures; Animals concentrated near towns and roads
Regulatory institutions	Traditional pastoral networks (little formal regulation)	Negdel	None (few newly emerged *hot ail* and new cooperatives)
Land-use regulation	No enforced formal regulation of movement; Neighborhood groups migrate together using animal cart	Negdel enforces seasonal moves and *otor*; Machinery provided by *negdel* for transportation and hay making; Species specialization by kind, age and sex	No formal regulation or enforcement; Little coordination of seasonal movements by *hot ail*; Diverse species composition
Land tenure and legal framework	Customary rights within administrative units	*Negdel* allocate pasture, often along customary lines; All property state-owned; Disputes resolved by brigades and *negdel*	Customary rights weak; Livestock, shelters and wells are privatized; Disputes are resolved by local governments (*bag* & *sum*)

The collectives made all decisions over the allocation of animals and specialization in tasks and species. They were also responsible for organizing labor, often drafting soldiers and students to help with more labor-intensive tasks such as hay making, shearing, and clipping. These factors created an atmosphere in which the cooperative functions of traditional households and other customary institutions either could not or had less need to function (Mearns 1996, Humphrey and Sneath 1999). All these changes led to a tacit policy of sedentarization under the socialist regime (Fernández-Giménez 1999). Although the traditional networks of movement may have been subsumed by the collectives, the customary institutions did not disappear altogether, as is demonstrated by the fact that many are now re-emerging (Schmidt et al. 2002, Janzen 2005, Schmidt 2006, Reading et al. 2006).

3. POLICY CHANGES AFTER DEMOCRATIZATION: EFFECTS ON LAND USE AND THE LANDSCAPE

3.1 Movement patterns since democratization

Since democratization in 1990, Mongolia has shifted to a free-market economy, which has led to changes in the livestock sector (see Table 8-1). With the privatization of livestock in the early 1990s collective farms were dissolved. Development interests have advocated the "privatization of public assets, price liberalization, cutting state subsidies and expenditure, currency convertibility, and the rapid introduction of markets" (Sneath 2003:441). The end of collective farms further reduced mobility because of less access to trucks, through the collapse of exports, and the reduction of incomes, public services, living standards, and food security (Sneath 2003).

Unemployment increased in the 1990s because of stagnation of jobs in the capital city and other civic centers, and as a result poverty increased. Accelerating inflation made things worse. Accompanying these trends has been a decline in social services available in the civic centers. Numerous small administrative units or villages have become less viable due to a lack of economic and resource support from the central government (Janzen 2005).

Political decisions in the early period of the Soviet influence during the 1930s were made to reduce the spatial extent of administrative units. This process has been accelerated since independence and has had a profound impact on pastoral movement patterns (Bawden 1968, Information Mongolia

1990). Smaller land-use areas have resulted in the destabilization of the pastoral system by decoupling the herders from the regional landscape and forcing them to utilize resources within fragmented units, which do not provide the diversity of settings needed to sustain their pastoral systems without external inputs.

3.2 Effects on livestock and the ecosystem

As livestock were privatized at the beginning of the 1990s and animals were distributed to the Mongolian population, the number of pastoral households increased dramatically (NSO 2003). An unintended effect of this transfer of livestock to the general public was that a number of households were given animals which they were not well-suited to manage. This creation of "new nomads" (Janzen 2005) led to poor grazing practices during the ealier period of transition in the 1990s and the resultant degradation of rangeland resources. In addition, a trend in pastoral movement toward less frequent and shorter distance moves has been observed due to the lack of subsidies to maintain transportation for long-distance travel and a preference to stay closer to settled areas and watering points (Janzen 2005, Fernández-Giménez 2006, Reading et al. 2006). The effect of this trend is for a higher concentration of livestock to be located near settled areas and year-round use of riparian zones, and has led to deterioration of the rangelands (Ojima et al. 2004, Janzen 2005, Chuluun et al. 2005). All this resulted in a declining livestock herd size and increased poverty, in general, at least until about 1993 when livestock numbers began to increase (Figure 8-1).

Livestock privatization eventually provided incentives for increasing livestock numbers. Without state subsidies, having greater numbers of animals in one's herd became the dominant risk management strategy as 'insurance' against future uncertainties and climate variability. The number of livestock increased from 25.8 million in 1990 to 33.6 million in 1999 (Figure 8-1). The increase in livestock numbers were driven primarily by the number of goats, which was a response to access to cashmere markets (World Bank 2003). Goat numbers doubled from 1990 to 1999, reaching 11 million in 1998. Associated with the overall increase in livestock numbers was a shift in livestock composition during this decade; the number of camels almost halved, cattle increased 34%, horses increased 40%, and sheep remained constant. The combined drought and *zuds* during the period from 1999 through 2002 resulted in severe livestock losses; livestock numbers were reduced up to 24 million in 2002. However, livestock numbers have increased back to 30.4 million in 2005, 13.3 million of which are goats (NSO 2006).

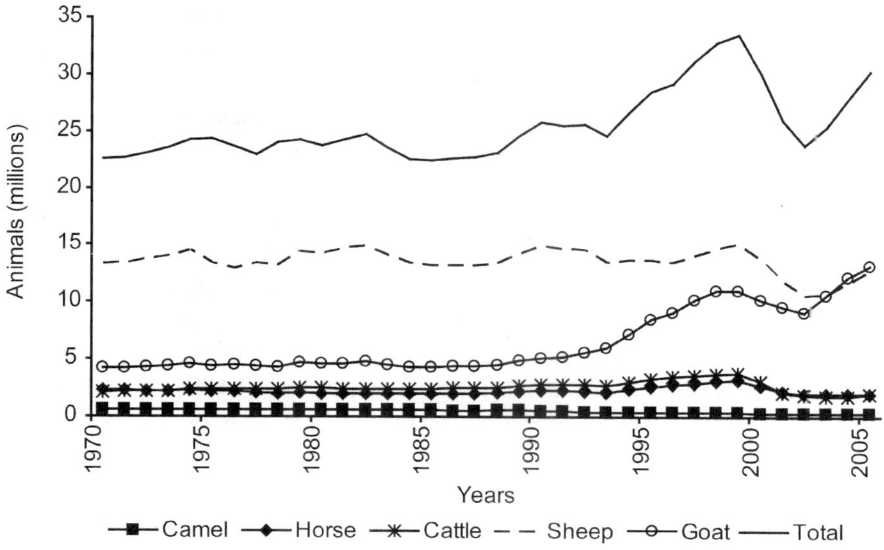

Figure 8-1. Livestock dynamics in Mongolia (NSO 2006).

Environmental degradation has increased markedly in Mongolia in the time that livestock numbers increasesd. Since about 1995, the area of highly degraded land increased 1.8 times (MNE 2001) and desertification in the arid and semi-arid region of Mongolia increased by 3.4% during 1990-2004 (MNE 2006). Acceleration in desertification has occurred in part because of human influences and in part from the changing climate. During the time that livestock numbers increased, it occurred without an expansion in grazing access outside the existing *bag* and *sum* administrative units resulting in inadequate pastures for seasonal utilization. As the number of livestock exceeded the pasture carrying capacity in these fragmented systems, pastureland degraded, output declined, and ecosystem breakdown occurred, causing grasslands to shift toward more desert-like conditions (Janzen 2005).

Similar patterns have emerged in Inner Mongolia in China and parts of Russia (Sneath 1998). Sneath found degradation, mostly caused by less mobile pastoral systems dependent on producing supplemental food for animals. This has implications for Mongolian rangelands where the number of livestock is rising yet there is a reduction of mobility.

3.3 Effects on water

In addition to grazing land, a key factor affecting spatial scale of livestock management is the availability and distribution of watering points.

Naturally occurring surface water in Mongolia is not abundant, so engineering related to well establishment and maintenance is an important component of pastoral management. Since 1990, there was failure to maintain water wells throughout Mongolia, with only 40 percent of wells constructed from 1960-1990 currently functioning (UNDP 2005).

There is also evidence that the volume of lakes and rivers has diminshed in recent years for various, interlinked reasons such as climate change, deforestation, land degradation, and other adverse human activities (UNDP 2005). According to the National Survey for Surface Water conducted in 2003 by the Ministry for Nature and Environment, 683 rivers (out of 5,565 rivers), 1,484 springs (out of 9,600 springs), and 760 lakes and ponds (out of 4,196 water bodies) disappeared since the last survey in 1995 (UNDP 2004). Because the number of water points decreased from year to year, movement patterns are constrained and this has resulted in concentrating the grazing on the remaining areas where water is available, that is, along rivers, springs, lakes, and in the villages where wells still operate. This further concentrated grazing pressure and has led to a rapid degradation of grassland ecosystems.

3.4 Climate effects

In addition to these anthropogenic effects, a warmer and drier climate has promoted the expansion of the Gobi desert region. From the beginning of the 20th century, global warming has intensified in northern latitudes and the temperature in Mongolia has increased by 1.9°C since 1940 (Ojima et al. 2004). The forage availability determined by the remote sensing data from 1982 through 2002 in the central parts of Mongolia was affected by these climate effects (Ojima et al. 2004). These climate effects extend beyond *sum* and *bag* areas (Ellis et al. 2002), and therefore result in conditions that are difficult to cope with since landscape units are not available to mitigate the loss of forage resources. Ellis et al. (2002) showed that the steppe area adjacent to the Gobi is especially vulnerable to climate change and, with increased grazing pressures during the past decade, has become more desertified (Report of the Government Committee on "Proposal on new administrative-territorial division of Mongolia" 2005).

3.5 Effects on the household

Under the legislation on privatization of livestock and the breakup of the *negdel* system, the former collectives were allowed a free hand in deciding how they should privatize (Mearns 1993). During the transition to capitalism, the livestock, shelters, and wells were privatized and customary rights to certain pasture lands became weak or unclear, especially in central

Mongolia. This has resulted in a variety of decision-making organizations determining grazing activites related to seasonal movements and access to resources seen today.

Since democratization, a rapid increase in pastoral households initially occurred (NSO 2006) followed by two additional trends during the past decade: first, an increased rural-to-urban migration (Janzen 2005) and second, an emergence of cooperatives (Schimdt et al. 2002, Schmidt 2006, Fernández-Giménez 2006). The Mongolian constitution of 1992 guarantees people the right to choose their residence (Fernández-Giménez and Batbuyan 2004, Janzen 2005), which has become the legal motivator of internal regional migrations to urban centers. In the central region, three major cities with good infrastructure and social services (Ulaanbaatar, Darhan and Erdenet) became attractors for the rural population, especially for households who lost their livestock between 1999-2002. Recent government policies are being developed to provide services in rural areas to reduce the migration to large urban centers and to promote a more even utilization of the natural resources (Report of the Government Committee on "Proposal on new administrative-territorial division of Mongolia" 2005).

In planning decollectivization, emphasis was placed on the transfer of assets from the state into the hands of private individuals. Many intermediate forms of organization between collective and private ownership of livestock have appeared (Schimdt et al. 2002). In some areas, they have persisted and will continue to persist; in others, they were only a short-term solution and have been dissolved (Fernández-Giménez and Batbuyan 2004). The most common form of livestock ownership is the livestock company. Of the 255 collectives in Mongolia which were privatized, 80 exist in the form of joint stock companies. In structure these are very similar to the former collectives, although the relationship between company and member herder is significantly different. Typically, companies retain ownership of large numbers of animals which are leased to their members, although the terms of the lease vary a great deal between companies. A small number of companies continue to pay their herders a salary and retain full ownership over animals.

Now, local governments at the *sum* and *bag* levels of administration responsible for resolving land-use disputes have replaced former policies of the collective (*negdel*) period (Fernández-Giménez and Batbuyan 2004). Yet at the same time, informal traditional pastoral institutions, particularly at the *hot ail* (a network of households) level, have re-emerged (Schimdt et al. 2002). Often the experienced elders within the region have become the leaders of a *hot ail* community. These networks of households are able to allocate use of rangelands during the different seasons, assist each other during long distance moves of the livestock, and facilitate re-stocking of the herds after calamitous events such as droughts or winter storms. This pattern of combined government and local responsibility of livestock management is what Fernández-Giménez (2002) calls co-management of formal state and regional regulations and local, more informal, pastoral controls.

4. POSSIBILITIES FOR FUTURE POLICY REFORM

4.1 National administrative boundary reform

The main elements for providing long-term sustainable development of pastoral animal husbandry are to use natural pastures and hay and to maintain ecosystem integrity. In Mongolia, almost half (159 out of 331) of the *sums* do not have the needed seasonal pasture areas due to their small sizes and the homogeneous nature of landscapes occupied by the current *sums* (Janzen 2005, Fernández-Giménez 2006). This has resulted in over-grazing of pasturelands during inappropriate seasons of the year, leading to reduced pasture carrying capacity and forage quality (Janzen 2005).

Recent policy changes have been proposed to modify the major administrative boundaries to allow greater access to natural resources and seasonal grazing lands to better sustain pastoral livelihoods (Report of the Government Committee on "Proposal on new administrative-territorial division of Mongolia" 2005, State Ikh Khural 2001). The policy is designed to develop a settlement pattern that reduces the concentration of population around major civic centers and to promote usage of resources associated with rural areas of the country. These policies are meant to reduce the extent of fragmentation of pastures and allow for more ecologically-based use of landscape features across a larger region (Report of the Government Committee on "Proposal on new administrative-territorial division of Mongolia" 2005).

These new administrative and territorial units have been proposed to enhance socio-economic optimality, environmental sustainability, and historical and cultural acceptability by citizens (Report of the Government Committee on "Proposal on new administrative-territorial division of Mongolia" 2005, Chuluun 2005). Reforming and enlarging administrative and territorial units may provide greater flexibility in managing livestock densities across a more diverse set of landscape types within a more comprehensive administrative unit. The overall result would be a greater utilization of the natural landscapes now restricted in the fragmented smaller *sums*. For this change to succeed, reinvestment in infrastructure such as transport and water resource developments - including well maintenance and reconstruction - to allow for longer movements within these larger administrative units will be necessary, as well as the establishment of rules for access and allocation of seasonal pastures within these larger administrative units.

4.2 Regional reforms

Balanced division of population, territory, natural wealth, and industries is of special significance to the self-sufficient development of the rural centers. There is still a strong adherence to traditional land use and nomadic pastoral systems. Development of regional rural policies which allow for greater flexibility of livestock movement along the more traditional routes is emerging. Support in government and among nomadic herders have led to development of a reorganization of regional government that encompasses territories of several ecological zones and restores culturally traditional landscapes similar to those existing in the early 1900s (Report of the Government Committee on "Proposal on new administrative-territorial division of Mongolia" 2005, Chuluun 2005). This system may provide greater flexibility of pastoral management, especially under high climate variability as experienced by herders in the Gobi region and other steppe areas where climate fluctuations are large and drought frequency is high.

4.3 Local reforms

Another trend is the emergence of cooperatives based on traditional pastoral networks such as *neg golynhon* (people in a one river area) or *neg nutgiinhan* (people from a single living place). Enhancing collective actions among herders through strengthening the traditional customary arrangements may be a key to achieving sustainable pastoral communities. The herders' interest in maximizing livestock in the current incentive structure is a primary challenge to building sustainable rangeland management (Enh-Amgalan 2002). Productivity improvement and alternative income-generation activities are crucial for changing the existing behavior and compensating for potential income losses from restriction of animal numbers.

Some grounds for cooperation include:

- Superseding the constraint of inadequate size (too small) households in terms of the number of members of the household needed for livestock management.
- Better access to services—veterinary, breeding, and social services.
- Strengthening of the traditional coping mechanisms for dealing with climate variability and extreme circumstances, sustainable use of rangeland

ecosystems, and increasing resiliency, while also decreasing the vulnerability of the pastoral communities.

The lessons learned from community-strengthening activities in Mongolia include (Schmidt et al. 2002, Fernández-Giménez 2006):

- Herders are eager to launch collective actions for reasons ranging from simple social gatherings to those aimed at developing further cooperation in production activities.
- Strengthening herding communities empowers initiatives for development thus encouraging equitable distribution of development resources.
- Because of the reduced requirements for establishment and lower transaction costs, herders prefer community-based networks.
- A sustainable community development approach, based on traditional pastoral networks, is one of the most cost-effective, adaptive strategies to deal with an uncertain climate and global changes.

5. CREATION OF A COMPLEX CULTURAL LANDSCAPE

Enforcing the policy of a 'complex social landscape' may be a way to cope more effectively with the impacts on society and environment of natural disasters such as droughts and *zud*. According to the definition by the UNESCO World Heritage Convention (2005), cultural landscapes are cultural properties and represent the integrated workings of nature and of humans. They are illustrative of the evolution of human society and settlement over time, under the influence of the physical constraints and/or opportunities presented by their natural environment and of successive social, economic, and cultural forces, both external and internal.

In the Mongolian context, the complex of seasonal pastures, long distance pastures, hay-making land, reserve pastures, and sacred lands would be a cultural landscape. The development of this cultural landscape in Mongolia would enhance the diversity of landscapes available to the herders through consolidation of *sums* into larger units.

The incorporation of this concept of cultural landscapes provides a landscape orientation to the management of resources and would require a minimum level of investment. This system would stabilize the existing biodiversity, if not enhance it, and the health of the system may improve as a result. Also, the sustainability of pastureland may be more attainable through adaptation of this concept so that improvements can be seen in the adaptability of pastoral nomadism, herd quality, and herder's living standards; herders may also have the opportunity to engage in processing and manufacturing of products from their pastoral systems. Opportunities for cultural and natural tourism could also be created.

On the other hand, if the landscape of arid regions is divided into parts, or if the seasonal pastures are not maintained in order to conserve specific portions of the landscape or ecosystem type, then management of these isolated portions of the cultural landscape would require higher amounts of investment. Because Mongolia's current administrative and territorial divisions do not always include complete access to the complex landscapes of the seasonal pastures, there are many areas where the complex cultural landscape have been lost. Therefore, reviving and enlarging the traditional cultural landscape by making the administrative and territorial divisions larger would improve the adaptability of pastoral nomadism to climate change and provide a positive effect on the livelihoods of rural people.

ACKNOWLEDGEMENTS

Funding for research conducted in Mongolia is supported by the NASA funded project entitled: "Northern Eurasian C-land use-climate interactions" (Contract No: NNG05GA33G) and is part of a coordinated research activity of the Northern Eurasian Earth System Partnership Initiative. Analysis and data sources were provided by the Mongolian Ministry of Construction and Urban Development. Review comments were provided by Dr. M. Glantz, National Center for Atmospheric Research and Dr. J. Belnap, US Geological Survey.

REFERENCES

Bawden, C.R. 1968. The modern history of Mongolia. Kegan Paul International, London and New York.

Bazargur, D., C. Shiirevad'ya, and B. Chinbat. 1993. Territorial organization of Mongolian pastoral livestock husbandry in the transition to a market economy, PALD Research Project No.1, Brighton, England. University of Sussex, Institute of Development Studies, Policy Alternatives for Livestock Development in Mongolia (PALD).

Bedunah, D. J. and S. M. Schmidt. 2004. Pastoralism and protected area management in Mongolia: the case of Gobi Gurvan Saikan National Park. Development and Change 35:167-191.

Chuluun, T. 2000. Climate variability: A Mongolian case study. Update, Newsletter of the International Human Dimensions Programme on Global Environmental Change, Nr. 1/2000.

Chuluun, T. 2005. A new administrative-territorial division of Mongolia as a mechanism to increase adaptive capacity to climate change. The 6th open meeting of the international human dimensions meeting of global environmental change on "Global environmental change, globalization and international security: new challenges for the 21st century", Bonn, Germany, Conference book, 436.

Chuluun, T. and D. S. Ojima, editors. 2002. Fundamental issues affecting sustainability of the Mongolian steppe. Proceedings of the open symposium on "Change and Sustainability of Pastoral Land Use Systems in Temperate and Central Asia," Ulaanbaatar, Mongolia, June 28-July 1, 2001, Interpress Publishing and Printing, Ulaanbaatar.

Chuluun, T., M. Altanbagana, and G. Sarantuya. 2005. Vulnerability and adaptation assessment of the Mongolian pastoral systems to climate and land use changes. The Proceedings of the Department of Biology, School of Natural Sciences, Mongolian Education University. 4:182-189.

Ellis, J. and T. Chuluun. 1993. Cross-country survey of climate, ecology and land-use among Mongolian pastoralists. Report to Project on Policy Alternatives for Livestock Development (PALD) in Mongolia, Institute of Development Studies at the University of Sussex, UK.

Ellis, J., K. Price, R. Boone, F. Yu, T. Chuluun, and M. Yu. 2002. Integrated assessment of climate change effects on vegetation in Mongolia and Inner Mongolia. Pages 26-34 *In* T. Chuluun and D. S. Ojima, editors. Fundamental issues affecting sustainability of the Mongolian steppe. Interpress Publishing and Printing, Ulaanbaatar.

Enh-Amgalan, A. 2002. Change and sustainability of pastoral land-use systems in Mongolia. Pages 228-231 *In* T. Chuluun and D. Ojima, editors. Fundamental issues affecting sustainability of the Mongolian steppe. Interpress Publishing and Printing, Ulaanbaatar.

Fernández-Giménez, M. E. 1999. Sustaining the steppes: A geographical history of pastoral land-use in Mongolia. Geographical Review 89:315-342.

Fernández-Giménez, M. E. 2002. Spatial and social boundaries and the paradox of pastoral land tenure: a case study from post socialist Mongolia. Human Ecology 30:49-78.

Fernández-Giménez, M. E. 2006. Land use and land tenure in Mongolia: A brief history and current issues Pages 30-36 *In* D. J. Bedunah, D. E. McArthur, and M. Fernández-Giménez, editors. Rangelands of Central Asia: Proceedings of the Conference on Tranformations, Issues, and Future Challenges. USDA Forest Service Proceedings RMRS-P-39.

Fernández-Giménez, M. E. and B. Batbuyan. 2004. Law and disorder: Local implementation of Mongolia's land law. Development and Change 35:141-165.

GISL. 1997. Strengthening of land use policies in Mongolia. ADB-TA No. 2458-MON, final report phases I and II. GISL, Hants, UK.

Humphrey, C. 1978. Pastoral nomadism in Mongolia: The role of herdsmen's cooperatives in the national economy. Development and Change 9:133-160.

Humphrey, C. and D. Sneath. 1999. The end of nomadism? Society, state, and the environment of Inner Asia. Duke University Press, Durham.

Information Mongolia: the comprehensive reference source of the People's Republic of Mongolia (MPR). 1990. Compiled and edited by the Mongolian Academy of Sciences, Pergamon Press plc, England.

Janzen, J. 2005. Changing political regime and mobile livestock keeping in Mongolia. Geography Research Forum 25:62-82.

Mearns, R. 1993. Territoriality and land tenure among Mongolian pastoralists: variation, continuity and change. Nomadic Peoples 33:73-103.

Mearns, R. 1996. Community, collective action and common grazing: The case of post-socialist Mongolia. Journal of Development Studies 32:297-339.

MNE (Ministry of Nature and Environment-Mongolia). 2001. State of the environment – Mongolia. United Nations Environment Programme. Klong Luang, Thailand.

MNE (Ministry of Nature and Environment). 2006. Mongolia: State of environment, 2004-2005 Report, Ulaanbaatar.

NSO (National Statistical Office – Mongolia). 2003. Mongolian Statistical Yearbook 2002. National Statisical Office, Ulaanbaatar, Mongolia.

NSO (National Statistical Office – Mongolia). 2006. Mongolian Statistical Yearbook 2005. National Statisical Office, Ulaanbaatar, Mongolia.

Ojima, D. S., T. Chuluun, B. Bolortsetseg, C. J. Tucker, and J. Hicke. 2004. Eurasian land use impacts on rangeland productivity. Pages 293-301 *In* R. DeFries and G. P. Asner, editors. Ecosystems and land use change. American Geophysical Union. Geophysical Monograph Series. Volume 153. Washington D.C.

Reading, R. P., D. J. Bedunah, and S. Amgalanbaatar. 2006. Conserving biodiversity on Mongolian rangelands: Implications for protected area development and pastoral uses. Pages 1-18 *In* D. J. Bedunah, D. E. McArthur, and M. Fernández-Giménez, editors. Rangelands of Central Asia: Proceedings of the conference on tranformations, issues, and future challenges. USDA Forest Service Proceedings RMRS-P-39.

Report of the Government Committee on "Proposal on new administrative-territorial division of Mongolia" to the President of Mongolia, April 1, 2005.

Schmidt, S. M. 2006. Pastoral community organization, livelihoods and biodiversity conservation in Mongolia's Southern Gobi Region. Pages 18-29 *In* D. J. Bedunah, D. E. McArthur, and M. Fernández-Giménez, editors. Rangelands of Central Asia: Proceedings of the conference on tranformations, issues, and future challenges. USDA Forest Service Proceedings RMRS-P-39.

Schmidt, S. M., G. Gansukh, K. Kamal, and K. Swenson. 2002. Comunity organization - a key step towards sustainable livelihoods and co-management of natural resources in Mongolia. Policy Matters 10:71-74.

Sneath, D. 1998. Ecology: State policy and pasture degradation in Inner Asia. Science 281:1147-1148.

Sneath, D. 2003. Land use, the environment and development in post-socialist Mongolia. Oxford Development Studies 31:441-459.

State Ikh Khural (Parliament) of Mongolia. 2001. Regional Development Concept, Resolution No. 57.

UNDP. 2004 . Access to water and sanitation services in Mongolia. UNDP, Ulaanbaatar, Mongolia.

UNDP. 2005. Economic and ecological vulnerabilities and human security in Mongolia. UNDP, Ulaanbaatar, Mongolia.

UNESCO WHC. 2005. Operational guidelines for the implementation of the World Heritage Convention. 2 /2005.

World Bank. 2003. From goats to goats: Institutional reform in Mongolia's cashmere sector. Report No. 26240-MOG, Washington D.C., 19 December 2003.

Chapter 9

FRAGMENTATION OF A PERI-URBAN SAVANNA, ATHI-KAPUTIEI PLAINS[1], KENYA

Robin S. Reid[1], Helen Gichohi[2], Mohammed Y. Said[1], David Nkedianye[1], Joseph O. Ogutu[1], Mrigesh Kshatriya[1], Patti Kristjanson[1], Shem C. Kifugo[1], Jasphat L. Agatsiva[3], Samuel A. Adanje[4], and Richard Bagine[4]
[1]*International Livestock Research Institute, P.O. Box 30709, Nairobi, Kenya;* [2]*African Wildlife Foundation, P.O. Box 48177, Nairobi, Kenya;* [3]*Department of Resource Surveys and Remote Sensing, P.O. Box 47146, Nairobi, Kenya;* [4]*Kenya Wildlife Service, P.O. Box 40241, Nairobi, Kenya*

1. INTRODUCTION

Many pastoral ecosystems around the globe are under pressure to produce more livestock or to make way for more intensive agricultural systems or new uses (Blench 2000). Some rangelands that used to be managed under communal land tenure are being privatized, with establishment of individual holdings; others are under state control (Galaty 1994). This is happening first in rangelands that receive more rainfall, are closer to urban centres, and/or contain significant key resources that are essential for successful crop cultivation (Galaty 1994). In these systems, pastoralists are either pushed onto more marginal lands for grazing or they begin to take up crop agriculture themselves, becoming agro-pastoralists (e.g., Campbell 1993, Campbell et al. 2003). One result is increased permanent settlement.

Pastoral people also choose to settle because they desire better education and health care for their families, their diets have changed and they have new needs for marketed goods and services (Little 1985, 1992, Fratkin and Smith 1995). In wetter, semi-arid savannas of East Africa, settled agro-pastoralists often build fences and take up cultivation to protect their access to forage and diversify their sources of food production (Rutten 1992,

K. A. Galvin et al. (eds.), Fragmentation in Semi-Arid and Arid Landscapes: Consequences for Human and Natural Systems, 195–224.

Kimani and Pickard 1998, Reid et al. 2004). These pressures (described generally in Chapters 1-2 and 13) fragment these rangelands into smaller holdings, which can have significant consequences for pastoral and agro-pastoral livelihoods and for biodiversity conservation (described generally in Chapters 1-3 and 15).

A few pastoral ecosystems are further along in this process of fragmentation than others, because of their history, their proximity to markets, human population pressure, policy, or other reasons. One such rangeland is the Athi-Kaputiei Plains (part of which is called the Kitengela) in Kenya. This area is unique because it continues to support migration of large wildlife over long distances despite its proximity to Kenya's capital of Nairobi, currently a city of over 2 million people (Figure 9-1). Only a fence separates wildlife from this bustling city. Nairobi National Park (117 km^2 in area), located at the northernmost tip of this 2,456 km^2 ecosystem, begins just 5 km from the central business district of Nairobi. South of the park stretches the rest of the ecosystem that is 21 times larger than the park itself. Twenty-four species of large mammals live on these rich plains, although not elephant, which was exterminated before 1962 (Stewart and Zaphiro 1963, Gichohi 1996). Migrating herds use the park during the dry season for its water and abundant grass and then move south into the open pastoral lands (the second and third of the three triangles shown in Figure 9-1) during the calving season when the rains begin. Here, the Kaputiei Maasai live along with a wide variety of other peoples. Together, they use the land for grazing their livestock, cultivation, horticulture, quarrying, settlement, local commerce, cement production, and export processing businesses.

The Athi-Kaputiei resembles many parts of the world—it is affected by processes that operate globally: urbanization, rapid in-migration, expansion of land use with little planning, high poverty rates, and shifts in systems of land tenure. However, this area is unusual because of its exceptional wildlife. We chose to describe this system as an example of the causes and consequences of fragmentation because this pastoral-wildlife system is one example of the ways that other rangelands in East Africa may change over the next few decades. If so, there is a great opportunity to learn from the issues and challenges in the Athi-Kaputiei as pastoral peoples struggle to understand and adapt to change, and decide, with others, whether or not to maintain viable (and potentially valuable) wildlife populations on their lands. We start the chapter with a description of fragmentation processes in the Athi-Kaputiei over many millennia, with a strong focus on the present.

We then present a synthesis conceptual model of the processes and feedbacks of change here. We next detail the current state of land use in this ecosystem and some of the consequences of fragmentation for people, livestock, and wildlife. We end with a brief description of some collaborative efforts to reverse fragmentation of the Kitengela part of the Athi-Kaputiei Plains to support movement of pastoral livestock and wildlife and finally, a discussion of future implications.

2. LOCATION AND ENVIRONMENT OF THE ATHI- KAPUTIEI PLAINS

Nairobi city bounds the Athi-Kaputiei ecosystem on the north, with the Lukenya hills to the east, the Rift escarpment to the west, and lower-lying rocky and hilly land to the south. The plains lie principally at the northern end of Kajiado District, but include a small piece of Machakos District; Nairobi National Park, at the northern tip of the ecosystem, though administratively falling in Nairobi District, is usually considered part of Maasailand. Today, many residents refer to the three 'triangles' that make up the pastoral part of the ecosystem: the first triangle bordering the park and the second and third triangles farther to the south (see Figure 9-1). Rainfall is moderate here, with 800 mm falling each year in the northwest and 500 mm in southeast (Norton-Griffiths 1977). Most precipitation occurs during two rainy seasons, but rains often fail; farmers say that crop production is generally successful only one year in five (Kristjanson et al. 2002). This ecosystem sits on very rich soils derived from phonolitic lava (Baker 1954) and thus is a nutrient-rich 'eutrophic' savanna, probably able to support 2-3 times more wildlife biomass than nutrient-poor 'dystrophic' savannas that are widespread elsewhere in Africa (Bell 1982, Huntley 1982, Fritz and Duncan 1994). The vegetation is principally wooded *Acacia/Balanites/Themeda* grassland, with gallery forests along rivers of *A. xanthophloea* and small forest patches of *Croton macrostachys* and *Olea africana*. Only two permanent rivers, the Kiserian and Empakasi, run through the northern part of the plains, and much of their flow is extracted by a pipeline running to Kajiado town (from the Kiserian River) or for irrigation and household consumption (Gichohi 1996). In the early 1990s, only 21% of the plains were within reach of permanent water for pastoral herders and their livestock within their normal grazing radii (Gichohi 1996), but this has likely changed with recent water development.

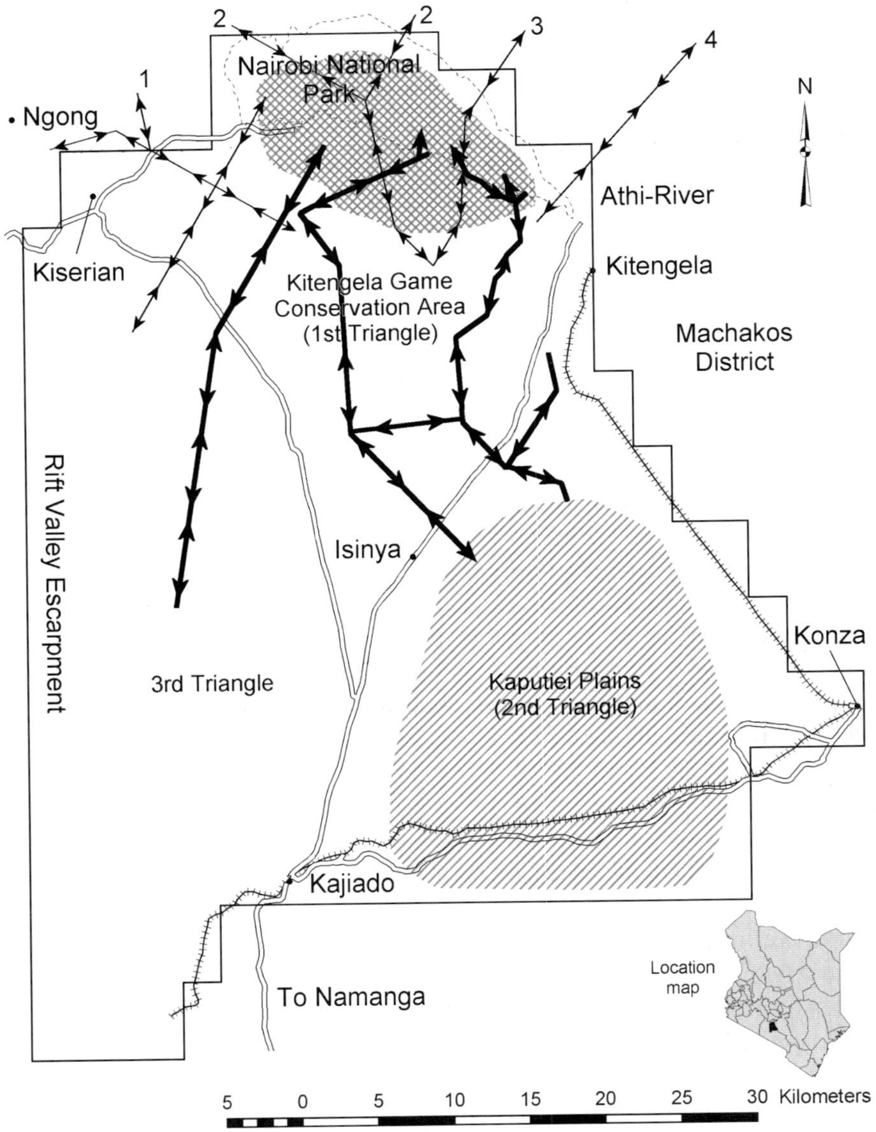

Figure 9-1. Map of Athi-Kaputiei ecosystem (outlined in light gray), showing the three triangles, Nairobi National Park, historical (thin solid lines and arrows, numbered) and current (bolded solid lines and arrows, not numbered) wildlife corridors and livestock grazing routes, and dry season (dark hatching) and wet season range (light diagonal striping) for wildebeest. The city of Nairobi is on the northern edge of the park, which is in south-central Kenya.

3. HISTORICAL FRAGMENTATION OF THE ATHI- KAPUTIEI PLAINS

Like other areas of East Africa (Leakey and Hay 1979), hominids and wildlife very likely lived together in the Athi-Kaputiei Plains ecosystem for millions of years; it seems unlikely that they actively fragmented this landscape in the distant past. Farmers began cultivating native sorghum, millet, and root crops in East Africa about 3,500 years ago, with crops from other continents like maize and cassava arriving much later (Robertshaw 1991). However, in East Africa, pastoral people with livestock arrived more than a millenium before crop cultivation, pushed south from the Sahara by a drying period that began about 5,500 years ago (Smith 1984, 1992, Marshall 1998, 2000, Marshall and Hildebrand 2002). The first pastoral people were likely Cushitic-speaking pastoral people, followed by Maa-speaking people thousands of years later, the latter migrating south from the Uganda-Sudan border region in the 1400s, probably reaching south to the Athi-Kaputiei Plains in the 1600s (Jacobs 1975, Robertshaw 1991, Sutton 1993). Over the last 400 years, the Maasai occupied much of the land in Kenya's southern Rift Valley and surrounding highlands (including Nairobi), defending this rich savanna and forest land from neighboring tribes (Rutten 1992).

At the end of the 1800s, some observers claimed that this ecosystem supported the "most spectacular concentration of wildlife in all of East Africa" (Simon 1962). In 1891, rinderpest reached this part of Maasailand, killing all but 5-10% of Maasai cattle herds and most of the grazing wildlife (Waller 1988). Human disease also took its toll. The Kaputiei Maasai in the Athi-Kaputiei were particularly hard-hit by smallpox (Rutten 1992). Wildlife counts in 1902 showed there were probably more wildlife than we see today in the Athi-Kaputiei Plains, despite the rinderpest epidemic about a decade earlier (Meinertzhagen 1957:58). The difference between then and now is that there were four times more wildlife than cattle in 1902/3, while nearly a century later, counts by the Kenyan Department of Resource Surveys and Remote Sensing show the reverse: livestock outnumber wildlife by 4:1.[2]

3.1 Policy

Over the last century, the Athi-Kaputiei pastoral-wildlife system became progressively compressed, bounded, and fragmented. British colonists appropriated land from pastoralists and brought private land ownership to East Africa, much as they did to eastern North America 150 years previously (Cronon 1983). Maasai gave up 60% of their best watered pastures in the early 1900s, and moved to two reserves in southern Kenya (Rutten 1992). Nairobi city grew next to the principal key water resource for people, livestock, and wildlife, at the border of the highland forest and lower and drier savannas, and along the Ugandan railway that runs along the eastern edge of this ecosystem today. Slowly, expansion of European and African settlement and farmland began to fragment this ecosystem that once stretched unbroken from just south of Mt. Kenya to Tanzania, progressively cutting off four of the known historical wildlife migration routes to the north (historic migration routes #1-4, thin solid lines and arrows in Figure 9-1) and to the east (Foster and Coe 1968, Gichohi 1996). The four historical routes for wildlife, livestock, and pastoral movement included: 1) to the Ngong Hills, 10 km from the current edge of the ecosystem; 2) to Nairobi, 5-10 km away; 3) to Ruiru-Thika, 40 km away; and 4) to Ol Doinyo Sabuk, 70 km.

In 1946, the colonial government excluded pastoral peoples from the wettest part of the existing grazing system (800 mm rainfall) by creating Nairobi National Park (dark cross-hatching, Figure 9-1). In the 1950s and 1960s, farmers and settlers gradually took up the land around the base of the Ngong Hills (migratory route #1 in Figure 9-1), until all the land north of the current Nairobi-Magadi road (the westbound road that goes through the town of Kiserian) was settled and unavailable for pastoral herders or wildlife by the 1970s. In 1963, the Royal Parks, the colonial park authority, built a fence around the western and northern sides of the park, between the park and the city, effectively ending migration of wildlife to the north from the park area, but also protecting the park wildlife from city residents. Ten years later, in 1976, there were still kongoni and Grant's gazelle in the highland areas of the Ngong Hills (northwest corner, Figure 9-1), an area that had been completely converted to housing (Foster and Coe 1968, Hillman and Hillman 1977).

In the late 1960s, development of group ranches was proposed as a way to help ensure Maasai ownership of land in Kenya, encourage development of rangelands, and solve the perceived degradation of rangelands (Njoka 1979). The first group ranches were formed in the Kaputiei section of Maasailand, in the Athi-Kaputiei ecosystem (Pasha 1986, Rutten 1992). In 1986, the Kaputiei Maasai again led the way in Kenyan Maasailand and began adopting individual private ownership of land. By 1990, forty of the original 52 group ranches in Kajiado District had subdivided or were in the process of doing so (Rutten 1992, Kimani and Pickard 1998). This meant splitting each group ranch into smaller plots: each member of the 15 former Kaputiei group ranches received title to private plots ranging in size from 51 to 298 acres (Rutten 1992). Kimani and Pickard (1998) found that the Kajiado group ranches with the smallest plot sizes were those that had subdivided first and/or those with the highest proportion of the plots sold to non-Maasai. They also found that those with the smallest plots were closest to Nairobi and received the most rainfall, although group ranches with many members at sub-division also have small plot sizes (J.S. Worden, pers. comm.). In the 1980s and 1990s, small towns, like Athi River and Kitengela (Figure 9-1), continued to grow, industries and the export processing zone (EPZ) were established nearby, and some pastoralists and farmers started to grow crops for the first time (Gichohi 2000). Land further fragmented as owners sold parts of original private plots or passed on plots to several inheritors. These changes are now having profound implications on how this landscape is used and how easily herders, livestock, and wildlife can move from one place to another in search of good pastures and water.

3.2 Human population

In addition, human populations in Kajiado District more than quadrupled from 4 to 19 people/km^2 in three decades from 1969-1999, with a slight slowing of growth recently (Katampoi et al. 1990, GoK 2001). Growth was four times faster than the district average in the Kitengela location within the first triangle, principally around the Kitengela shopping centre and other smaller villages (GoK 2001). Throughout the district, rapid population increase has led to more settlements, which, in this area, brought more fencing (Figure 9-2). Many of the new residents are non-Maasai farmers and townspeople who, unlike the pastoral Maasai, have a long history of eating wild meat (Nkedianye 2003).

4. CURRENT PROCESSES OF LOSS AND FRAGMENTATION IN THE ATHI-KAPUTIEI PLAINS

As described above, these historical events, and other cultural, natural resource, economic, and political conditions set the context for the wide range of ultimate (underlying) and proximate (nearby) forces that cause this landscape to fragment into smaller patches of different land uses and change the access of people and grazing animals to key forage and water resources (Figure 9-3). It is important to recognize that these same forces also initiate a range of changes beyond fragmentation, like improvement in crop production with the expansion of cropland, but we focus on fragmentation processes for the purpose of this book. We propose here that the most important of these causes are, as described above, land tenure, settlement and protected area policy, inheritance by multiple inheritors and land sales, urbanization (particularly expansion of settlements and industrial activities), and human population growth. In addition to these, high access and use of

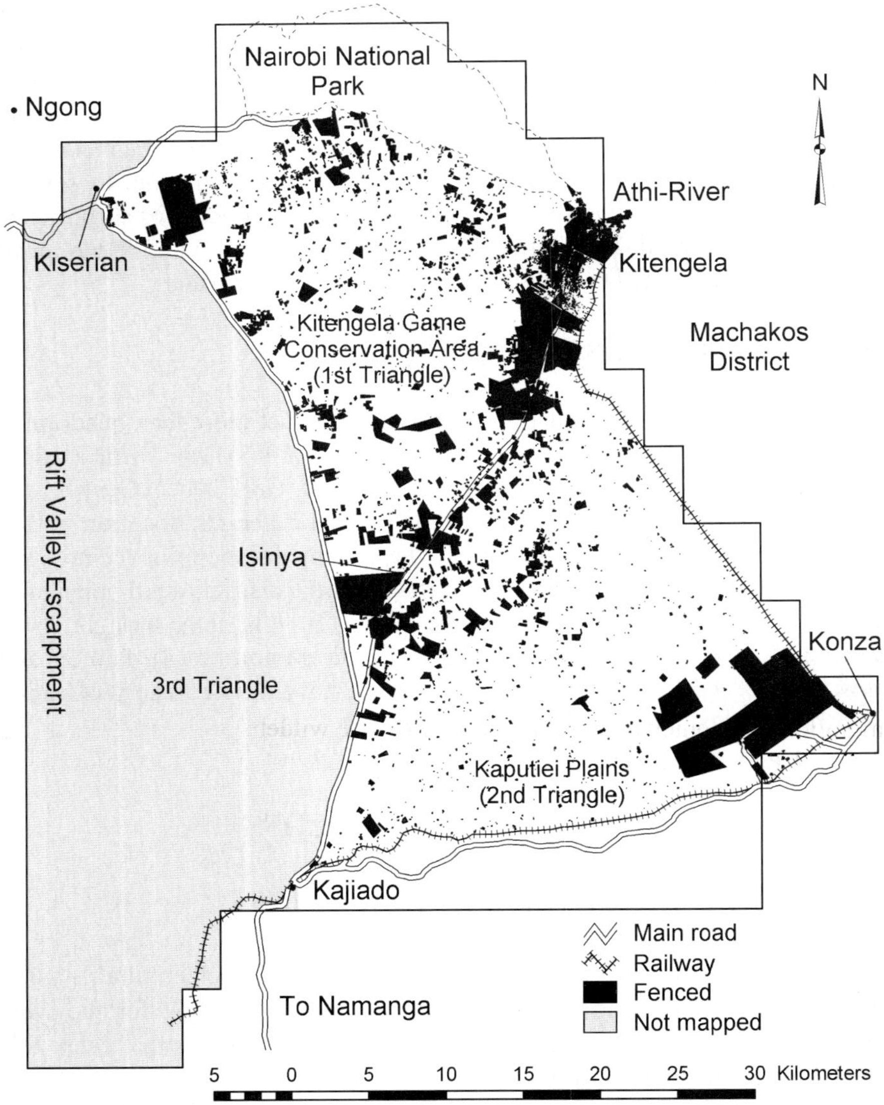

Figure 9-2. Fences and land use in first two triangles of the Athi-Kaputiei Plains, July-October 2004.

markets (e.g., for flowers), commoditization of livelihoods in response to these markets, and good access and use of educational opportunities also contribute to fragmentation.

Today, we see a strongly truncated and fragmented landscape in response to these forces, with a fragmented pastoral-wildlife savanna bounded on the north and east by towns and the city, and rapid demarcation of private plots through fence building by pastoral people themselves and by subsistence farmers, commercial (flower) farmers, city dwellers, non-governmental organizations (NGOs), the export processing businesses, and others (Figure 9-2). These changes affect the amount and spatial arrangement of rangeland open for wildlife and livestock grazing. Pastoral families sometimes fence their land to keep wildlife away from their homesteads, forage, and water (Mwangi and Warinda 1999). In 1999, nearly all the families in the first triangle had a small fence around their homes, 83% around their small cultivated plots next to their homes, but only 16% around any of their grazing land.

5. EFFECTS OF LOSS AND FRAGMENTATION ON WILDLIFE MOVEMENTS AND POPULATIONS

Historically, wildlife (wildebeest, zebra, and probably others) as well as pastoral people and their livestock accessed water and forage in the dry seasons and droughts at higher elevations near the footslopes of Mt. Kenya and in the Ngong hills (Figure 9-1). They then likely migrated back into the drier rangelands in the wet season to reach high quality forage (Gichohi 1996) and salt licks. Since the 1940s, loss of corridors restricted this migration to a somewhat circular pattern for the wildebeest between Nairobi National Park and their calving grounds in the drier 'second triangle' (see Figure 9-1) to the south. Zebra move widely, spending the wet season in particular areas in each of the three triangles, while other species like eland can move as far south as Amboseli (Hillman and Hillman 1977). Pastoral people and livestock cannot access the park legally, but often do so at night. Herders also have sole daytime access to pastures crowded with people (although wildlife may graze in these areas at night). During the 1999-2000 drought, like other recent droughts, it was common to see Maasai herders grazing cattle on the verges of highways and roads deep in the city of Nairobi.

In the rest of this section, we look at more recent trends in wildlife and livestock populations based on ground counts in Nairobi National Park from 1961-2004, and aerial survey data from the rest of the Athi-Kaputiei Plains from 1977-2002. In Nairobi National Park, counting teams completed total ground counts of wildlife from vehicles in 15 blocks, from 1961-1979, resuming again in 1990 to 2004, about six times a year (Gichohi 1996). In the three triangles in the Athi-Kaputiei Plains, the Department of Resource

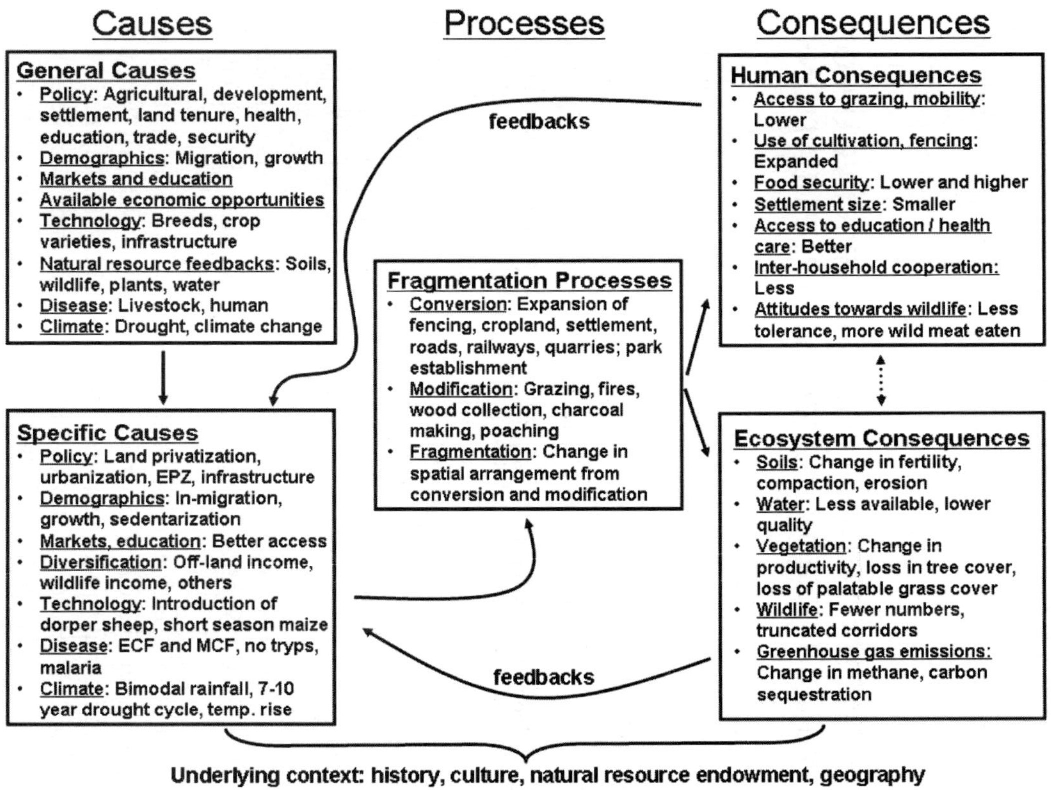

Figure 9-3. Conceptual model of the causes and consequences of fragmentation in the Athi-Kaputiei ecosystem. Arrows indicate directional connections between boxes. ECF = East Coast Fever, MCF = Malignant Catarrhal Fever, EPZ = Export Processing Zone, tryps = trypanosomosis.

Surveys and Remote Sensing counts used systematic reconnaissance flights from the air. Wet season counts were generally conducted between April and June and dry season counts between October and March (Gichohi 1996). In this study we used the wet season aerial counts to analyze animal trends for the Athi-Kaputiei. The trend analysis was based on the 5 x 5 km transects and covered the period 1977-2002. We used a polynomial regression of the log-transformed animal counts for each year and accounted for temporal autocorrelation in the counts using continuous-time generalizations of the first order autoregressive model. Model selection based on the corrected Akaike information criterion was then used to select the appropriate model from a set of candidates comprising linear, quadratic, and cubic polynomial trend models.

These counts show that from 1977-2002, wildlife populations fell precipitously by 72%, or an average of 5% per year, in the three triangles outside Nairobi National Park (Figure 9-4), nearly identical to the rate of loss of resident wildlife in the Mara ecosystem over a similar time period (Ottichilo et al. 2000). More than 90% of the eland, giraffe, and wildebeest disappeared over this 25-year period, twice the average wildlife loss. Impala and Thomson's gazelle declined by 78% overall, while Grant's gazelle populations halved. Much of these changes are probably due to mortality of animals, but some could be due to movement of animals out of the ecosystem.

The total density of migratory wildlife species (wildebeest, eland, and zebra) declined faster than the non-migrants (Grant's gazelle, Thomson's gazelle, kongoni, impala, giraffe, and ostrich, 76% vs. 63% loss). Only zebra numbers showed no overall change, with a humped distribution. Even browsers and mixed feeders (giraffe, eland, impala, gazelles), species likely to compete only with goats for forage, declined strongly. Wildlife populations declined dramatically during the droughts of 1960/1 (Foster and Coe 1968, Hillman and Hillman 1977), 1973/4, 1983/4, 1994 and 1999/2000. Loss during drought may indicate livestock or wildlife deaths, but also indicates movements of animals outside the system, which are usually temporary (Hillman and Hillman 1977, Nkedianye 2003).

With declining wildlife populations, one might expect livestock populations to rise in this pastoral part of the ecosystem, as more forage and water become available, and wildlife-livestock disease transmission presumably might decrease in some parts. Remarkably, sheep and goat populations dropped by 63% in the last 25 years, at the same rate as the small-bodied wildlife. Donkeys nearly disappeared altogether. Our key informants suggest that recent losses in sheep and goats are caused by the increased susceptibility of improved dorper sheep (introduced during the last decade or so) to diseases like blue tongue, which was widespread after a prolonged period of drought followed by unusually high rainfall. Cattle populations were stable except for heavy declines during the more recent droughts, between the periods 1994-96 and 1998-2000. Families in the first triangle of Kitengela lost, on average, 54% of their cattle herds during the most recent (2000) drought (Nkedianye 2003). The total biomass of wildlife and livestock together was almost halved in the pastoral part of the Athi-Kaputiei system in the last 25 years (Figure 9-4). It is possible that free-ranging wildlife and livestock decline for some of the same reasons, as the savanna fragments.

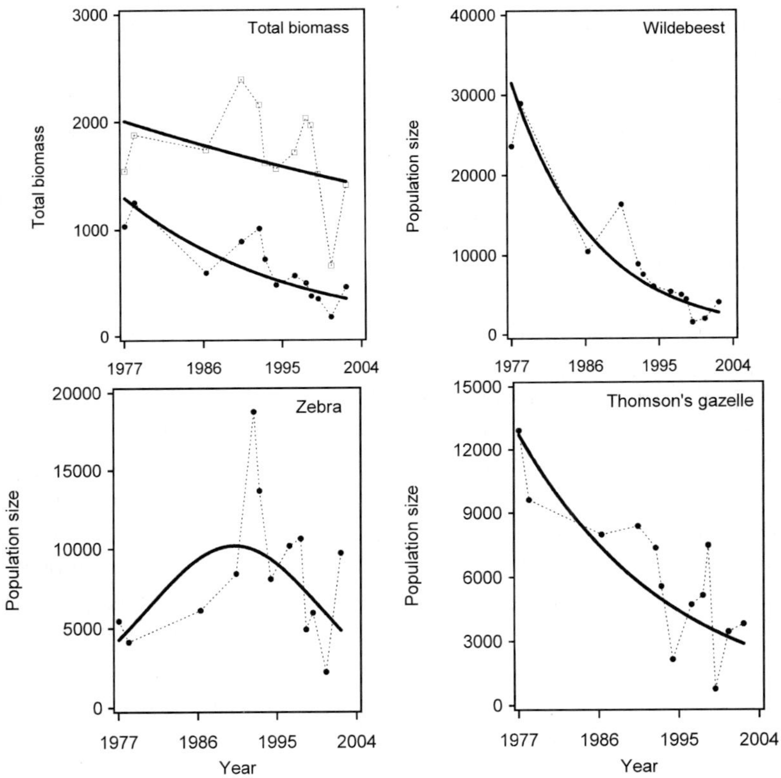

Figure 9-4. Wet season trajectories and trends of wildlife and livestock biomass (kg/km^2) and selected wildlife species numbers from 1977-2002 in the three triangles of pastoral lands of the Athi-Kaputiei Plains south of the park. Total biomass shows wildlife biomass (lower line) and livestock biomass (upper line). Dotted lines with markers show actual data, solid lines show trends.

We expected fewer losses of wildlife inside Nairobi National Park than in the three triangles outside the park, because of differences in land use. Our data support this. Before the national ban on wildlife hunting in 1977, wildlife in the park was in decline. Since 1977, while total wildlife biomass dropped strongly outside the park, there was no perceptible change, over the same period, inside the park, with some indication of a slight increase in total biomass (1977-2002, Figure 9-5). Wildebeest in the park increased during the late 1980s and then declined strongly in the late 1990s. Populations of zebra, also a migrant, grew strongly from 1977 to 2002 in the park, as did rhino. Thomson's gazelles changed little in the park, like buffalo and eland. Note that buffalo were introduced into the park in 1966. Kongoni increased to a peak of 3,323 by 1973 then declined to only 179 following the 1974 drought and have since stabilized around 380 individuals. Giraffe and ostrich consistently declined between 1990 and 2004. These trends suggest that there were only weak links between wildlife inside and outside the park in the period 1977-2002, except for wildebeest. Wildebeest, eland, and zebra populations in the park fluctuate strongly between the wet and dry seasons, suggesting significant movement of wildlife inside and outside the park (Hillman and Hillman 1977, Gichohi 2000), a phenomenon often observed by local people. Other species of wildlife varied less strongly between seasons during the 1961 to 2004 period, implying that some animals do stay relatively permanently within the park boundaries.

In the Machakos commercial ranches to the southeast of the Athi-Kaputiei, there was no decline in overall numbers of large mammals between 1991-2000 (Parker 2003). Fencing here prevents most movement between these ranches and the surrounding farming land to the east and the pastoral land to the west. Despite the stable populations, ranchers commonly find poaching snares on their properties on these ranches.

5.1 Causes of wildlife decline

Why are wildlife in decline in some places and not others? We cannot definitively establish the causes, but we can suggest likely candidates and their relative importance. Poaching of wildlife by people is probably the primary cause, with strong secondary causes. We assume poaching rates are rising in the Athi-Kaputiei Plains, caused by a rapid influx of outsiders who historically hunt, sell, and/or consume wild meat (Barnett 2000), but we know of no data that shows how fast poaching is growing. However, today, 61% of pastoral families in the Kitengela triangle currently consume some wild meat, especially when food is scarce, shifting away from their traditional prohibition on consuming non-domestic meat (Nkedianye 2003).

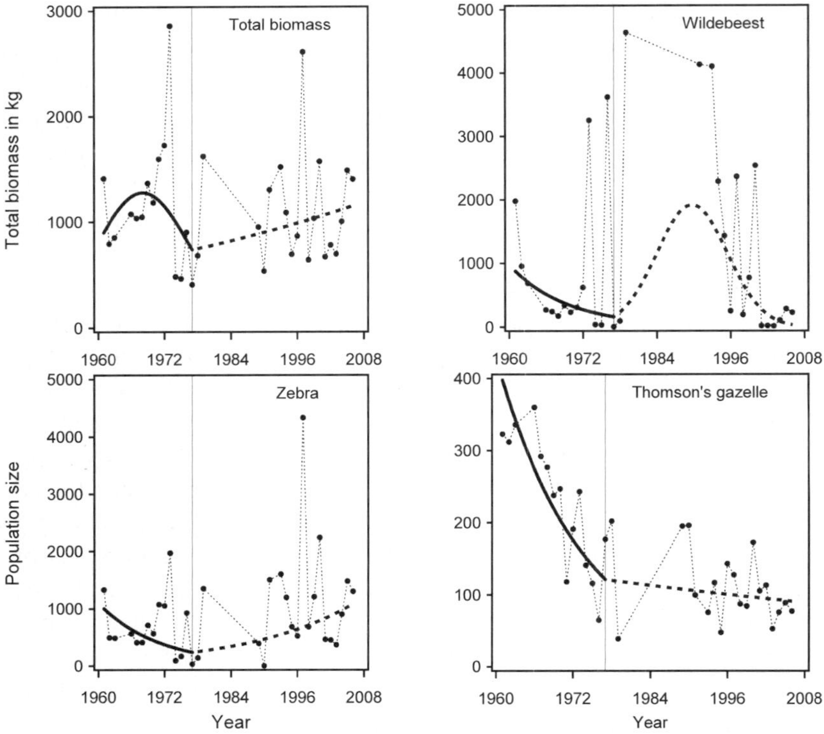

Figure 9-5. Trajectories and trends in numbers and biomass of selected wildlife species from 1961-2004 in Nairobi National Park based on averages of monthly counts conducted within each year. Dotted lines with markers show actual data. Solid lines without markers show trends before 1977, while dotted lines without markers show trends during 1977-2004 to facilitate comparison with the data from the pastoral lands in Figure 9-4. Trend lines were calculated separately for these two time periods.

Maasai respondents in the first Kitengela triangle say that people prefer the taste of eland and wildebeest meat and find that zebra is unpalatable and hard to catch, preferences that match the decline in numbers by species. However, non-Maasai poachers sell meat by weight and will kill almost any type of grazer. There are few controls on poaching because anti-poaching efforts are weak across Kenya (Barnett 2000). We speculate that increasing poverty also leads pastoral and non-pastoral people to eat more wild meat. On the other hand, increased education and diversification of incomes seems to reduce dependence on wild meat among pastoral families in this system (Nkedianye pers. obs.) Consumption of wild meat by Kamba farmers/hunters in Kitui District to the east of the study area was 67 grams/person/day or 14 kg/month/family in the late 1990s (Barnett 2000). Poachers cut fencing to make snares and burn pastures to create patches of green grass that attract wildlife and then ring these areas with snares (Parker 2003). Local organizations find large numbers of snares along the southern, and unfenced, edge of the park, so animals passing from the park to the pastoral land are at risk of injury or death. Ranchers in Machakos fence most of their properties, which presumably reduces the rates of poaching inside these large fenced properties.

As in southern Africa, loss and fragmentation of habitat (forage, water) by fencing and some cultivation is probably also a major cause of the decline in migratory wildlife (Whyte and Joubert 1988, Spinage 1992, Perkins 1996, Boone and Hobbs 2004) and likely elsewhere. The incidence of poaching may be related to land use and fragmentation because poachers often corner wildlife by driving them from open rangelands into fencelines to trap and kill them (Nkedianye pers. obs.). On the other hand, well-fenced land may deter poaching inside the fence, as is probably the case in the Machakos ranches. But it is unlikely that fencing is the sole cause of wildlife loss in the Athi-Kaputiei because the loss in wildlife (72%) is far higher than the proportion of land fenced (14%) from 1977 to 2004 (assuming no fences in 1977, although the relationship may not be linear). Even though the amount of fenced land is relatively low, there were a total of 6,741 parcels with fencing. Many scattered fences probably disrupt wildlife behavior and movement, even if their areal coverage is still low. In Kitengela, the fenced parcels are spread throughout the range of wildlife, suggesting that wildlife are almost always in visible distance of people, wet season or dry, whenever they are outside the park.

Fences may differentially enclose wildlife habitat of high value (good grazing lands, water points). Changes in the distribution of wildlife across the ecosystem show that few wildlife still use areas around roads and towns, where fencing and human population growth are highest (Figure 9-3). Furthermore, fences may cause a disproportionate loss in wildlife if there is a threshold of habitat area needed to sustain healthy populations (caused, for example, by fencing key resources first). Gardner et al. (1987) and Stauffer (1985) predicted that the ease of movement of animals through a connected landscape is rapidly lost when 30-50% of the landscape is converted to uses incompatible with animal movement.

Fences may also reduce the number of animals particular parts of the landscape can support (Boone and Hobbs 2004, Boone et al. 2005). Using a model of the Amboseli ecosystem, just 70 km south of the Athi-Kaputiei (BurnSilver et al., Chapter 10), they found that the diversity of types of patches that wildlife and livestock can access declines as their access to the landscape becomes restricted to smaller and smaller areas. For example, the amount of variation in green forage (measured by greenness) accessible to a herd of livestock declines by 12% when the parcel they can access halves from 20 to 10 km^2, similar to their findings for cattle in northwest South Africa.

But fencing may represent more than mere fragmentation; if herders, farmers, and townspeople exclude wildlife entirely from fenced areas, there is less wildlife habitat altogether. This is sometimes the case in the Athi-Kaputiei, particularly around settlements. So far, expansion of subsistence cultivation is limited, and thus probably has limited impact on wildlife, similar to Ngorongoro to date (Boone et al. 2002, McCabe 2003). Currently, expansion of commercial cultivation (in this case flower farms) seems to be more of a threat to wildlife, as it is elsewhere (e.g., Homewood et al. 2001, Serneels and Lambin 2001).

Recurrent droughts can cause up to a 50% loss in wildlife populations, as happened between 1958-1962 because of the 1961 drought (Stewart and Zaphiro 1963). From Figures 9-4 and 9-5, it appears that the 1999-2000 drought had more effect on animal populations than any other drought since the early 1960s. Some of this loss is through starvation, but animals also move out of the ecosystem in the hardest times. Drought is probably a less important cause of long-term wildlife loss, unless droughts are becoming more frequent or more severe because of climate change. Or, other changes, like fragmentation, may make wildlife (and livestock) populations more vulnerable to drought or make recovery after drought more difficult (e.g., Holling and Meffe 1996). In southern Africa, a quarter to half of selected mammal species are predicted to go extinct by 2050 because of climate change (Thomas et al. 2003), principally because of decreased rainfall. Predictions of climate change near the equator are uncertain, with a good possibility of increased rather than decreased rainfall (as measured by length of growing period) in 50 years (Jones and Thornton 2003). But temperatures are also increasing (Altmann et al. 2002, Hemp 2005), which will likely negate the impact of increased rainfall by increasing evapotranspiration.

Pastoral Maasai in the Kitengela often observe that wildlife cluster just outside the park on the short, 'grazing lawns' created by livestock grazing and avoid the coarse, tall grasses in the park to access better food and avoid predators (Nkedianye, Reid, pers. obs.). Park management burned and mowed park grasslands to attract wildlife into the park from the late 1950s to 1963 and from 1968 to the mid-1970s, but from then until the late 1990s, no burning was done (Gichohi 1990). The Kenya Wildlife Service recently resumed burning, and wildlife are clearly more abundant on burnt, short grass than in unburnt, tall grass. However, significant numbers of wildlife still cluster outside the park in the areas grazed by livestock (Nkedianye, Reid, pers. obs.).

6. EFFECTS OF LOSS AND FRAGMENTATION ON LIVESTOCK POPULATIONS AND MOVEMENTS

At the time of sub-division of the Kaputiei group ranches in 1986/7, there were few fences on the communally owned land (Nkedianye, pers. obs.). Livestock moved freely among the three triangles, from Empakasi (northeast) to Oloosirkon (northwest) to Enkirgirri (southeast), depending on where the pastures were better, similar to wildlife. By 2004, herding cattle on foot from Isinya to Oloosirkon took at least twice as long as it did in the late 1980s (Nkedianye, pers. obs.). Landowners have fenced pastures, salt licks, and water, making it difficult for the majority of the pastoral residents and their livestock, as well as wildlife, to access these resources, thus magnifying the effect of fencing beyond the area the fences enclose.

To better understand the effects of fencing on livestock movements, we contrasted the movements of herds of cattle in open rangeland with little fencing and congested ('closed') rangeland with abundant fencing. We collected data at a temporal scale of one minute to capture feeding behavior at three scales: the feeding station, micro-patch, and plant community scales (Senft et al. 1987). Future work will capture regional-scale herd movements over time through interviews. Herders carried a GPS unit which logged the position of the herd they were following every minute automatically throughout the day; in addition, every ten minutes the herders recorded the distance to the nearest fence from the cattle herd. Observations were made 15 times between March 2003 and April 2004.

Fencing strongly changed the speed, pattern, and area grazed by cattle at fine scales. In the open rangeland, cattle grazed, on average, 200 m from the nearest fence; in the congested rangeland, this fell to 50 m. The total area grazed was significantly smaller for the herd in the area with many fences than in the open area. Although the sizes of the herds were relatively similar in the two areas (45-54 animals), the grazing orbits for the herd in the congested rangeland were more convoluted than in the open rangeland, where fencing did not hinder cattle choices of where to feed (Figure 9-6).

Cattle moved more slowly in the landscape with few fences. On average, the herd in the "open", less fenced area, walked 35% more slowly than the other herd in the "closed" area ($\overline{U}_{open} = 0.227$ m/s vs. $\overline{U}_{close} = 0.308$ m/s; $p < 0.001$). Cattle in the unfenced landscape walked quickly from place to place and then lingered to feed and rest throughout the day (Figure 9-7). Cattle in the fenced areas did stop to feed and rest but were more constantly on

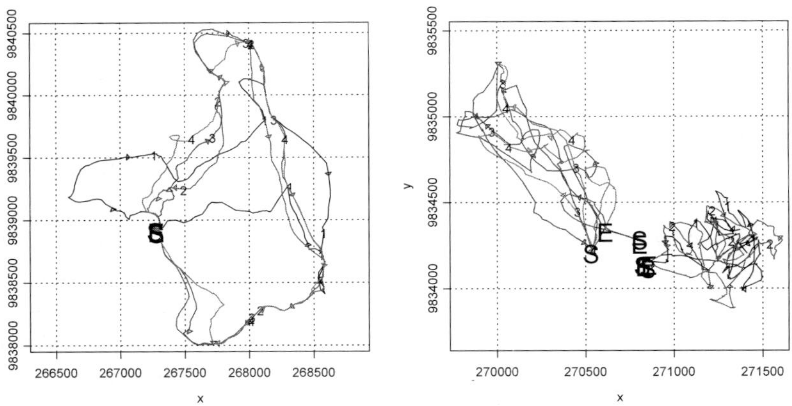

Figure 9-6. Four daily grazing orbits in April 2003 in a relatively enclosed area with a herd size of 54 (right) and in an open area with a herd size of 45 (left). The letter "S" and "E" designate the start and end of the track respectively. In both maps, each grid square represents 500 meters in length and the axes are in UTM coordinates.

the move from place to place than the unfenced herds. It is possible that food quality and quantity are different in the two areas, but we did not measure this.

We hypothesize that fences affect livestock foraging by limiting the number and diversity of plant patches and communities that livestock have access to at moderate scales (e.g., Senft et al. 1987). This will be particularly true where good quality plant patches are clustered in certain locations on the landscape. At a finer scale, fencing likely has little effect on the choice livestock make about which plant part to eat, because these choices are made once the animal chooses to stop at a feeding location which should be independent of the presence or absence of fencing. Fences may also affect the quantity and quality of food available at each feeding station if the intensity of grazing is different in fenced compared to unfenced areas, which we think is likely. This may explain the greater velocity of the cattle herd in our fenced landscape. Fencing will also increase travel costs for short or long-distance movements because travel paths will need to be more convoluted to avoid fences. Indeed, there is some evidence that higher walking speed can incur higher energy expenditures, which, in turn can affect milk yields (Homewood and Rodgers 1991, Figure 9-7).

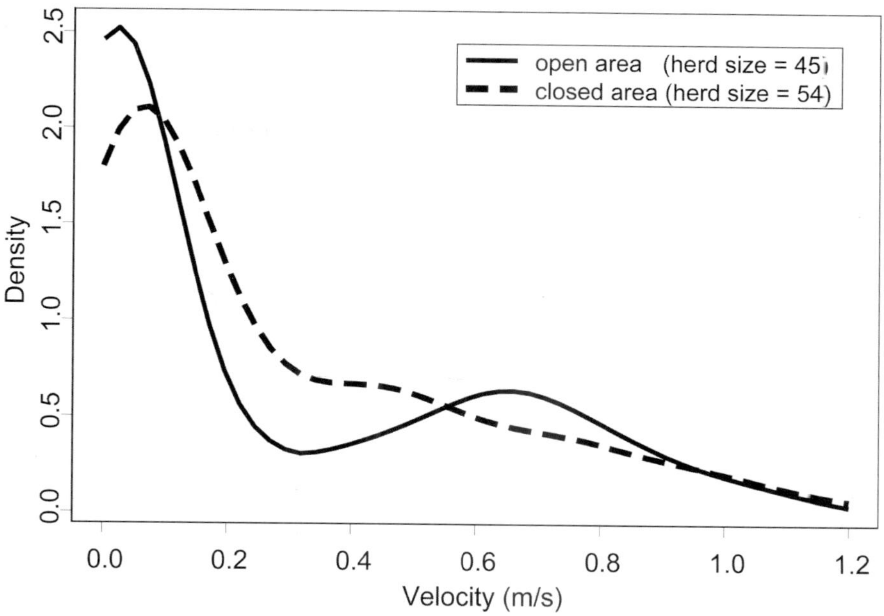

Figure 9-7. Herd velocity (m/s) profile (probability density function) based on four grazing orbits for two herds in a relatively open and closed area.

7. LIKELY EFFECTS OF LAND SUB-DIVISION ON PASTORAL LIVELIHOODS

We did not design data collection to assess the consequences of fragmentation through land sub-division on pastoral livelihoods. However, several recent surveys, in and near the Athi-Kaputiei ecosystem, allow us to briefly summarize some of these impacts as well as the linked processes of sedentarization and diversification of livelihoods (BurnSilver et al., Chapter 10).

Land privatization and sub-division initiates a suite of processes in the pastoral communities of Kajiado. The number of families that share the same homestead declines after sub-division as families move to their own piece of land (Njoka 1979). This and other changes mean that family members have less leisure time (Rutten 1992). Kaputiei Maasai, only a few years after sub-division, said that cooperation among herders was about the same, but they made fewer cooperative decisions on where cattle should graze, because the land was privately owned (Rutten 1992). Herd movement can become more difficult as people settle, as we saw above, but this was not the case for a sample of sedentary households in northern Kenya, because herds moved but households did not (McPeak and Little 2005). During the drought of 2000, herders started moving in search of pastures a bit earlier in the Athi-Kaputiei Plains than in previous droughts partly due to the loss of grazing lands as a result of land sales and fencing (Nkedianye 2003). Marcel Rutten (1992) calculated that only 10% of all households in Kaputiei would have enough grass to support their livestock on their own plots following sub-division in 1986. Thornton and colleagues (2006), using a linked savanna ecosystem-pastoral household model (SAVANNA-PHEWS) found that sub-division into smaller parcels can have significant consequences for long-term food security in southern Kajiado. If households only have access to their own parcels, even if those parcels are relatively large, the former group ranch supports fewer households at the same level of well-being after sub-division than before sub-division. In addition, there are some particular disadvantages to having a homestead fixed in some locations; for example, Maasai with plot allocations in the wildebeest calving area in the second triangle have to move their cattle for three months each year from their plots to avoid contracting malignant catarrhal fever between March and May (Nkedianye 2003, Bedelian et al. 2007).

Once land is privately held, herders can sell land for the first time and they often do. Particularly important to fragmentation, Maasai often sell land to non-Maasai who come from nearby farming cultures. For example, Maasai land owners had sold 30% of their plots in Kisaju Group Ranch, mostly to non-Maasai, only six years after privatization (Rutten 1992). While the sales bring in much-needed cash, these cash gains can be short-lived, followed by decreases in income from the remaining, smaller parcels. In the Kitengela triangle, sales will continue in the future because a third of the landowners plan to sell an average of 22 acres of land in the next three years (Nkedianye 2003). However, Nkedianye (2003:52) notes that there is widespread "disillusionment among the landowners as a result of the poor performance of those who rushed to sell land, most of whom ended up poorer".

Nearly all residents now fence their homesteads and adjacent gardens, but most leave their grazing lands open (Mwangi and Warinda 1999). However, new landowners with a farming tradition put up more fences than the Maasai families who were given plots when the group ranch was sub-divided (Kimani and Pickard 1998, Nkedianye 2003). Three-quarters of all farmers involved in cultivation in Kajiado District were non-Maasai only a few years after sub-division (Rutten 1992). Non-Maasai have more need to fence than Maasai because non-Maasai have smaller plots, they cultivate most of their land and thus must protect crops from wildlife, and they are also more familiar with fencing (Kimani and Pickard 1998). By so doing, they deny herders and wildlife access to water, pastures, salt licks, and routes that they have used over the years (Nkedianye 2003). Interestingly, rules restricting grazing access on private lands that are not fenced have not yet arisen in Kitengela.

But land sub-division does have its advantages. From a Maasai perspective, the biggest advantages are security of land ownership, easier access to credit, access to land ownership by younger Maasai who were too young to acquire land when group ranches were formed, and ending their frustration with group ranch management (Grandin 1986, Pasha 1986). However, the Kenyan government initially did not support sub-division because it was concerned about the ecological and economic viability of individual parcels (Bekure et al. 1991). In addition, sometimes the poor fare better once they are allocated their share of land and the associated resources (Nkedianye, pers. obs.). Kristjanson et al. (2002) found that younger, more educated households with diverse income sources who were more market-oriented typically had overall incomes that were significantly higher than more traditional, older and less educated households. During droughts, the wealthy may pay to access the pastures of the poor who have fewer livestock, a common practice today in Kima, Arroi, and Nkama areas of Mashuuru division of Kajiado district (Nkedianye, pers. obs.). However, Kimani and Pickard (1998) suggest that both pastoral viability, tenure security, and ecological integrity will be met better by maintaining group ranches rather than privatizing land. All the same, privatization is already with us, so the key question is how to soften the disadvantages of this process and magnify the advantages. Although we know something about the more immediate advantages and disadvantages of sub-division over the short term, we have little idea of the wider and longer-term consequences. Most Maasai feel that they had to sub-divide to gain secure tenure of their land and water; they are acutely aware that sub-division will be injurious to their long-term interests and well-being (Nkedianye, pers. obs.).

Land privatization or sub-division usually results in increased sedentarization and is often linked to livelihood diversification and/or intensification, although it is not a necessary pre-condition for either process. In the Kitengela survey, households that had smaller land holdings earned more income per acre from farming and livestock than the more traditional pastoral households, suggesting a move towards more intensive crop and livestock production with shrinking landholdings for some households (Kristjanson et al. 2002). But, when wildlife incomes are available, sedentarization may be incompatible with maintenance of wildlife populations, and thus sedentarization and fencing removes this potential source of income for pastoral families.

Although sedentarization implies a loss in herd mobility, it is not always so. In northern Kenya, McPeak and Little (2005) found that pastoral households use paired sedentary homesites and satellite livestock camps to gain the advantages of sedentary life around towns, like access to wage employment, education, and health care, while maintaining herds that can access water and pasture far from town and can move in response to a variable environment. Sedentarization also can provide income-earning opportunities for poorer women (Little et al. 2001, Nduma et al. 2001).

Diversification is widely seen as a response to declining livestock holdings per household (Little et al. 2001). Sedentary households tend to diversify the ways that they earn their incomes beyond pastoralism (McPeak and Little 2005). Pastoral households often use crop agriculture to support pastoralism, by reducing the need for the family to sell livestock to buy grains during dry periods. Crop growing is particularly important after droughts to provide a food source if herds are decimated in northern Kenya (Little 1985, 1992), but is only advantageous when cropping occurs in wetter areas where returns to cropping are reliable (Little et al. 2001). Similarly, in Kajiado, PHEWS model results suggest that rainfed cropping has an adverse impact on household food security because farming is so risky in this region (Thornton et al. 2006) and (even small) crop input costs are incurred in years when yields are very low or zero. These findings are supported by survey data from Kitengela showing negative net crop incomes even in good rainfall years (Kristjanson et al. 2002). The range of off-land income-earning activities was very large in Kitengela, with 31% of households obtaining over 30% of their total income from income-earning activities other than crops and livestock, 34% saying that off-land income made up 10-30% of their total household income, while 34% of households had no off-land income. In Kitengela and northern Kenya, more educated households were more diversified (Kristjanson et al. 2002, McPeak and Little 2005). PHEWS model results also suggest that more diversified households are better off in southern Kajiado (Thornton et al. 2006), particularly poor households with few livestock. This was not always the case in northern Kenya; richer households often benefited from diversification, but poor households often did not (Little et al. 2001).

8. THE KITENGELA LEASE PROGRAM: UN-DOING SOME OF THE NEGATIVE ASPECTS OF SUB-DIVISION

As we have seen, one disadvantage of sub-division is the fragmentation of land, loss of wildlife habitat, and restriction of the movement of both domestic and wild animals. Initiated in April 2000, the Wildlife Conservation Lease Program was created to ensure that wildlife in the Athi-Kaputiei

Plains could move freely to their traditional habitats. The program requires participants to allow free movement of wildlife on their land, refrain from poaching themselves, report poaching by others, protect natural vegetation, and avoid fencing or sub-dividing their land. In return, they receive Ksh 300/acre/year (US $4.25 in late 2006). The program started by leasing 214 acres from two participants in 2000 and grew to leasing 8,600 acres from 118 families in the first triangle of Athi-Kaputiei Plains by late 2004. In late 2004, the project disbursed approximately Ksh 3,000,000/year. The installments of Ksh 1,000,000 per school term to the 118 families ensured many local parents found school fees for their children, particularly those in secondary school. Participants in the leasing program have more positive attitudes towards wildlife, are more willing to share water and pastures with wildlife, and strongly support keeping the range open without fencing (Table 9-1).

Table 9-1. Attitudes of pastoral households who do and do not participate in the conservation leasing program about wildlife and conservation, measured by percentage of respondents who strongly agree with statements posed by the interview team (extracted from Nkedianye 2003:106-107). NNP = Nairobi National Park. N = 104 respondents, 52 participants and 52 non-participants.

Statement	% of non-participants who strongly agree	% of participants who strongly agree
Wildlife is important to you	24	62
Wildlife conservation is important to society and future generations	33	56
Area be left open for livestock and wildlife with benefits	42	59
Area be left open for livestock and wildlife without benefits	10	10
All landowners to fence their land to keep away wildlife	28	6
Livestock and wildlife to share basic resources (water and pasture)	12	38
Development of tourist related activities be encouraged	51	67
Government to plough back revenue from NNP to the area	71	77
Government policy re: human-wildlife conflict resolution fair	4	6
Government policy re: wildlife revenue sharing with communities fair	4	0
Lease Program an adequate method for saving wildlife	20	50
Fenced Nairobi National Park would be more beneficial	17	6

Many participants say that the lease program allows them to choose not to sell land because the strongest motivation for the sale of land among most Kitengela households is the need for school fees (Nkedianye 2003). In a bad rainfall year (when the long rains fail), lease program payments double the income of the poorest households (Kristjanson et al. 2002). There is some indication that lease payments are allowing parents to afford to send more girls to school (Nkedianye, pers. obs.).

9. IMPLICATIONS: ALTERNATIVE FUTURES FOR PASTORALISM AND WILDLIFE IN ATHI-KAPUTIEI ECOSYSTEM?

It is clear that the lease program is a success in the eyes of the participants, but is this effort too limited currently to allow continued pastoral and wildlife use of the Athi-Kaputiei ecosystem? Urbanization of the Athi-Kaputiei is so rapid that the lease program will need to be expanded significantly, which is the focus of current efforts. Land prices are rising in desirable areas next to the all weather road, near the national park, and in areas contiguous to shopping centers, reducing incentives for landowners to participate in the lease program in these areas. Such a program also requires strong collective action and community support since a few individuals can spoil the efforts of many (e.g., by putting up fences along key migration routes). New strategies of land purchase, permanent conservation easements, tax incentives, implementation of land zoning, and others will be needed if the massive wildlife losses in this area are to be reversed. There also has been no study, to our knowledge, that looks at the effects of sedentarization, intensification, and diversification on the attitudes of pastoral/agro-pastoral families towards wildlife, nor on their incentives to participate in different conservation initiatives.

Also critical is a strong government policy on land use and enforcement of current anti-poaching regulations. Significant progress in policy has been made in the last 50 years, so that wildlife conservation is not just focused on protected areas, and communities have started to receive some returns from conservation (Hulme and Murphree 2001). However, it is still the case that government policy tends to favor farmers, since administrators come predominately from agricultural backgrounds (Horowitz and Little 1987). There is also a strong assumption that food security is only gained through production of crops, rather than livestock products. There is deep irony in this prejudice. In late 2004, tourism was the biggest foreign exchange earner for Kenya, with 42 billion Kenya shillings (US $560 million) in earnings from 1.4 million visitors, supporting thousands of livelihoods throughout the country (Mugambi 2005). The existence of the wildlife-related returns rests almost entirely on the long history of compatibility between wildlife and pastoral land use in savanna regions of the country; the future viability of these returns also depends, in part, on the continuing good will of these same pastoral communities towards wildlife. It is ironic that government policy does not support pastoralism: pastoral families are responsible for maintaining the livelihoods of many people outside pastoral lands through conserving the wildlife that forms the base of employment in the tourism sector. Development and implementation of strong land-use policies that consider pastoral livelihood needs and wildlife conservation on equal ground with other development needs, will allow these communities to meet these responsibilities. This policy would ensure, for example, that sub-division below a certain acreage is illegal (and enforced). Just as important is effective education, an appropriate legal framework, and enforcement of anti-poaching by a wide range of actors. Political and financial support of pastoralism from individuals and businesses supported by the tourism will help also.

10. EPILOGUE: AND WHAT OF THE ROLE OF RESEARCH?

This research just begins the collection of information needed to allow actors to fully assess the societal trade-offs of different futures for the Athi-Kaputiei. Such research is interesting, but not particularly useful, unless it gets beyond academia into the hands of actors (communities, policymakers). In cases like this, researchers can strengthen management of natural resources by communities and improve policies affecting the sustainability of pastoral livelihoods and their ecosystems by listening to policy makers and community members and designing research to address their pressing questions (e.g., Tomich et al. 2004). Our research group is attempting to do exactly this in the Athi-Kaputiei Plains. The research group consists of a united researcher – community facilitator team that attempts to knit the needs of communities and policymakers throughout the research process and strengthen researcher-community-policymaker networks. One key to this approach is identification of the salient, policy-relevant issues for research with local community members and leaders and also with national-level research and management institutions (e.g., Cash et al. 2003). The team attempts to strengthen the legitimacy of the research for different stakeholder groups by including and addressing the concerns of a wide range of actors (individuals, institutions) that focus on agricultural development, land-use planning, water resources, and wildlife conservation. Another effective strategy is for the core research – communication team to act as a convener and catalyst for other national and international researchers working in the same ecosystems to communicate with communities and policymakers. Specific activities to strengthen these links include community involvement in all data collection, interpretation and feedback with the wider local communities, meetings with policymakers to revise policy acts on wildlife and pastoral development, grants to national and international students to report their PhD results back to communities and discuss policy and management options, and meetings for researcher – policymakers to discuss salient issues. The goal of this engagement is to help stakeholders to better evaluate the trade-offs of alternative ways of using the Athi-Kaputiei Plains landscape, so that they can create more viable and vibrant futures for themselves, their communities, and wildlife.

ACKNOWLEDGEMENTS

Foremost, we thank the communities in the Athi-Kaputiei for helping us understand their issues. Most of the data was collected in collaboration with the Kitengela Landowners' Association (KILA) and local community members. We thank Simon Ole Mula, Daniel Ole Issa, Lugard Ole Makui, Joseph Ole Matanta, Simon Ole Peria, Joseph Ole Tuletto, Nelson Ole Oiputa, Mark Ole Koikai, Joseph Ole Kimiti, Michael Arunga, Vincent Odour, James Ole Turere, Nathaniel Ole Sinkeet, and Ogeli Ole Makui for collecting data on 6,741 fence lines in Kitengela. The mapping of cattle movements was assisted by Nelson Ole Oiputa, Mark Ole Koikai, Walter Ole Sinkeet, John Ole Sayiore, Edward Ole Sinkeet, and Paul Ole Sinkeet. We acknowledge the contribution of the members of the East Africa Natural History Society, Kenya Wildlife Service (KWS), the Nairobi National Park staff members, Friends of Nairobi National Park (FoNNaP), and the former Wildlife Conservation International for collecting valuable historical data on wildlife numbers in NNP. We are grateful to the Wildlife Conservation Society for supporting H. Gichohi to collect aerial surveys of wildlife from 1992 to 1998. We highly appreciate the support of the directors of both the Department of Resource Surveys and Remote Sensing (DRSRS) and KWS for enabling us to access the datasets on animal population counts. Finally, we thank Phil Thornton, Jeff Worden, Jill Lackett, and Tom Hobbs for thoughtful reviews of this chapter.

ENDNOTES

[1] These plains are marked on maps as the Athi Kapiti Plains. Kapiti is a shortened spelling of the name of the sectional tribe of Maasai, the Kaputiei, after which the plains are named. We choose to use the latter spelling here.

[2] Calculations made by R. Reid based on Meinertzhagen (1957), Gichohi (1996).

REFERENCES

Altmann, J., S. C. Alberts, S. A. Altmann, and S. B. Roy. 2002. Dramatic change in local climate patterns in the Amboseli basin, Kenya. African Journal of Ecology 40:248-251.

Baker, B. H. 1954. Geology of South Machakos District. Geological Survey of Kenya, Nairobi, Kenya.

Barnett, R. 2000. Food for thought: The utilization of wild meat in Eastern and Southern Africa. TRAFFIC East/Southern Africa, Nairobi, Kenya.

Bedelian, C., D. Nkedianye, and M. Herrero. 2007. Maasai perception of the impact and incidence of malignant catarrhal fever (MCF) in southern Kenya. Preventive Veterinary Medicine 78:296-316.

Bekure, S., P. N. De Leeuw, B. E. Grandin, and P. J. H. Neate. 1991. Maasai herding: An investigation of the livestock production system of Maasai pastoralists in eastern Kajiado District, Kenya. International Livestock Centre for Africa, Nairobi, Kenya.

Bell, R. H. V. 1982. The effect of soil nutrient availability on community structure of African ecosystems. Pages 191-216 *In* B. J. Huntley and B. H. Walker, editors. Ecology of tropical savannas. Springer-Verlag, Berlin, Germany.

Blench, R. 2000. 'You can't go home again', extensive pastoral livestock systems: issues and options for the future. ODI/FAO, London, UK.

Boone, R. B., M. B. Coughenour, K. A. Galvin, and J. E. Ellis. 2002. Addressing management questions for Ngorongoro Conservation Area, Tanzania, using the SAVANNA modelling system. African Journal of Ecology 40:138-150.

Boone, R. B. and N. T. Hobbs. 2004. Lines around fragments: Effects of fencing on large herbivores. South African Journal of Grass and Forage Science 21:147-158.

Boone, R. B., S. B. BurnSilver, P. K. Thornton, J. S. Worden, and K. A. Galvin. 2005. Quantifying declines in livestock due to land subdivision. Rangeland Ecology & Management 58:523-532.

Campbell, D. J. 1993. Land as ours, land as mine: Economic, political and ecological marginalization in Kajiado District. Pages 258-272 *In* T. Spear and R. Waller, editors. Being Maasai. James Currey, London, UK.

Campbell, D. J., D. P. Lusch, T. Smucker, and E. E. Wangui. 2003. Land use change patterns and root causes in the Loitokitok Area, Kajiado District, Kenya. LUCID report #19, International Livestock Research Institute, Nairobi, Kenya.

Cash, D. W., W. C. Clark, F. Alcock, N. M. Dickson, N. Eckley, D. H. Guston, J. Jager, and R. B. Mitchell. 2003. Knowledge systems for sustainable development. Proceedings of the National Academy of Sciences 100:8086-8091.

Cronon, W. 1983. Changes in the land. Hill and Wang, New York.

Foster, J. B. and M. J. Coe. 1968. The biomass of game animals in Nairobi National Park (1960-1966). Journal of Zoology, London 155:413-425.

Fratkin, E. and K. Smith. 1995. Women's changing economic roles and pastoral sedentarization: varying strategies in alternative Rendille communities. Human Ecology 23:4233-4454.

Fritz, H. and P. Duncan. 1994. On the carrying capacity for large ungulates of African savanna ecosystems. Proceedings of the Royal Society of London Series B Biological Sciences 256:77-82.

Galaty, J. G. 1994. Rangeland tenure and pastoralism in Africa. Pages 185-204 *In* E. Fratkin, K. A. Galvin, and E. A. Roth, editors. African pastoralist systems: An integrated approach. Lynne Reiner Publishers, Boulder, Colorado.

Gardner, R. H., B. T. Milne, M. G. Turner, and R. V. O'Neill. 1987. Neutral models for the analysis of broad-scale landscape patterns. Landscape Ecology 1:19-28.

Gichohi, H. 1990. The effects of fire and grazing on grasslands of Nairobi National Park. MSc. University of Nairobi, Nairobi, Kenya.

Gichohi, H. 2000. Functional relationships between parks and agricultural areas in East Africa: The case of Nairobi National Park. *In* H. H. T. Prins, J. G. Grootenhuis, and T. T. Dolan, editors. Wildlife conservation and sustainable use. Kluwer Academic Publishers, Dordrecht, Netherlands.

Gichohi, H. W. 1996. The ecology of a truncated ecosystem - The Athi-Kapiti Plains. Ph.D. University of Leicester, U.K., Leicester.

GoK. 2001. 1999. Population and housing census. Volume 1. Counting our people for development. Population distribution by administrative areas and urban centres. Ministry of Finance and Planning, Nairobi, Kenya.

Grandin, B. E. 1986. Land tenure, subdivision, and residential change on a Maasai group ranch. IDA Development Anthropology Network 4:9-13.

Hemp, A. 2005. Climate change-driven forest fires marginalize the impact of ice cap wasting on Kilimanjaro. Global Change Biology 11:1013-1023.

Hillman, J. C. and A. K. K. Hillman. 1977. Mortality of wildlife in Nairobi National Park during the drought of 1973-74. East African Wildlife Journal 15:1-18.

Holling, C. S. and G. K. Meffe. 1996. Command and control and the pathology of natural-resource management. Conservation Biology 10:328-337.

Homewood, K., E. F. Lambin, E. Coast, A. Kariuki, I. Kikula, J. Kivelia, M. Y. Said, S. Serneels, and M. Thompson. 2001. Long-term changes in Serengeti-Mara wildebeest and land cover: pastoralism, population or policies. Proceedings of the National Academy of Sciences 98:12544-12549.

Homewood, K. M. and W. A. Rodgers. 1991. Maasailand ecology: pastoralist development and wildlife conservation in Ngorongoro, Tanzania. Cambridge University Press, Cambridge, UK.

Horowitz, M. M. and P. D. Little. 1987. African pastoralism and poverty: some implications for drought and famine. Pages 59-82 *In* M. Glantz, editor. Drought and famine in Africa: denying drought a future. Cambridge University Press, Cambridge, UK.

Hulme, D. and M. Murphree, editors. 2001. African wildlife and livelihoods: The promise and performance of community conservation. East African Educational Publishers, Nairobi, Kenya.

Huntley, B. J. 1982. Southern African savannas. *In* B. J. Huntley and B. H. Walker, editors. Ecology of tropical savannas. Springer-Verlag, Berlin, Germany.

Jacobs, A. H. 1975. Maasai pastoralism in an historical perspective. Pages 406-425 *In* T. Monod, editor. Pastoralism in tropical Africa. International African Institute, Oxford, UK.

Jones, P. G. and P. K. Thornton. 2003. The potential impact of climate change on maize production in Africa and Latin America in 2055. Global Environmental Change 13:51-59.

Katampoi, K. O., G. O. Genga, M. Mwangi, J. Kipkan, J. Ole Seitah, M. K. Van Klinken, and M. S. Mwangi. 1990. Kajiado District Atlas. Arid and Semi-arid Lands Programme, Kajiado, Kenya.

Kimani, K. and J. Pickard. 1998. Recent trends and implications of group ranch sub-division and fragmentation in Kajiado District, Kenya. The Geographical Journal 164:202-213.

Kristjanson, P. M., M. Radeny, D. Nkedianye, R. L. Kruska, R. S. Reid, H. Gichohi, F. Atieno, and R. Sanford. 2002. Valuing alternative land-use options in the Kitengela wildlife dispersal area of Kenya. International Livestock Research Institute, Nairobi, Kenya.

Leakey, M. D. and R. L. Hay. 1979. Pliocene footprints in the Laetolil beds at Laetoli, northern Tanzania. Nature 278:317-323.

Little, P. D. 1985. Social differentiation and pastoralist sedentarization in northern Kenya. Africa 55:243-261.

Little, P. D. 1992. The elusive granary: Herder, farmer and state in northern Kenya. Cambridge University Press, Cambridge, UK.

Little, P. D., K. Smith, B. A. Cellarius, L. D. Coppock, and C. B. Barrett. 2001. Avoiding disaster: Diversification and risk management among East African herders. Development and Change 32:401-433.

Marshall, F. 1998. Early food production in Africa. The Review of Archaeology 19:47-58.

Marshall, F. 2000. The origins of domesticated animals in Eastern Africa. *In* K. C. McDonald and R. M. Blench, editors. The origins and development of African livestock: Archaeology, genetics, linguistics and ethnography.

Marshall, F. and E. Hildebrand. 2002. Cattle before crops: The beginnings of food production in Africa. Journal of World Prehistory 16:99-143.

McCabe, J. T. 2003. Sustainability and livelihood diversification among the Maasai of northern Tanzania. Human Organization 62:100-111.

McPeak, J. and P. D. Little. 2005. Cursed if you do, cursed if you don't: the contradictory processes of sedentarization in northern Kenya. Pages 87-104 *In* E. Fratkin and E. A. Roth, editors. As pastoralists settle. Kluwer Academic Publishers, Dordrecht, Netherlands.

Meinertzhagen, R. 1957. Kenya Diary 1902-1906. Oliver and Boyd, London, UK.

Mugambi, K. 2005. Tourism reclaims its place as the leading foreign exchange earner. Sunday Nation, Nairobi, January 16, 2005, p. 19.

Mwangi, A. and E. Warinda. 1999. Socio-economic dimensions of sustainable wildlife conservation in the Kitengela Dispersal Area. African Conservation Center, Nairobi, Kenya.

Nduma, I., P. Kristjanson, and J. McPeak. 2001. Diversity in income-generating activities for sedentarized pastoral women in northern Kenya. Human Organization 60:319-325.

Njoka, T. J. 1979. Ecological and socio-cultural trends of Kaputiei group ranches in Kenya. Ph.D. University of California, Berkeley, California, USA.

Nkedianye, D. 2003. Testing the attitudinal impact of a conservation tool outside a protected area: the case for the Kitengela Wildlife Conservation Lease Programme for Nairobi National Park. MSc. University of Nairobi, Nairobi, Kenya.

Norton-Griffiths, M. 1977. Aspects of climate of Kajiado District. FAO Project DP/KEN/71/ 526, Nairobi, Kenya.

Ottichilo, W. K., J. de Leeuw, A. K. Skidmore, H. H. T. Prins, and M. Y. Said. 2000. Population trends of large non-migratory wild herbivores and livestock in the Masai Mara ecosystem, Kenya, between 1977 and 1997. African Journal of Ecology 38:202-216.

Parker, I. 2003. The Machakos experience. Swara 26:48-50.

Pasha, I. K. O. 1986. Evolution of individuation of group ranches in Maasailand. Pages 303-317 *In* R. M. Hansen, B. M. Woie, and R. D. Child, editors. Range Development and Research in Kenya. Winrock International Institute for Agricultural Development, Morrilton, Arkansas.

Perkins, J. S. 1996. Botswana: fencing out the equity issue. Cattleposts and cattle ranching in the Kalahari Desert. Journal of Arid Environments 33:503-517.

Reid, R. S., P. K. Thornton, and R. L. Kruska. 2004. Loss and fragmentation of habitat for pastoral people and wildlife in East Africa: Concepts and issues. South African Journal of Grass and Forage Science 21:171-181.

Robertshaw, P. 1991. Early pastoralists of south western Kenya. British Institute of East Africa, Nairobi, Kenya.

Rutten, M. M. E. M. 1992. Selling wealth to buy poverty: The process of individualisation of land ownership among the Maasai pastoralists of Kajiado District, Kenya, 1890-1990. Breitenbach Publishers, Saarbrucken, Germany.

Senft, R. L., M. B. Coughenour, D. W. Bailey, R. W. Rittenhouse, O. E. Sala, and D. M. Swift. 1987. Large herbivore foraging and ecological hierarchies. BioScience 37:789-795.

Serneels, S. and E. F. Lambin. 2001. Impact of land-use changes on the wildebeest migration in the northern part of the Serengeti-Mara ecosystem. Journal of Biogeography 28:391-407.

Simon, N. 1962. Between the sunlight and the thunder. Collins, London, UK.

Smith, A. B. 1984. The origins of food production in North East Africa. Palaeoecology of Africa 16:317-324.

Smith, A. B. 1992. Pastoralism in Africa: Origins and development ecology. Hurst & Company, London; Ohio University Press, Athens; and Witwatersrand University Press, Johannesburg.

Spinage, C. A. 1992. The decline of the Kalahari wildebeest. Oryx 26:147-150.

Stauffer, D. 1985. Introduction to Percolation Theory. Taylor and Francis, London, Kenya.

Stewart, D. R. M. and D. R. P. Zaphiro. 1963. Biomass and density of wild herbivores in different East African habitats. Mammalia 27:483-496.

Sutton, J. E. G. 1993. Becoming Maasailand. Pages 38-60 *In* T. Spear and R. Waller, editors. Being Maasai. James Currey, London, UK.

Thomas, C. D., A. Cameron, R. E. Green, M. Bakkenes, L. J. Beaumont, Y. C. Collingham, B. F. N. Erasmus, M. F. d. Siqueira, A. Grainger, L. Hannah, L. Hughes, B. Huntley, A. S. v. Jaarsveld, G. F. Midgley, L. Miles, M. A. Ortega-Huerta, A. T. Peterson, O. L. Phillips, and S. E. Williams. 2003. Extinction risk from climate change. Nature 427:145-148.

Thornton, P. K., S. B. BurnSilver, R. B. Boone, and K. A. Galvin. 2006. Modelling the impacts of group ranch subdivision on agro-pastoral households in Kajiado, Kenya. Agricultural Systems 87:331-356.

Tomich, T. P., K. Chomitz, H. Francisco, A.-M. N. Izac, D. Murdiyarso, B. D. Ratner, D. E. Thomas, and M. van Noordwijk. 2004. Policy analysis and environmental problems at 3 different scales: asking the right questions. Agricultural Ecosystems and Environment 104:5-18.

Waller, R. 1988. Emutai: crisis and response in Maasailand 1883-1902. Pages 73-112 *In* D. H. Johnson and D. M. Anderson, editors. The ecology of survival: Case studies from northeast African history. Lester Crook Academic Publishing, London, UK and Westview Press, Boulder, Colorado, USA.

Whyte, I. J., and S. C. J. Joubert. 1988. Blue wildebeest population trends in the Kruger National Park and the effects of fencing. South African Journal of Wildlife Research 18:78-87.

Chapter 10

PROCESSES OF FRAGMENTATION IN THE AMBOSELI ECOSYSTEM, SOUTHERN KAJIADO DISTRICT, KENYA

Shauna B. BurnSilver[1,2], Jeffrey Worden[1,3], and Randall B. Boone[1]
[1]Natural Resource Ecology Laboratory, Colorado State University, Fort Collins, CO 80523, USA; [2]Department of Anthropology, Colorado State University, Fort Collins, CO 80523, USA; [3]International Livestock Research Institute, Nairobi, Kenya

1. INTRODUCTION

The Amboseli ecosystem is known worldwide as one of Kenya's "conservation jewels," and is recognized as a landscape where humans, livestock, and wildlife have co-existed for centuries. However, there is a long-term shift underway, pushed by a transition in human land-use from extensive pastoralism by Maasai to intensive pastoralism carried out within legally-prescribed private parcels of land. In the face of this transition, the region's wildlife populations and its system of seasonal livestock and wildlife movements appear increasingly fragile, and Maasai pastoralists themselves are facing significant challenges to their economic and cultural well-being.

Fragmentation of the resource base, from communally-managed rangelands down to subdivided individual parcels is an ongoing process with far-reaching implications. The antecedents of this process are alternately historical, demographic, political, and economic in origin, and have originated exogenously at national and international levels, and endogenously from within the dynamics of Maasai society itself. The effects of these changes on Maasai society and culture have been emphasized by a variety of researchers (White and Meadows 1981, Evangelou 1984, Campbell 1984,

K. A. Galvin et al. (eds.), Fragmentation in Semi-Arid and Arid Landscapes: Consequences for Human and Natural Systems, 225–253.

Grandin 1986, Grandin et al. 1989, Bekure et al. 1991, Galaty 1992, Kerven 1992, Rutten 1992, Campbell 1993, Zaal 1998, Desta and Coppock 2004). However, the linkages between economic and land tenure change and the ecological implications of declining scale of resource use have not been clearly addressed.

Ninety-two percent of the Amboseli ecosystem is categorized as arid and semi-arid (Ole Katampoi et al. 1990). Resource availability for pastoralists is highly variable across space and time. Herders traditionally offset temporal variability in resources by using mobility to access ecological heterogeneity in the form of discontinuous grazing resources varying in quality and quantity across space. Amboseli therefore conforms to a system where connectivity between resources is "movement mediated," as described by Hobbs et al. (Chapter 2). There are two patterns of fragmentation currently discernible on the Amboseli landscape (Figure 10-1). Subdivision occurs as communal rangelands are divided into private parcels. Alternately, sedentarization describes a process by which herders settle permanently in one location. These patterns are linked and are self-reinforcing, in that sedentarization occurs prior to or as a direct effect of subdivision, or, households settle permanently for reasons of economic opportunity or survival (Little et al. 2001a), without subdivision being a deciding factor. Both patterns, however, imply a decline in the mobility of households. Fragmentation of the pastoral resource base occurs as a result of both of these patterns, as key resource areas (e.g., prime settlement and grazing zones) are excised from use, and cultural and ecological systems are disrupted. The implications of this fragmentation are three-fold: a reduction in pastoral mobility (BurnSilver et al. 2003), an implied reduction in the spatial scale of interactions – down to scales which fail to include a full range of key resources or ecological heterogeneity (Hobbs et al., Chapter 2), and a general erosion of ecological heterogeneity itself. For wildlife, subdivision and sedentarization means extensive habitat modification and increasing conflict and competition with pastoralists (Campbell et al. 2003a, Worden et al. 2003, Worden in prep.). There is a gain in security of tenure and investment for households in subdivided areas, but there are costs in terms of lost flexibility and increased risk. The companion assumption made by policymakers is that declines in pastoral mobility will be offset by better returns from newly intensified livestock raising and greater access to productive inputs. Our discussion questions the appropriateness of this assumption in the case of the infra-structure-poor Amboseli system.

Our goal is to integrate the historical, political, and economic threads of the Amboseli story with the key ecological concepts of fragmentation and

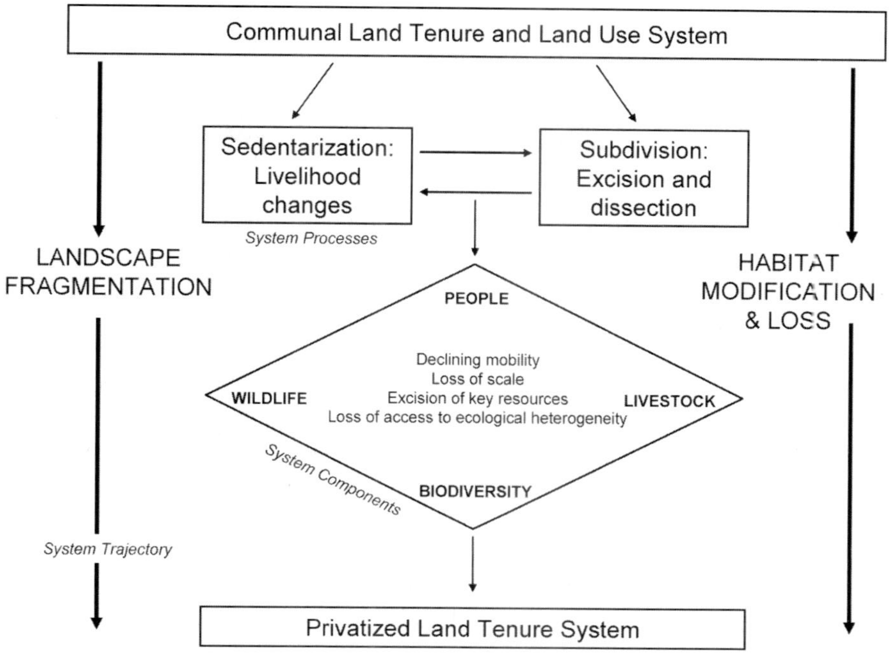

Figure 10-1. Conceptual diagram of fragmentation processes in Amboseli.

scale. Researchers have highlighted the critical role of mobility for wildlife and pastoral livestock in arid ecosystems (Ellis and Swift 1988, Swift et al. 1996, Niamir-Fuller and Turner 1999, Fratkin and Mearns 2003). But the Amboseli ecosystem provides an example of the cascading effects of losses in spatial scale of land use on human economies, wildlife populations, and rangeland ecology – effects that currently are being played out in other pastoral areas on a global scale (Blench 2001).

2. THE GREATER AMBOSELI ECOSYSTEM

2.1 Biophysical description

Kajiado District (21,852 km^2) represents the north-eastern corner of Greater Maasailand. The Greater Amboseli Ecosystem (approximately 8,500 km^2) is defined as a core area encompassing the Amboseli Basin and swamps along the northern foot of Kilimanjaro, but includes the dry season dispersal movements of herbivores between the swamps of Amboseli National Park and neighbouring rangelands (Western 1973). The system is dominated by two topographic gradients running north-south and east-west. Soils range from a complex of Luvisols and Cambisols in the north, to the recent volcanics of the Chyulu Hills in the east and the rocky basement derived Central hills, to the saline and sodic lacustrine plains of the Amboseli Basin and the well drained Pleistocene volcanics of the Kilimanjaro foothills (Touber 1983). Dominant vegetation communities are broad leaf, dry tropical forests and woodlands on the Kilimanjaro and Chyulu slopes; open grasslands and seasonally flooded plains, riverine forests, halophytic grass and scrubland in the Amboseli Basin; and scattered *Commiphora* and *Acacia* woodlands. Rainfall is both spatially and temporally heterogeneous. Annual rainfall ranges from 500-600 mm in the north to 250-300 mm in Amboseli National Park. Localized areas of higher rainfall occur along the northern slopes of Kilmanjaro and the Chyulu Hills (>800 mm and 500-600 mm respectively). While rain typically falls in two seasons (Nov-Jan and April-May) with two intervening dry periods, it is not unusual for either or both of these rainy seasons to fail in some areas, or altogether.

Surface water is scarce throughout the region with permanent water in the form of rivers and a line of swamps occurring only in the southeast. Seasonal rivers such as the Eselenkei and Olkejuado in the north provide important surface water during the rains. Dry season water sources are hand dug wells and boreholes.

The Greater Amboseli Ecosystem currently encompasses multiple land-use types and land tenure systems. This chapter focuses on the changing dynamics between land-use and tenure systems in six study areas (Osilalei, Eselenkei, Lenkisim, Emeshenani, North Imbirikani and South Imbirikani) on four Maasai Group Ranches (Osilalei, Eselenkei, Olgulului/Lolarashi and Imbirikani) (Figure 10-2). Socio-economic and ecological data were collected by BurnSilver and Worden during the period 1999-2002.

2.2 Maasailand

Kajiado and Narok Districts currently form the core territories of the Kenya Maasai. The Maasai are best known as transhumant pastoralists who moved historically between seasonally wet and dry season pastures, dependent for their livelihoods on animal herds that were a combination of cattle, sheep, and goats. Individual herds were privately owned, while land was held communally, and livestock movements were arranged by elders' consensus according to seasonal climatic conditions. Large, multi-generational households managed livestock herds, and layered relationships between households based on clan membership, blood ties, marriage and stock-friendships formed the basis of a pastoral society that was at once

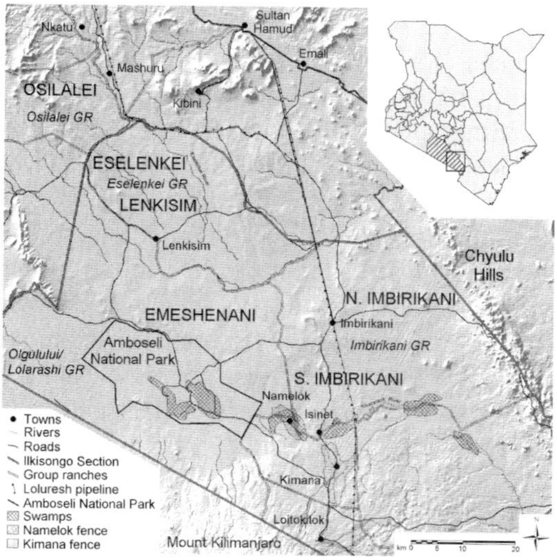

Figure 10-2. The Amboseli study area.

cohesive and flexible in the face of climatic uncertainty and risk (Bekure et al. 1991). A majority of Maasai nutritional energy came from milk (primarily cows' milk), but blood and meat from opportunistic slaughter also contributed significantly to household diets (Nestle 1985).

Six sub-tribe designations of Maasai, called sections or *oloshon*, are located within Kajiado District. The Greater Amboseli Ecosystem overlaps portions of three Maasai sections: Ilkisonko, which lies in the southeast corner of the district, Kaputei in the north, and Matapaato, which extends north and west of Ilkisonko (Figure 10-2). Maasai sections have exclusive claims to rangeland territories, follow their own cultural calendars, and have unique variations in dress and language (Spear and Waller 1993). However, economic and cultural interactions take place across sectional boundaries, and in times of severe drought access to grazing in other sections is negotiated on a reciprocal basis (Galaty 1993).

The preceding paragraphs briefly describe the traditional, cultural, and productive milieu for Maasai pastoralism in Kajiado District. However, change is widespread in this system and where extensive pastoralism was the norm, a gradient of land use now exists, ranging from transhumant pastoralism in some areas to sedentary agropastoralism in others. Alternatively setting the stage for emergence, or constraining new strategies from surfacing, is the constellation of physical infrastructure and ecological features characterizing areas of the landscape (Little et al. 2001a). Key productive features of the Amboseli region include the higher rainfall areas on the slopes of Kilimanjaro and Chyulu Hills (e.g., as reserve grazing "banks"), as well as the network of swamps and irrigation canals arranged east-west along the base of Kilimanjaro (Figure 10-2). The swamps are important for humans (for agricultural and domestic water use), livestock (as grazing reserves), and wildlife (for forage and water). The access of households to physical infrastructure (e.g., roads, boreholes and water pipelines) and services (e.g., schools, medical facilities, shops, and livestock markets) differs dramatically depending on settlement location (Figure 10-2). Core rangeland zones north of Amboseli National Park are significantly more isolated from services and infrastructure. In contrast, other areas lie at the crossroads of livelihood opportunities provided by the swamps, and the main north-south transportation routes and services, for example the area around Isinet town (Campbell 1999). However, access to other productive infrastructure (e.g., banks and veterinary services) is generally low across the Amboseli ecosystem (Rutten 1992, Zaal 1998, Boone et al., Chapter 14).

3. FRAGMENTATION PROCESSES

Historical precedents, formal policy alterations in land tenure laws, and land use change linked to a series of ultimate and proximate drivers have transformed the political, economic, demographic, and cultural conditions facing Maasai pastoralists. Ultimate drivers represent fundamental system characteristics, sometimes originating from outside the system, that set the stage for subsequent interactions between system components. Proximate drivers arise from the interactions between these fundamental features of the system and local conditions. The iterative effect in Amboseli has been a gradual fragmentation of communal rangelands, and this process has implications for the system's resident human, livestock, and wildlife populations.

3.1 Ultimate drivers

A variable and dry climate, resource heterogeneity, rising human populations, pastoral policy, and limited market access have functioned as ultimate drivers within the Amboseli ecosystem.

The climate in Amboseli is highly variable temporally and spatially, resulting in a resource base that is heterogeneous, or *patchy,* from the perspective of pastoralists and their animals. Maasai herders maximized flexibility and minimized risk through mobility. Pastoral households timed their seasonal migrations to take advantage of diverse vegetation communities and key resource zones (e.g., swamps, riparian areas), and actively managed other areas as grazing reserves (e.g., highlands).

However, human population in Kajiado climbed consistently throughout the 20th century as a function of both intrinsic growth within the Maasai population and immigration of agriculturalist non-Maasai (Ole Katampoi et al. 1990, Rutten 1992). Additionally, while human population increased through time, livestock (cattle) populations fluctuated dramatically in the short-term, but overall livestock numbers have remained consistent over the long-term (Bekure et al. 1991). This has translated into a steady decline in livestock available per capita. Most researchers agree (and Maasai themselves point out) that pastoralists have become poorer in recent decades (Rutten 1992, Desta and Coppock 2004, BurnSilver et al. 2005).

Historical land-use policies and priorities are also critical drivers to fragmentation in Maasailand. At the advent of British rule Maasai territory extended over 60,000 km^2, but by 1911 the Maasai had signed treaties agreeing to remain within the boundaries of a 38,000 km^2 southern reserve (Kerven 1992). Losses of Maasai territory in the Amboseli region continued from 1930 to 1960 as successive influxes of non-Maasai cultivators to the Kilimanjaro highlands and swamps excised valuable grazing areas and habitat from use by herders and wildlife. Wildlife conservation priorities also led to the initial designation of the 3,260 km^2 Amboseli National Reserve in 1947, and finally the gazetting of a smaller Amboseli National Park in 1974 (390 km^2) (Lindsay 1987), which nonetheless represented the permanent loss of access to key forage and water resources for local herders.

The economic policies of the British towards the Maasai in the early part of the century were equal parts "benign neglect" and "obstructionist" (Kerven 1992:40). However, policymakers in the 1950s began to emphasize the economic importance of pastoral areas to the Kenyan economy and charted the transformation of subsistence, milk-based pastoralism to a system of intensive beef production based on the assumptions of private property, enforced grazing controls, and intensified use of production inputs (Oxby 1982, Rutten 1992, Fratkin and Wu 1997).

Market access, or the *lack thereof*, has been a key system driver that defines production options available to pastoral households. Historically, livestock markets were the most vital to Maasai livelihoods, but access to agricultural markets and a growing demand by Maasai for consumer goods are increasingly important market features. Livestock marketing by Maasai was blocked during the colonial period to limit competition with white settlers (Kerven 1992), but was then later mandated by government intervention to decrease stocking rates (Rutten 1992). The Maasai have been criticized historically as market-averse (Herskovitz 1926, Lamprey 1983). Currently, selling of livestock is critically important for pastoralists, although it remains largely need-driven rather than timed to take advantage of price competitiveness (Evangelou 1984, Zaal 1999). Marketing of livestock in Kajiado is frequent, but largely on an *ad hoc* basis, limited by distance, lack of information, and unstable prices (Kerven 1992, Holtzman and Kulibaba 1995, Zaal 1998).

Marketing and consumption of agricultural products has also emerged as a critical livelihood strategy in the swamps and the Kilimanjaro highlands since the droughts of the 1960s-80s (Southgate and Hulme 1996, Campbell 1999). Large swamp areas have been converted to agriculture and significant conflicts are emerging over water management and reserve grazing areas for livestock and wildlife. Overall, the area of highland rainfed agriculture has increased 177% over the time period 1973-2000, while irrigation in the swamp areas increased by 45.2% (Campbell et al. 2003b). Market access is limited by poor transport infrastructure and high costs, and crop prices are highly variable (BurnSilver in prep.). So while demand for agricultural products is expanding, agricultural returns to producers are often highly unstable (Norton-Griffiths and Butt 2003).

The trend is therefore towards greater articulation between Maasailand and the Kenyan economy. There is greater availability of and demand for services (e.g., education), foodstuffs, and consumer goods in pastoral areas. Pastoral households thus have an increased need for cash to support broadening education and lifestyle goals, but instability in livestock and agricultural markets limits efforts to satisfy these needs.

3.2 Proximate drivers

Proximate system drivers arise from interactions between fundamental system characteristics and local conditions. Proximate drivers in Amboseli are recent drought history, land tenure change, changes in settlement patterns, pressure on pastoralists to diversify and intensify, and evolving livelihood expectations.

Droughts involving failure of either or both the long and short rains occur regularly in Amboseli. The severe droughts of 1977 and 1984 signalled a period of sedentarization for many households as they settled in swamp areas to pursue agriculture as a survival mechanism and as a "short-term" strategy to build up their herds, with the goal of eventually returning to the system as pastoralists (Southgate and Hulme 1996, Campbell et al. 2003b). However, many of these households have not re-transitioned into extensive pastoralism, and zones of sedentarized agropastoralism continued to grow.

Economic policies based on the premise that private lands would be managed more productively were the basis for formal changes in land tenure rules in Kajiado from the 1950s to the mid-1980s. Colonial administrators in the 1950s allowed county councils to grant *ad hoc* private title of communal trust lands to influential Maasai for individual ranches (Galaty 1992, Fratkin and Wu 1997). In the late 1960s, the newly independent Kenyan Government adjudicated over 38,000 ha of agricultural land on the Kilimanjaro slopes and individual ranches in Kaputei Maasai section (Rutten 1992). Then beginning in 1968 the government, with funding from the World Bank, pushed to adjudicate Maasailand into community-leasehold Group Ranches. There is currently a rich history of research illustrating that the intensive production goals of the group ranch concept were not attained. However, they did set the stage for a series of cascading changes to both formal land tenure rules and pastoral land use, the fragmentation repercussions of which continue to play out in the system.

One effect of the group ranch initiative was the installation of livestock infrastructure (e.g., stock dip tanks and water points), and other government-supported services (e.g., schools and health centers) in central locations. When group ranch boundaries disrupted traditional grazing arrangements in the Amboseli area, pastoral elders created a system of land use based on phased and enforced migrations between permanent (*emparnati*) and seasonal grazing (*enkaroni*) settlement areas (BurnSilver in prep., Worden in prep.). *Emparnati* settlement zones evolved adjacent to newly installed local infrastructure, services, and/or other key resources (e.g., roads and swamps) and these permanent settlement areas attracted additional services (e.g., local shops and grain mills). Thus infrastructure development and changes in settlement patterns were precursors to a process of land use change and additional sedentarization in these core areas.

Group ranches were an initial step in a formal effort to 'rationalize' pastoral production. For policymakers, they were an intermediate step towards privatization of the rangelands, but many pastoralists saw group ranches foremost as a means of protecting their lands from further encroachment. However, from the outset, internal dissatisfaction with group ranch management and external agitation for their final subdivision into private parcels was supported by emergence of national level policy in 1983 in favor of individualized land tenure (e.g., private property). Privatization was considered to be a precursor to economic development (Fratkin and Mearns 2003). By 1990, forty of the original fifty-two group ranches in Kajiado had subdivided (Kimani and Pickard 1998). Within the study area, only Osilalei is officially subdivided (since 1990). But, the membership of the three other ranches is currently debating how and when to subdivide their grazing lands. Division of these ranches into privately deeded parcels is now considered to be "inevitable" by many, although not necessarily desirable by all (Ntiati 2002).

At this stage, successive droughts, population pressure, and a mixture of economic opportunity and need have pushed and pulled Maasai households to diversify livelihoods and intensify livestock production strategies (Galaty and Johnson 1990, Little et al. 2001a, Desta and Coppock 2004). As well, life expectations are changing and the recognition that being a 'pure' pastoralist is becoming more, not less, difficult is expressed through an increasing emphasis on schooling Maasai children.

The preceding discussion underscores the linkages between ultimate and proximate drivers in the Amboseli system and their physical manifestation on the landscape in the form of increasing sedentarization and subdivision. Drought, or changing livelihoods may then lead to sedentarization whereby household grazing is circumscribed within a 'reachable' radius around settlements. Alternately, subdivision may be the starting point, as households become more sedentary, curtailing larger-scale seasonal mobility as they take possession of their parcels and use them individually. Regardless of the catalyst, and whether stemming from land tenure or land use change, the trend is increased exclusivity of use by pastoral households and ultimately a more fragmented landscape for livestock, wildlife, and pastoralists.

4. THE SHAPE OF LANDSCAPE FRAGMENTATION

The process of fragmentation in Amboseli linked to subdivision and sedentarization leaves a spatial signature on the landscape. The following section illustrates important patterns of fragmentation.

4.1 Patterns of subdivision

Figure 10-3 depicts conditions of gradual compression of pastoral and wildlife rangeland areas linked primarily to changes in policy and land tenure in the study area. Prior to the 1960s, extensive pastoral movements occurred seasonally within and across Maasai sectional boundaries – indicated by black boundary lines in Figure 10-3i. Subsequently, Maasai sections were adjudicated into group ranches and initial areas along the slopes of Mount Kilimanjaro were divided into private parcels (Figure 10-3ii). Key resource areas for irrigated agriculture (e.g., Kimana and Namelok Swamps) were then informally divided and made available to Olgulului/ Lolarashi and Imbirikani ranch members beginning in the late 1970s (dark grey shading in Figure 10-3iii). Additional dark shading in Figure 10-3iv shows, 1) the 1990 subdivision of Osilalei Group Ranch, whereupon group ranch members moved onto private parcels gradually over the following decade, 2) the subdivision and distribution of additional highland rainfed agricultural (e.g., an area called Emurutot) to Olgulului/Lolarashi members in 2002, and 3) the beginning of the subdivision process in Imbirikani group ranch with the official division of swamp areas into private parcels.

Figure 10-3. Subdivision patterns in Amboseli. Medium gray shading corresponds to the study area. Dark grey shading indicates subdivided rangelands and agricultural zones.

4.2 Patterns of sedentarization

While often identified only as a result of fragmentation, sedentarization is an underlying and parallel process of land-use change in the Greater Amboseli Ecosystem. Patterns and processes of sedentarization for four study areas in Amboseli – Emeshenani, Eselenkei, Lenkisim, and Osilalei - are characterized in Figure 10-4. Distinct spatial patterns of settlement emerge as a result of both subdivision and other factors linked to sedentarization. At the household level, production strategies become more individualized and the number of houses per settlement declines (Osilalei and northern Eselenkei; Figure 10-4i). However, where movement remains extensive, larger settlements reflect larger traditional social and labor-sharing units (e.g., Lenkisim, Emeshenani and southern Eselenkei). Another indicator of sedentarization is a change in construction materials from mud/dung to tin or grass roofs (Figure 10-4ii). Permanent structures exist in subdivided Osilalei, but also in as yet unsubdivided areas, linked to infrastructure availability and the rise of permanent settlements.

Moving up from the settlement to the landscape level, distinct configurations of settlements emerge. Traditional Maasai settlement patterns were based on sectional boundaries and loose associations of elders, with the

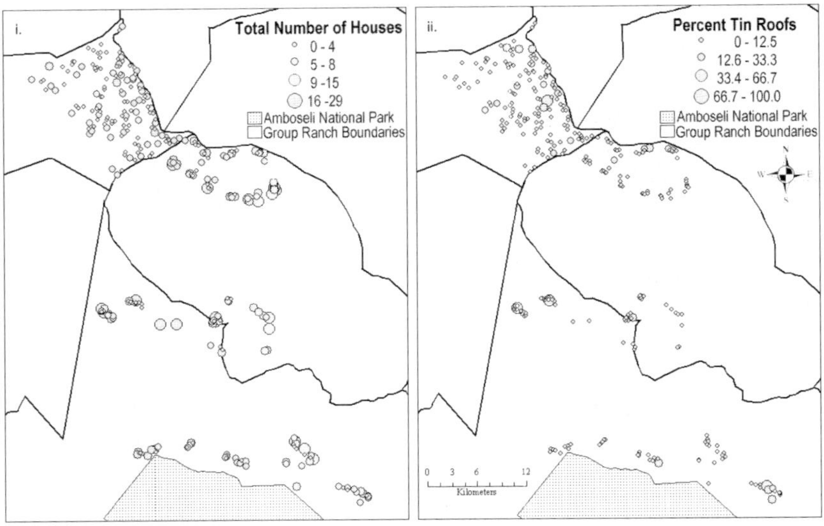

Figure 10-4. The impacts of subdivision and sedentarization on settlement patterns. i) Number of houses per settlement, ii) Percentage of houses in the settlement that are grass or tin roofed. Modified from Worden (in prep.).

ultimate decision-making power resting with the household head (Bekure et al. 1991, Mwangi 2003). However, currently groups of households, organized as neighborhoods of permanent settlement are clustered near infrastructure (e.g., water sources, schools, shops, health centers) reflecting a process of sedentarization not linked directly to subdivision. In contrast, subdivision in Osilalei has resulted in the even distribution of individual parcels across the landscape (Worden in prep.).

5. IMPLICATIONS FOR PASTORAL RESOURCE UTILIZATION

The patterns and trajectory of ongoing fragmentation in the Amboseli region have clear implications for the mobility of pastoralists and their ability to access ecological resources at the right time. Similarly, subdivision and sedentarization in tandem with other system drivers compresses the range of traditional economic choices households perceive as being practicable, and mandates new coping strategies. The following section highlights the responses of pastoral households in the face of fragmentation – specifically their observed mobility patterns, their access to resources in critical time periods, and their emerging economic choices.

5.1 Quantifying pastoral mobility

We hypothesized that fragmentation translates into a decrease in system connectivity, a decrease in scale of resource use, and a decline in access to ecological heterogeneity for pastoralists and their livestock. We looked at how mobile households are within the Amboseli system in subdivided and sedentary areas, compared to zones where household movements are still unrestricted. We also explored scenarios of forage resource availability for pastoral households pre- and post-subdivision for a 24-month period extending from 1999 (a good/average year) to 2000 (a significant drought year).

Results indicate that S. Imbirikani and Osilalei households are less mobile on average than households from other areas. Osilalei is a subdivided group ranch and S. Imbirikani is a center of agropastoral activities where households are increasingly sedentary. These households migrated seasonally less often from permanent to dry grazing settlements (e.g., to access 1 or more *enkaroni*), and they moved fewer times per year overall (Table 10-1). This was true both in an average year (1999) and a drought year (2000). Similarly, average daily distances traveled by herds in Osilalei, S. Imbirikani, and Eselenkei in the wet season were low; but dry season orbits rebounded in length everywhere but Osilalei, where households grazed inside their parcels. Thus, sedentary areas maintained some seasonal flexibility, but subdivided households did not.

However, looking at the differences in migration patterns across years, we find that even Osilalei and S. Imbirikani followed the general pattern of increasing mobility in response to drought pressure (Table 10-1). Sixty-four percent of households overall migrated in 1999, while fully 91.8% of households moved away from their permanent settlements in 2000. A majority of those not migrating were Osilalei and S. Imbirikani households, however,

Table 10-1. Indicators of pastoral mobility across the study areas in 1999 and 2000.

Study Areas	Mobile Households (%) *		Number of Moves (Mean no./yr) *		Daily Mean Distance Traveled (km)**	
	99	00	99	00	Wet	Dry
Osilalei	8.3	66.7	0.2	1.5	4.7	***
S. Imbirikani	44.0	56.0	1.0	1.4	5.9	10.1
Eselenkei	87.5	91.7	1.8	3.0	4.7	9.9
Lenkisim	62.5	100.0	1.3	2.9	8.4	10.5
Emeshenani	100.0	95.7	1.7	1.9	9.9	8.7
N. Imbirikani	87.5	100.0	3.3	5.1	10.4	10.7

*Calculated based on n=146 households. ** Calculated based on n=62 grazing orbits of 38 sub-sampled households. *** Households graze in private parcels.

it is clear that some sedentary households do maintain the ability to migrate when circumstances demand it. As well, of the 44 households that moved their herds out of their home group ranches in 2000, ten were from Osilalei. Off-ranch moves are long-distance, implying substantial labor investments and added disease risk (Bekure et al. 1991). In the case of the 2000 drought, most moves were to Imbirikani group ranch – the only area in the region with forage still available after two failed rainy seasons. Thus, the trend in 2000 was greater mobility in both subdivided and communal group ranches, but movement was *towards* unfragmented areas. This highlights the importance of maintaining these "intact" areas for people and livestock within the system. It also poses the question, would the same flexibility exist if the locus of the drought changed and forage was available only in subdivided areas?

5.2 Access to green forage: NDVI analyses

The four group ranches included in this study are diverse – they exhibit different levels of ecological productivity and contain a variety of vegetation types. Only Osilalei is currently subdivided, but we used Normalized Difference Vegetation Indices (NDVI) analyses to illustrate conceptually the potential effects of subdivision on forage access and ecological heterogeneity for herders across the four group ranches.

Trends in NDVI were averaged for 10-day periods throughout the mid-1990s (USGS 2002). We calculated average annual greenness profiles for 1 km^2 pixels on the ground and linked these forage profiles to the progression in land adjudication and subdivision in Ilkisonko Maasailand for the four group ranches, as pastoral access decreased from an entire section (Figure 10-5i) to within group ranch boundaries (Figure 10-5ii) and ultimately to subdivided individual parcels (Figure 10-5iii). Taking vertical slices through the profiles during any time of the year shows that the diversity of forage responses available to herders decreases dramatically as lands are subdivided. Figure 10-5iii reflects profiles for five adjacent 1 km^2 patches (1 km^2 = 100 ha or 247 acres), while column Figure 10-5iv reflects NDVI access for five non-contiguous 1 km^2 parcels, mimicking a situation in which individual herders cooperate in a post-subdivision environment to share parcels located in different areas of the group ranches. The resolution of the NDVI data precluded taking the analyses below 1 km^2 resolution, but note that the reported range of actual and potential sizes of subdivided parcels in the group ranches is significantly below the 247 acre figure used here (e.g., Osilalei = +/- 100 acres/member, Imbirikani = 64 acres, Olgulului/Lolarashi = 33 acres, Eselenkei = 148 acres) (Ntiati 2002). These analyses therefore *over* estimate the choices actually available to herders post-subdivision.

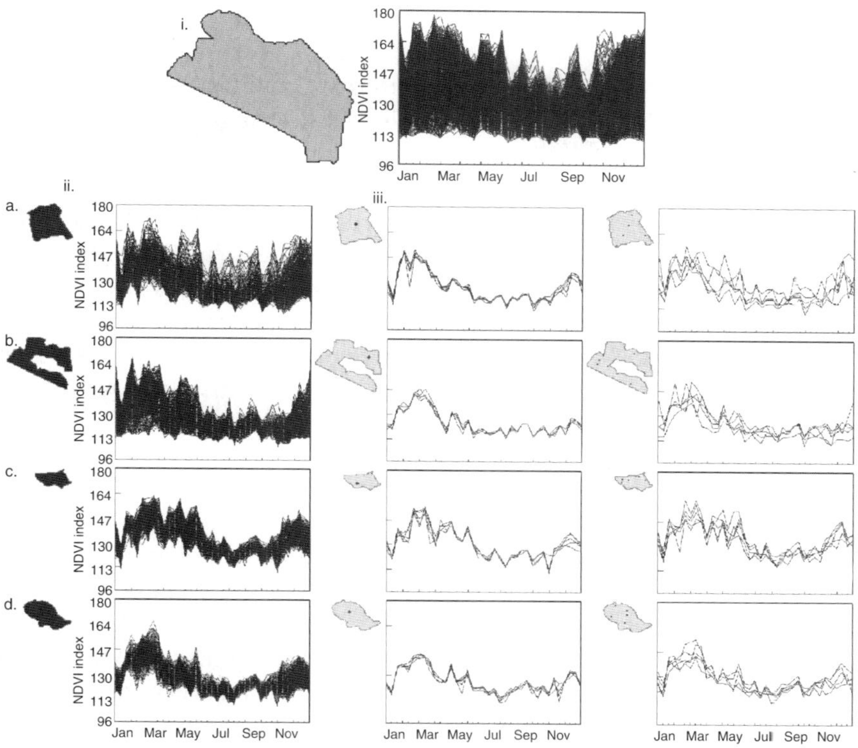

Figure 10-5. Access to green forage in four group ranches (a-Imbirkani, b-Olgulului/Lolarashi, c-Osilalei and d-Eselenkei) under different fragmentation scenarios; i) Ilkisonko Maasai section, ii) Group ranches, iii) five adjacent individual parcels, and iv) five non-contiguous shared parcels.

Confining herders to progressively smaller areas thus reduces access to ecological heterogeneity and limits pastoral flexibility (Boone et al. 2005). Larger areas contain patches with a variety of vegetation responses to variable rainfall levels that herders can access through movement, while smaller areas severely limit access to forage options through time. However, if households share parcels and rotate their herds between locations, the range of options widens once again, suggesting the potential importance of maintaining some degree of flexibility and collaboration following subdivision. We include this latter scenario in the discussion because although subdivision may be imminent in Eselenkei, Olgulului/Lolarashi and Imbirikani group ranches, how the physical division of the ranch will occur is not yet decided. Pastoralists are actively debating their options and the ramifications of subdividing the rangelands (BurnSilver et al. 2005).

5.3 Pastoral responses to fragmentation

As discussed previously, the expectations and abilities of pastoral households to depend solely on their livestock for subsistence is declining. Kajiado households therefore are pursuing a combination of both economic diversification and intensification strategies in response to systemic economic change and fragmentation Results are based on socio-economic surveys of 183 households conducted across the six study areas by BurnSilver from 1999-2001. Household wealth was identified using Grandin's (1988) wealth ranking technique yielding the following distribution: poor 34.7%, medium 37.3% and rich 28.3%. Mean tropical livestock units (for both cattle and smallstock) held by rich, medium and poor sampled households were 126.4, 47.0, and 23.5 TLUs, respectively.

5.3.1 Economic Diversification

Kajiado households are diversifying their productive strategies, and this is true to a level beyond that found in prior studies of Maasai and other pastoral groups (Bekure et al. 1991, Rutten 1992, Zaal 1998, McPeak and Little 2005). While a majority of households are still dependent on livestock as their base economic activity, the proportion of income generated by additional activities increasingly represents a substantial component of household income (Table 10-2).

Livestock revenue and in-kind consumption combined represent an average of 64% of household income for the study sample, but relative importance ranges from 45-84% depending on location. Proportion of income from livestock sales alone ranges from a high of 0.84 in isolated Emeshenani, to a low of 0.45 in centrally-located, agropastoral S. Imbirikani, where sold and consumed agriculture accounts for much of the cash and in-kind income generated by households. Business, salary/wages, wildlife-based income, and petty trade also account for a high proportion of household income in different locations.

We also compared mean annual gross income by study area for households. A majority of households combine livestock with multiple other livelihood streams, particularly in areas where fragmentation is pronounced, subdivided Osilalei and sedentary S. Imbirikani. Average gross income from cash and in-kind sources is greatest for the most diversified households (e.g., those combining livestock + agriculture + milk sales + off-farm activities), but even livestock combined with just off-farm activities (e.g., business, petty trade and wage/salary incomes), or households adding

Table 10-2. Mean proportion of gross income from activity types by study area.

Study Areas	Livestock Revenue*		Agricultural Revenue**		Off-land Revenue***		N
	%	No.	%	No.	%	No.	
Osilalei	.62	28	.18	22	.34	19	29
Eselenkei	.67	30	.05	8	.52	17	30
Lenkisim	.63	28	.23	4	.55	21	30
Emeshenani	.84	29	.23	11	.15	12	29
S. Imbirikani	.45	33	.44	32	.35	13	34
N. Imbirikani	.64	32	.20	13	.40	22	32
Mean Totals	**.64**	**180**	**.28**	**90**	**.40**	**104**	**184**

* Livestock revenue calculated based on combined income from sales of animals, hides and skins, milk and in-kind slaughter/consumption. **Agricultural income combines the value of agricultural products sold and consumed. *** Off-land revenue calculated based on income from salary/wages, petty trade and business activities.

agriculture to off-farm activities have gross incomes 33-50% greater than less diversified households. Household modeling results for Kajiado households under various subdivision scenarios also support these findings (Thornton et al. 2006). It might therefore be tempting to predict a gradual decline in the economic importance of livestock activities within pastoral households. However, herders themselves point out that livestock are the capital base for the diversification occurring (BurnSilver in prep.), and animals are the preferred investment and savings strategy even when these other activities are successful (McPeak and Little 2005).

5.3.2 Intensification of Livestock Production Strategies

Intensification of livestock production is also an emerging strategy in Amboseli as households struggle to increase outputs under conditions of declining mobility (e.g., sedentarization), or on private parcels of land with limited or no seasonal mobility (e.g., subdivision *and* sedentarization).

"Improving" livestock has been a goal of Maasai herders in the Amboseli region since the 1960s (White and Meadows 1981), but the pace and extent of the practice has intensified while the end points of the process remain undefined in the minds of producers and in terms of its ultimate effects on land use. Preferred cattle crosses are the local Zebu cattle with either the Sahiwal or Boran zebu sub-breeds, known respectively for higher milk and meat production (Scarpa et al. 2003). Maasai are extremely aware of the high-risk trade-offs implied in the transition to raising hybrid animals in an arid environment. However, the benefits associated with the 45-65% higher prices these animals bring in the marketplace (Bekure et al. 1991, Rutten

1992, KARI 2001) seem to outweigh the costs associated with their greater forage and water requirements, increased veterinary costs, inability to walk long distances, and drought intolerance (Rutten 1992, BurnSilver et al. 2005).

Only 38% of households (n=70) had no improved animals in their herds in 2000-2001. Rutten (1992) identified a link between low forage productivity and lack of improved animals in dry area herds, and this pattern still holds true, as 79%, 40% and 50% of households in drier Emeshenani, Lenkisim and S. Imbirikani areas were maintaining their cattle herds as pure Zebu. Alternately, N. Imbirikani, Osilalei and Eselenkei show the lowest maintenance rate of zebu animals in herds by households (21%, 24%, and 16.7%, respectively). When the mean percentage of improvement for each cattle age and sex category is weighted by the number of improved animals per herd, results show some degree of improvement for between 50-75% of bulls and 50-58% of calves in Osilalei, Eselenkei, Lenkisim, and N. Imbirikani (BurnSilver in prep.). These areas are either wetter or closer to the Kaputei Maasai section where there has been a long history of crossbreeding (Rutten 1992). The process begins by crossbreeding improved bulls with local cows, so that calves, heifers, and immature steers will show initially higher levels of improvement than mature cows and mature steers. However, even heifers and mature cows in these wetter areas are showing substantial indications of breed improvement, and the process has begun in drier zones, although it is not yet advanced (BurnSilver in prep.).

In a series of 17 focus groups organized across the six study sites (BurnSilver et al. 2005), participants voiced that the perceived inevitability of subdivision was contributing to their decision to crossbreed their animals *now*, given that they would be required to curtail their mobility under privatization. However, respondents expressed concern that even though these larger animals are more valuable in the marketplace, would do 'better' walking less, and are therefore better suited perhaps to smaller, private parcels, drought was not going to 'go away' with subdivision. This cross-breeding effort was therefore a very risky undertaking for herders who would have few compensation mechanisms in place other than mobility for supplementing the hearty appetites of their large crossbred cattle in future droughts.

6. ECOLOGICAL RESPONSES TO FRAGMENTATION

As with pastoral responses to fragmentation, ecological responses result from the interaction of ultimate and proximate drivers through time and across space. In this section we consider the ecological responses to fragmentation in the Greater Amboseli Ecosystem at three levels. First, we address the initial separation of the ecosystem into domestic and wild components, and ultimately protected and non-protected areas, with reference to the broad scale "segregation effects" of conservation policy (Western and Gichohi 1993). Secondly, within this broader context, pastoral rangelands are also experiencing a spatial and temporal fragmentation of ecosystem processes within landscapes. For example, the fragmentation of pastoral areas through subdivision, sedentarization, and the reduced scale of pastoral land-use has resulted in a spatial separation of ecosystem processes and the removal of livestock grazing and settlement creation from certain areas of the landscape. These changes in pastoral land-use interact to create a landscape which is *structurally* and *functionally* fragmented through habitat modification and loss, a gradual reduction in spatial extent, and the separation of fundamental ecosystem processes. Finally, we will consider the implications of these interacting processes for wildlife at the ecosystem level – a critical concern in the heretofore integrated human-livestock-wildlife Amboseli system.

6.1 Wildlife responses

The segregation of wildlife and pastoralists in Amboseli began with the creation of the Amboseli National Reserve and culminated with the demarcation of the Amboseli National Park in 1974. While pastoralists and wildlife had both depended on diverse resources throughout the ecosystem, these new boundaries began a process of separation that continues today. For example, elephants (*Loxodanta Africana*) have long been a key component of the Amboseli ecosystem, but this large-scale fragmentation of the system into two opposing landscapes, protected versus human-dominated, and the subsequent collapse of seasonal elephant and livestock movements, created the potential for a polarization of influences and the disintegration of the shifting landscape mosaic created historically by the dynamic interaction of pastoralists and elephants (Western and Maitumo 2004).

At the time of demarcation in 1974, the Amboseli elephant population was estimated at a minimum of 584 animals. This population continued to decline throughout the 1970s due to poaching and drought to 480 animals in 1978. With protection the elephant population increased rapidly over the next 20 years to 1087 in 1999 (Moss 2001). Rapid intrinsic rates of growth combined with population compression due to an intensification of human activities outside of protected areas resulted in extremely high elephant densities inside the park (Western 1989, Worden et al. 2003). While the porous boundaries of the park meant that these animals could migrate, the disruption of two important historical processes: traditional elephant dispersal patterns (Western and Lindsay 1984) and the removal of Maasai and their livestock from within the park, led to significant vegetation change on both sides of the protected area boundary. Field observations and personal accounts from local Maasai suggest that the *Commiphora* and *Acacia* woodlands surrounding the park increased as a result of livestock grazing in the absence of elephants (Worden, pers. obs., Western and Maitumo 2004, Western, pers. comm.). In contrast, the woodlands within Amboseli National Park, and particularly *Acacia xanthoploea*, collapsed. While it is still not clear what caused the initial decline in *Acacia xanthophloea* woodlands within the protected area, it is now apparent that elephants alone are preventing their regeneration (Western and Maitumo 2004).

The concentration of elephants inside, and humans and their livestock outside, the protected area has implications for other wildlife as well as vegetation. The collapse of the woodlands has resulted in a loss of biodiversity in Amboseli National Park (Western 1989, Worden et al. 2003). An aerial count of wildlife and livestock in and around the Amboseli, Namelok and Kimana Swamps suggested clear differences in species composition and abundance between the protected and unprotected areas (Worden et al. 2003). The protected areas contained far greater numbers of elephant, wildebeest (*Connochaetes taurinus*), and zebra (*Equus burchelli*) as well as locally important populations of eland (*Tragelaphus oryx*), oryx (*Oryx gazella callotis*), and warthog (*Phacochoerus aethiopicus*). Hippopotamus (*Hippopotamus amphibius*) and buffalo (*Cyncerus caffer*) were only observed in the protected areas. In contrast, Coke's Hartebeest (*Alcelaphus buselaphuscokei*) and Gerenuk (*Litocranius walleri*) were only found in the unprotected areas.

A survey of large herbivores conducted simultaneously by the Kenya Wildlife Service recorded 1090 elephants for the core of the ecosystem, as well as evidence supporting the assertion that elephant and buffalo favor protected areas while species such as eland and giraffe are gradually being pushed to the woodlands outside (Omondi et al. 2002). The impacts of woodland decline documented within Amboseli National Park have also led to significant shifts in primate distributions (Struhsaker 1976), and while yet to be documented, the declines almost certainly have far reaching implications for understudied taxonomic groups such as ants, butterflies, and birds (cf. Cumming et al. 1997).

The three western swamps at the base of Kilimanjaro highlight the effects of segregation compounded by land-use intensification (e.g., elephant compression and agricultural conversion) for Amboseli National Park and its immediate surroundings. But this is only part of the story of fragmentation in the Amboseli Ecosystem; the more subtle effects of changes in settlement distribution and livestock grazing arising from subdivision and sedentarization processes (Figures 10-4 and 10-5), and changes in movement patterns for pastoral rangelands, can also have significant ecological consequences. In an effort to document the impacts of these processes on the distributions of livestock and wildlife, we conducted a dry season aerial count along a fragmentation gradient from Emeshanani (least fragmented) to subdivided Osilalei (most fragmented). Osilalei has by far the highest density of domestic biomass (5,130 kg/km^2), followed by Eselenkei (2,592 kg/km^2) then Emeshenani (841 kg/km^2). While these differences may partly result from the more mobile nature of pastoral herds in Eselenkei and Emeshanani, they are also indicative of the intensification of pastoral production associated with fragmentation.

Fragmentation through changes in settlement and grazing patterns has important impacts on vegetation and wildlife at local (Western and Dunne 1979, Muchiru 1992, Muchiru et al. in press, Worden in prep.) and landscape levels (Western and Maitumo 2004, Worden in prep.). Vegetation impacts include local decreases in woody vegetation cover (Western and Dunne 1979, Muchiru 1992, Worden in prep.), changes in woody composition through selective utilization (Jensen 1983, Worden in prep.), a pronounced decrease in herbaceous biomass adjacent to sedentary settlements (Muchiru 1992, Worden in prep.), and enhanced grass and forb biomass on abandoned settlements and in calf grazing reserves (known as *olopololi*).

The implications of *de facto* rangeland fragmentation for meso-scale wildlife distributions are clearest in a comparison of the adjacent Eselenkei and Osilalei areas. Generally, Osilalei supports far less wildlife biomass (116 kg/km^2 vs. 339 kg/km^2). In particular, while Thomson's gazelle and ostrich appear to be doing well in this subdivided area, we observed no wildebeest or impala; and zebra, Grant's gazelle, and giraffe densities were respectively 43%, 51%, and 27% lower than those found in Eselenkei.

Similar trends emerge if we consider the Greater Amboseli Ecosystem as a whole. Long-term aerial counts (1977-2001) suggest that impala have declined significantly while eland, kongoni and giraffe suggest possible, though not significant, population declines (Western and Nightingale 2003). In contrast to our dry season count results for Osilalei and Eselenkei, at the larger-scale Thomson's gazelle are possibly declining, while zebra have increased significantly. Hansen et al. (2005) present evidence for increasing zebra populations at the district level as well, but suggest that giraffe are declining. This implies that the lack of giraffe in Amboseli National Park during our high resolution count in 2002 (Worden et al. 2003) may be part of a larger decline.

These results indicate that broad scale fragmentation of the Amboseli Ecosystem into domestic and wild (non-protected and protected) and subsequent changes in patterns of settlement and grazing associated with subdivision and sedentarization on the Amboseli rangelands, has variably impacted wildlife and vegetation at multiple scales. The somewhat surprising increase in zebra throughout the ecosystem reiterates that responses to fragmentation are species dependent and that while many species may decline in the face of widespread subdivision and sedentarization, some species may initially benefit from intensification (e.g., from leaky water pipes and irrigated agriculture). However, evidence from ecosystems that are farther along the fragmentation and intensification trajectory (e.g., Reid et al., Chapter 9) suggest that these peripheral benefits may be short-lived and that widespread declines in wildlife are the end result.

7. DISCUSSION AND CONCLUSIONS

Ultimate and proximate drivers are interacting to facilitate the fragmentation of the Greater Amboseli Ecosystem at multiple scales. Fragmentation, through the linked processes of subdivision and sedentarization, is evident in the gradual individualization and dissection of the rangelands, changes in the size and distribution of settlements, as well as landscape scale changes in wildlife populations. As a result, pastoralists, their livestock, and wildlife are faced with increasing constraints on flexibility and movement. The disruption of socio-cultural and land-use systems, as well as the segregation and fragmentation of underlying ecological patterns and processes have important implications for this mixed human-livestock-wildlife system.

While seasonal movement is still the norm in unfragmented core areas, the linked processes of sedentarization and subdivision in the Greater Amboseli Ecosystem have resulted in a decline in pastoral mobility for some households. Yet, results from the 2,000 drought indicate that some sedentary and subdivided households still maintain the ability to move when it is critical. Subdivision and sedentarization therefore do not necessarily restrict *all* movement, but rather may imply more extreme movement in the face of necessity. However, indications are that the trajectory of the system in these areas is towards less mobility and less flexibility overall. If and how the core rangelands of the Amboseli system will be subdivided is also being debated as this chapter is being written. Subdivision and sedentarization mean that a system historically based on extensive and multi-staged movements by pastoralists and predicated on punctuated and short-term use of successive areas on the landscape, is now trending overall towards intensive grazing of private or localized areas over extended periods of time. The result of subdivision and sedentarization is the imposition of gradual constraints on traditional flexibility to respond to ecological heterogeneity.

The institutionalization and spread of fragmentation with increasing subdivision threatens to erode traditional coping strategies and increase pastoral vulnerability. We find that households are engaged in a series of activities in an effort to respond to ongoing uncertainty. The case could be made that livelihood diversification and intensification are occurring as responses to a variety of drivers, fragmentation being only one of these system constraints; however, our results indicate that households use these strategies to both alleviate risks associated with changing economic realities and the looming uncertainty represented by fragmentation and declining mobility. While livestock are clearly still the economic base for most pastoral households, diversification is widespread, both as a contribution to alleviating basic needs and as a means of investment in addition to animals. Wealth level as a determinant of diversification pathways is not addressed in these analyses, however, how wealthy vs. poor pastoralists diversify remains a critical question (Little et al. 2001b). Livelihood strategies are also linked with an intensification of livestock production through an investment in 'improved' animal breeds. The realization that future resource constraints through subdivision are 'inevitable' has led increasing numbers of Maasai to embrace the uncertain trade-offs associated with crossbreeding: greater potential outputs from livestock, but higher risk to improved livestock associated with drought. Mobility has been arguably *the* critical pastoral response to system variability in the past (Little et al. 2001b), however it remains to be seen how new economic and productive strategies buffer pastoral households from the loss of mobility in the future.

Fragmentation of the Amboseli rangelands has had important ecological implications for vegetation patterns as well as the distribution and abundance of wildlife at local and landscape scales. The broad scale separation of the ecosystem into its domestic and wild components, and the institutionalization of this separation through conservation policy, has led to widespread segregation effects (Western and Gichohi 1993). Fragmentation at this ecosystem scale has resulted in a polarization of vegetation communities with woodlands declining inside the protected areas and increasing in the pastoral rangelands (Western 1989, Western 2002). A similar pattern emerges at a landscape level where sedentarization and subdivision create patches of intense utilization around settlements and also create areas of less intense utilization, therefore increasing woody vegetation, where settlement construction is limited (Worden in prep.). This disruption of dynamic settlement regimes and the historical interaction of pastoralists and elephants (Western and Maitumo 2004) have resulted in a general simplification and polarization of habitats (Western and Nightingale 2003, Worden in prep.).

Wildlife populations show a similar multi-scale response to fragmentation. A consideration of wildlife distributions along a gradient of land use in the Amboseli, Namelok, and Kimana Swamps suggested that the interaction of segregation effects and land-use intensification (e.g., agriculture and sedentarization) results in clear differences in species composition and abundance inside and outside protected areas. While some species do disappear as elephants eliminate woodlands inside Amboseli National Park, protected areas also provide refuges for local populations of large herbivores such as buffalo, eland, elephant, and hippopotamus. In the pastoral rangelands, we presented evidence that subdivision results in a reduction of herbivore density and richness. We suggest that the interaction of these local and broad scale processes of fragmentation may also be an important factor in the declining population trends for a number of species at the ecosystem level.

While the political-economic and cultural environment in Maasailand is changing rapidly, the ecological exigencies of the system – those mandating flexibility - remain unchanged. Access to compensating infrastructure and productive inputs in Amboseli remains very low (see Boone et al., Chapter 14) and pastoral households are caught between the proverbial 'rock and a hard place'. Over previous decades, many researchers and policymakers have debated the future viability of mobile pastoralism in general, and of Maasai pastoralism in particular (Evangelou 1984, UNDP 2003); but, there are indications that pastoralists are now utilizing a range of both traditional and emergent strategies to adapt to fragmentation and change. In order to maintain mobility as an option in the face of shrinking household size and increased emphasis on schooling, flexible labor-sharing arrangements between households, the use of stop-gap parental labor, and hiring of herders are emerging household coping mechanisms (Sikana and Kerven 1991). There is limited scope to save or 'bank' grass on individual parcels, however, parcel exchanges, pasture rental, and grazing networks or associations are emergent strategies in subdivided areas (Rutten 1992, Mwangi 2003, BurnSilver in prep.). These grazing arrangements take place on the basis of traditional kinship, stock associations or new norms of economic exchange. Thus, there is potential to reintegrate flexibility and access to both scale and heterogeneity back into an ostensibly fragmented pastoral grazing system (BurnSilver and Mwangi 2006). There are positive implications for wildlife here as well, as these options may facilitate maintaining rangelands in an open state, without fencing. These strategies are not addressed in-depth here, but as emergent trends they speak to the ability of pastoralism – in Maasailand and elsewhere – to remain a relevant and productive land-use strategy in the face of ongoing fragmentation and system change.

ACKNOWLEDGEMENTS

We would like to thank the Maasai of Southern Kajiado who gave us endless time and related their stories during this crucial time of transition. Our field research assistants were also critical in this work. Our sincere thanks to: Richard Solonga Supeet and Justus Lekimankusi Supeet, Jacob Mayiani Loorimirim, and Sauna Joseph Lemiruni. The work of Randall Boone was instrumental in the NDVI analyses. BurnSilver's fieldwork was supported by the Global Livestock CRSP (Collaborative Research Support Program of the office of Agriculture and Food Security Global Bureau USAID under grant no. PCE-G-98-0036-000), a supplementary research grant from ILRI, and the ILRI Ereto O Ereto project supported by the Belgian Ministry of Foreign Affairs, Foreign Trade and International Co-operation under their program for Belgian Support of International Agricultural Research for Development. Worden's fieldwork was supported by the Global Livestock CRSP (Collaborative Research Support Program of the office of Agriculture and Food Security Global Bureau USAID under grant no. PCE-G-98-0036-000), as well as a Dissertation Enhancement Award from the National Science Foundation (grant no. 0096706). Additional support for writing and analysis was provided by ILRI through the Ereto O Ereto project, and the National Science Foundation through the SCALE project of the Biocomplexity Program (grant no. 0119618).

REFERENCES

Bekure, S., P. N. de Leeuw, B. E. Grandin, and P. J. H. Neate, editors. 1991. Maasai herding: An analysis of the livestock production system of Maasai pastoralists in Eastern Kajiado District, Kenya. ILCA Systems Study 4. ILCA (International Livestock Centre for Africa), Addis Ababa, Ethiopia. 172.

Blench, R. 2001. You can't go home again: Pastoralism in the new millennium. London, Overseas Development Institute: 103.

Boone, R. B., S. B. BurnSilver, P. K. Thornton, J. S. Worden, and K. A. Galvin. 2005. Quantifying declines in livestock due to land subdivision. Rangeland Ecology & Management 58:523-532.

BurnSilver, S. B. In prep. Economic strategies of diversification and intensification among Maasai pastoralists: Changes in landscape use and movement patterns, Kajiado District, Kenya. Graduate Degree Program in Ecology. Fort Collins, Colorado State University.

BurnSilver, S., R. Boone, and K. A. Galvin. 2003. Linking pastoralists to a heterogeneous landscape: The case of four Maasai group ranches in Kajiado District, Kenya. Pages 119-134 *In* J. Fox, R. Rindfuss, S. Walsh and V. Mishra, editors. People and the environment: Approaches for linking household and community surveys to remote sensing and GIS. Kluwer Academic Publishers, New York.

BurnSilver, S. B., L. Onetu, and R. Ole Supeet. 2005. Amboseli Group Ranches Focus Group Proceedings. ILRI (International Livestock Research Institute), Nairobi.

BurnSilver, S. B. and E. Mwangi. 2006. Beyond group ranch subdivision: Collective action for livestock mobility, ecological viability and livelihoods. Paper presented at: Pastoralism and Poverty Reduction in East Africa: A Policy Research Conference. International Livestock Research Institute. June 27-28. Safari Park Hotel, Nairobi, Kenya.

Campbell, D. 1984. Response to drought among farmers and herders in southern Kajiado District, Kenya. Human Ecology 12:35-64.

Campbell, D. J. 1993. Land as ours, land as mine: economic, political and ecological marginalization in Kajiado District. Pages 258-272 *In* T. Spear and R. Waller, editors. Being Maasai. Ohio University Press, Athens.

Campbell, D. J. 1999. Response to drought among farmers and herders in southern Kajiado District, Kenya: A comparison of 1972-1976. Human Ecology 27:377-416.

Campbell, D. J., H. Gichohi, R. S. Reid, A. Mwangi, L. Chege, and T. Sawin. 2003a. Interactions between people and wildlife in SE Kajiado District, Kenya. Nairobi, ILRI.

Campbell, D. J., D. P. Lusch, T. A. Smucker, and E. E. Wangui. 2003b. Root causes of land use change in the Loitokitok Area, Kajiado District, Kenya. Nairobi, LUCID (Land Use Change Impacts and Dynamics Project): International Livestock Research Institute, Nairobi.

Cumming, D. H. M., M. B. Fenton, I. L. Rautenbach, R. D. Taylor, G. S. Cumming, M. S. Cumming, J. M. Dunlop, A. G. Ford, and M. D. Hovorka. 1997. Elephants, woodlands, and biodiversity in sourthern Africa. South African Journal of Science 93:231-236.

Desta, S. and D. L. Coppock. 2004. Pastoralism under pressure: tracking system change in southern Ethiopia. Human Ecology 32:465-486.

Ellis, J. and D. Swift. 1988. Stability of African pastoral ecosystems: alternate paradigms and implications for development. Journal of Range Management 41:450-459.

Evangelou, P. 1984. Livestock development in Kenya's Maasailand: Pastoralists' transition to a market economy. Westview Press, Boulder.

Fratkin, E. and T. S. Wu. 1997. Maasai and Barabaig herders struggle for land rights in Kenya and Tanzania. Cultural Survival Quarterly Fall:55-61.

Fratkin, E. and R. Mearns. 2003. Sustainability and pastoral livelihoods: lessons from East African Maasai and Mongolia. Human Organization 62:112-122.

Galaty, J. G. 1992. The land is yours: social and economic factors in the privatization, sub-division and sale of Maasai ranches. Nomadic Peoples 30:26-40.

Galaty, J. G. 1993. Maasai expansion and the new East African pastoralism. Pages 61-86 *In* T. Spear and R. Waller, editors. Being Maasai. James Currey LTD, Athens.

Galaty, J. G. and D. L. Johnson. 1990. Pastoral systems in global perspective. Pages 1-31 *In* J. G. Galaty and D. L. Johnson, editors. The world of pastoralism: Herding systems in comparative perspective. Guilford Press, New York.

Grandin, B. E. 1986. Land Tenure, Subdivision, and Residential Change on a Maasai Group Ranch. Development Anthropology Network 4(2):9-13.

Grandin, B. E. 1988. Wealth ranking in smallholder communities: A field manual. Intermediate Technology Development Group, Nairobi: 47 pp.

Grandin, B. E., P. N. de Leeuw, and P. Lembuya. 1989. Drought, resource distribution, and mobility in two Maasai group ranches, southeastern Kajiado District. Pages 245-263 *In* T. E. Downing, K. W. Gitu and C. M. Kamau, editors. Coping with drought in Kenya: national and local strategies. Lenne Rienner Publishers, Boulder.

Hansen, A. J., M. Said, J. O. Ogutu, R. S. Reid, R. Dufries, J. Dempewolfe, J. Robinson-Cox, S. A. R. Mduma, and G. Hopcraft. 2005. Land-use change around protected areas: effects on large mammal populations in Maasailand of East Africa. ILRI. Unpublished report.

Herskovits, M. J. 1926. The cattle complex in East Africa. American Anthropologist 28: 230-72, 361-88, 494-528, 633-64.

Holtzman, J. S. and N. P. Kulibaba. 1995. Livestock marketing in pastoral Africa: Policies to increase competitiveness, efficiency and flexibility. Pages 79-95 *In* I. Scoones, editor. Living With uncertainty. Intermediate Technology Publications, Ltd, London.

Jensen, C. L. 1983. The effect of settlement on shrub vegetation by the Amboseli Maasai. Unpublished report. African Conservation Centre. 1-34. pp.

KARI (Kenya Agricultural Research Institute). 2001. Socio-economics of the current livestock production, utilization and marketing in arid and semi-arid lands. Kajiado District Stakeholder's Workshop, Kiboko, Kenya. 67 pp.

Kerven, C. 1992. Customary commerce: A historical reassessment of pastoral livestock marketing in Africa. ODI Agricultural Occassional Paper 15. London, Overseas Development Institute: 119.

Kimani, K. and J. Pickard. 1998. Recent trends and implications of group ranch subdivision and fragmentation in Kajiado District, Kenya. The Geographical Journal 164:202-213.

Lamprey, H. F. 1983. Pastoralism yesterday and today: the over-grazing problem. Pages 643-666 *In* D. W. Goodall, editor. Ecosystems of the world - Tropical savannas. Elsevier Scientific, Amsterdam.

Lindsay, K. 1987. Integrating parks and pastoralists: some lessons from Amboseli. Pages 149-168 *In* D. M. Anderson and R. Grove, editors. Conservation in Africa: people, policies and practice. Cambridge University Press, New York.

Little, P., K. Smith, B. A. Cellarius, D. L. Coppock, and C. B. Barrett. 2001a. Avoiding disaster: diversification and risk management among East African Herders. Development and Change 32:387-419.

Little, P. D., H. Mahmoud, and D. L. Coppock. 2001b. When deserts flood: risk management and climatic processes among East African pastoralists. Climate Research 19: 149-159.

McPeak, J. and P. D. Little. 2005. Cursed if you do, cursed if you don't: the contradictory processes of pastoral sedentarization in northern Kenya. Pages 87-105 *In* E. A. Roth and E. Fratkin, editors. As pastoralists settle: Social, health, and economic consequences of pastoral sedentarization in Marsabit District, Kenya. Kluwer Academic Publishers, New York.

Moss, C. J. 2001. The demography of an African elephant (Loxodonta africana) population in Amboseli, Kenya. Journal of Zoology 255:145-156.

Muchiru, A. N. 1992. The ecological impact of abandoned Maasai settlements on savanna vegetation and its herbivores in the Amboseli Ecosystem. Biology of Conservation. Nairobi, University of Nairobi: 129.

Muchiru, A. N., D. Western, and R. S. Reid. In press. The impact of abandoned pastoral settlements on plant and nutrient succession in the African savannas. Journal of Arid Environments.

Mwangi, E. N. 2003. Institutional change and politics: The transformation of property rights in Kenya's Maasailand. Ph.D., Indiana University.

Nestle, P. S. 1985. Nutrition of Maasai women and children in relation to subsistence food production. Nutrition. London, Queen Elizabeth College, University of London: 261.

Niamir-Fuller, M. and M. D. Turner. 1999. A review of recent literature on pastoralism and transhumance in Africa. Pages 18-46 *In* M. Naimir-Fuller, editor. Managing mobility in African rangelands: The legitimization of transhumance. IT Publications, London.

Norton-Griffiths, M. and B. Butt. 2003. The economics of land use change in Loitokitok Division of Kajiado District, Kenya. Land Use Change Impacts and Dynamics Project: LUCID Working Paper Series, 34. International Livestock Research Institute, Nairobi.

Ntiati, P. 2002. Group ranches subdivision study in Loitokitok Division, Kajiado District, Kenya. Nairobi, LUCID (Land Use Change Impacts and Dynamics Project): International Livestock Research Institute, Nairobi. 26.

Ole Katampoi, K. O., G. O. Genga, M. Mwangi, J. Kipkan, J. O. Seitah, M. K. van Klinken, and M. S. Mwangi. 1990. Kajiado District Atlas. Nairobi, ASAL Programme Kajiado: 120.

Omondi, P., P. Muruthi, R. Mayienda, and E. Bitok. 2002. Total aerial count of elephants in Amboseli-Longido Ecosystem, Kenya Wildlife Service: 42.

Oxby, C. 1982. Group ranches in Africa. London, Overseas Development Institute. 13 pp.

Rutten, M. M. 1992. Selling wealth to buy poverty. Verlag breitenbach Publishers, Saarbrucken.

Scarpa, R., P. Kristjanson, E. Ruto, M. Radeny, A. Drucker, and J. E. O. Rege. 2003. Valuing indigenous cattle breeds in Kenya: An empirical comparison of stated and revealed preference value estimates. Ecological Economics 45:409-426.

Sikana, P. M. and C. K. Kerven. 1991. The impact of commercialisation on the role of labour in African pastoral societies. Overseas Development Institute, London: 31.

Southgate, C. and D. Hulme. 1996. Land, water and local governance in a Kenyan wetland in dryland: The Kimana Group Ranch and its environs. Institute for Development Policy and Management, Manchester: 63.

Spear, T. and R. Waller. 1993. Being Maasai. Athens, James Currey LTD.

Struhsaker, T. T. 1976. A further decline in numbers of Amboseli vervet monkeys. Biotropica 8:211-214.

Swift, D. M., M. B. Coughenour, and M. Astedu. 1996. Arid and semi-arid ecosystems. Pages 243-299 *In* T. R. McClanahan and T. P. Young, editors. East African ecosystems and their conservation. Oxford University Press, New York.

Thornton, P. K., S. B. BurnSilver, and K.A. Galvin. 2006. Modelling the impacts of group ranch subdivision on agro-pastoral households in Kajiado, Kenya. Agricultural Systems 87: 331-356.

Touber, L. 1983. Soils and vegetation of the Amboseli-Kibwezi Area: (quarter degree sheets 173, 174, 181, and 182). Nairobi, Kenya Soil Survey: 17.

United Nations Development Programme (UNDP). 2003. Pastoralism and mobility in the drylands. Global Drylands Imperative (GDI) Challenge Paper Series. Drylands Development Centre, Nairobi, Kenya.

USGS. 2002. Global Land 1-km AVHRR Project. Sioux Falls, SD: U.S. Geological Survey, EROS Data Center. http://edcwww.cr.usgs.gov/landdaac/1KM/1kmhomepage.htmnl.

Western, D. 1973. The structure, dynamics and changes of the Amboseli ecosystem. Nairobi, Kenya, University of Nairobi: 222.

Western, D. 1989. Conservation without parks: wildlife in the rural landscape. Pages 158-165 In D. Western and M. C. Pearl, editors. Conservation for the 21st century. Oxford University Press, New York.

Western, D. 2002. In the dust of Kilimanjaro. Island Press, Washington.

Western, D. and T. Dunne. 1979. Environmental aspects of settlement site decisions among pastoral Maasai. Human Ecology **7**:75-97.

Western, D. and W. K. Lindsay. 1984. Seasonal herd dynamics of a savanna elephant population. African Journal of Ecology 22:229-244.

Western, D. and H. Gichohi. 1993. Segregation effects and the impoverishment of savanna parks: the case for ecosystem viability analysis. African Journal of Ecology 31:269-281.

Western, D. and M. D. L. Nightingale. 2003. Environmental change and the vulnerability of pastoralists to drought: A case study of the Maasai in Amboseli Kenya. UNEP. 35 pp.

Western, D. and D. Maitumo. 2004. Woodland loss and restoration in a savanna park: a 20-year experiment. African Journal of Ecology 42:111-121.

White, J. M. and S. J. Meadows. 1981. Evaluation of the contribution of group and individual ranches in Kajiado District, Kenya, to economic development and pastoral production strategies. Ministry of Livestock Development, Nairobi. 167 pp.

Worden, J. S., R. S. Reid, and H. Gichohi. 2003. Land-use impacts on large wildlife and livestock in the swamps of the Greater Amboseli Ecosystem, Kajiado District, Kenya. International Livestock Research Institute, Nairobi, Kenya. 50 pp.

Worden, J. S. In prep. Maasai settlement and land-use, landscape mosaics, and the spatial patterning of vegetation and wildlife in East African Savannas. Ph.D., Dissertation. Graduate Degree Program in Ecology. Colorado State University, Fort Collins, CO.

Zaal, F. 1998. Pastoralism in a global age: livestock marketing and pastoral commercial activities in Kenya and Burkina Faso. University of Amsterdam, Amsterdam.

Zaal, F. 1999. Economic integration in pastoral areas: commercialisation and social change among Kenya's Maasai. Nomadic Peoples 3:97-115.

Chapter 11

NGORONGORO CONSERVATION AREA, TANZANIA: FRAGMENTATION OF A UNIQUE REGION OF THE GREATER SERENGETI ECOSYSTEM

Kathleen A. Galvin[1,2], Philip K. Thornton[3,4], Randall B. Boone[2], and Linda M. Knapp[1,2]

[1]*Department of Anthropology, Colorado State University, Fort Collins, CO 80523, USA;* [2]*Natural Resource Ecology Laboratory, Colorado State University, Fort Collins, CO 80523, USA;* [3]*International Livestock Research Institute, Nairobi, Kenya;* [4]*Institute of Atmospheric and Environmental Sciences, School of Geosciences, University of Edinburgh, Edinburgh, UK*

1. INTRODUCTION

The Ngorongoro Conservation Area (NCA), occupying 8,292 km^2 of northern Tanzania, is unique among East Africa's protected areas because of its multiple land-use status. This distinction includes the explicit mandate of conserving wildlife and other natural resources while also serving the needs of the resident Maasai pastoralists and promoting tourism. It is an interesting case of highly productive land being protected from fragmentation into privately-owned parcels, yet it is being fragmented from pastoral use through conservation policy.

In this chapter we briefly describe the natural resource base of the region followed by a discussion of East African protected areas and tourism. This is followed by a detailed discussion of land tenure in the NCA. These sections set the stage for an understanding of the processes of fragmentation of the Greater Serengeti Ecosystem, and place the NCA within a regional perspective. We then look at the specific sources of resource fragmentation

K. A. Galvin et al. (eds.), Fragmentation in Semi-Arid and Arid Landscapes: Consequences for Human and Natural Systems, 255–279.

in the NCA. Consequences of diminished access to resources are then described. Finally, we conclude with a discussion of the implications of fragmentation for the Ngorongoro Conservation Area and the Greater Serengeti Ecosystem.

The NCA is part of the Greater Serengeti Ecosystem (GSE) through which the great African wildebeest migrations occur and in which numerous other species of wildlife make their home (Figure 11-1). There are several types of protected areas in the GSE including a national park, game reserves, a game controlled area, open areas, and the NCA. Fragmentation of the GSE has occurred through historical processes of gazetting the protected areas. In addition human population increases and cultivation have reduced the amount of land available to the movement of livestock and wildlife. The areas to the west and south of Serengeti National Park are populated by agriculturalists or agro-pastoralists whose livelihoods are based on cultivation (Galvin et al. in press). Human population growth in that region has increasingly put pressure on wildlife populations (Polasky et al. in press).

In the NCA and the Loliondo Game Controlled Area live the pastoral Maasai. It can be argued that living in protected areas allow the Maasai to have a pastoral and agro-pastoral existence. This is true at one level, in that processes that are affecting other African pastoral areas, such as land privatization and fence construction, do not occur in the NCA. However, the process of chopping up the GSE into different administrative units has resulted in fewer options for responding to temporal variability in production of vegetation and in availability of water. Conservation/protected lands policy within the NCA is having the same effect.

The NCA is bordered to the west by the Serengeti National Park, to the east and south by the Lake Eyasi basin and agricultural communities, and to the north by semi-arid rangelands making up the Salei Plains and the Lake Natron basin (Perkin 1997) (Figure 11-2). NCA can be divided into three broad ecological zones: a relatively cool and wet volcanic highland massif characterized by tropical montane forests, where annual rainfall ranges between 800 and 1,200 mm per year; semi-arid lowlands or short-grass plains, where rainfall averages between 400 and 600 mm per year; and between the highlands and plains are slopes of woodlands, bushlands, and grasslands forming a transition zone between the two areas (Galvin et al. 2002).

Present land uses within NCA include the conservation of natural resources, livestock herding, cultivation, tourism, and research. The main trends that are limiting Maasai access to natural resources include climate, exclusion from key resources (water and forage), human population growth and disease.

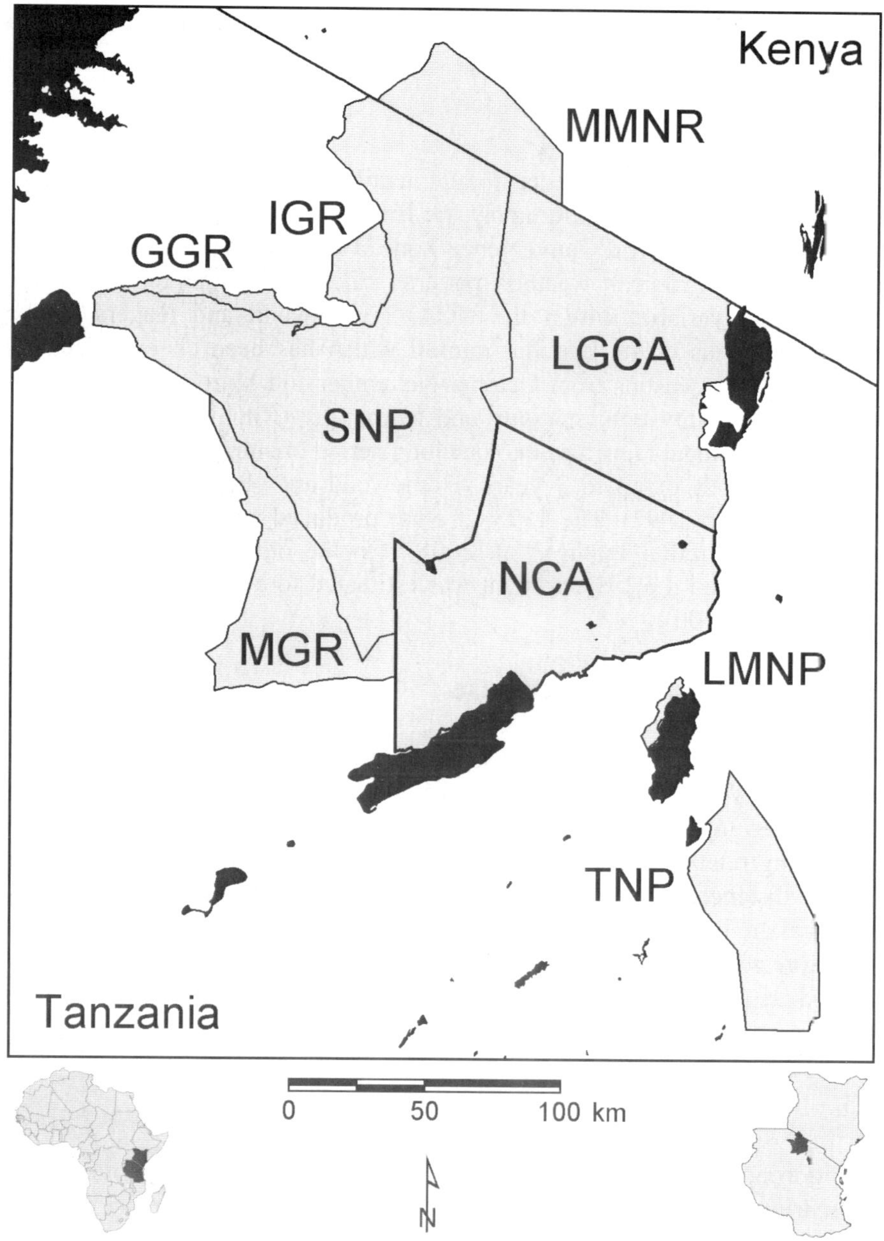

Figure 11-1. The Greater Serengeti Ecosystem which includes Ngorongoro Conservation Area (NCA), Loliondo Game Controlled Area (LGCA), Serengeti National Park (SNP), Maswa Game Reserve (MGR), Grumeti Game Reserve (GGR), Ikoronga Game Reserve (IGR) and Masai Mara National Reserve (MMNR) in Kenya. Other national parks in the region include Lake Manyara National Park (LMNP) and Tarangire National Park (TNP).

2. CLIMATE AND NATURAL RESOURCES

2.1 Climate

Rainfall seasonality affects forage availability, livestock production, availability of water, and ultimately the livelihoods of pastoralists. Located within the Inter Tropical Convergence Zone, Tanzania's climate is governed by large-scale tropical weather patterns and local topography (which is extremely variable within the NCA) (Homewood and Rogers 1991). Prevailing winds create bimodal rainfall which has been characterized by two annual wet seasons from October-November and March-April although there is uncertainty both spatially and temporally (Ritchie et al. in press). Sometimes bimodal rains fuse into one long period or simply fail altogether—even for periods of up to 3 years (Homewood and Rogers 1991, Sinclair 1995, Potkanski 1997). The 1998 El Niño produced an estimated five-fold increase in rainfall (Galvin et al. 2001). On the other hand, 1997 was a drought year and the 1999 drought was estimated to be one of the worst on record (WFP 2000).

2.2 Natural resource base

East African pastoralist ecosystems tend to be heterogeneous in nature, unlike their West African systems (Ellis and Galvin 1994). Thus pastoralists have tended to be able to acquire key wet season and dry season resources by moving relatively short (tens and hundreds of km rather than thousands of km) distances. As the natural systems become fragmented or lost, key resources are harder to acquire. Vegetation in the NCA is obtained at different elevation zones and different water points and its heterogeneity can be defined as course-grained, meaning access to a relatively large area is necessary to obtain all resource types (Hobbs et al., Chapter 2).

2.3 Water

Ngorongoro Maasai depend primarily on surface water sources for their drinking water. Lack of water development is illustrated by research showing that during the dry season only 2.4% of NCA Maasai use tap sources compared with 20% of Kenyan Maasai (Coast 2002). Of this surface water, much is alkaline. Aikman and Cobb (1997) found significant numbers of NCA residents using water sources that contain unsafe levels of saline and fluoride according to the World Health Organization.

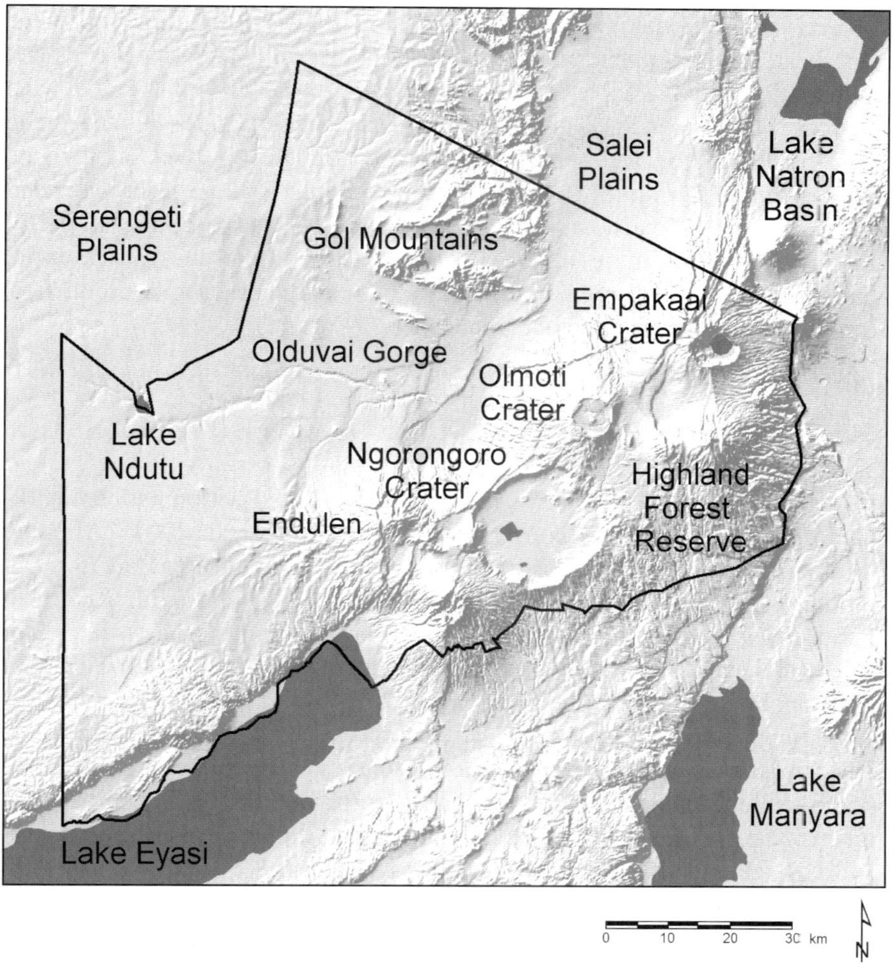

Figure 11-2. Map of Ngorongoro Conservation Area with elevation and place names.

The NCA Crater Highlands catchment is an important source of water for wildlife, livestock, and people from the Serengeti plains and Gol Mountains (west of the highlands) and people from the south and east of NCA (Homewood and Rogers 1991). The Highlands became an important center of Maasai immigration during the drought years of the 1970s when pastoralists from the surrounding lowlands moved to the higher rainfall areas of the NCA in search of pasture and water (Perkin 1997). There are 23 permanent streams in NCA and the watershed is such that all streams flow into crater lakes or depressions (Homewood and Rogers 1991).

While it is clear that the highlands maintain important water sources for the area, water acts as a limiting resource within NCA rangeland dynamics.

Maasai decision-making involves choosing between high-quality forage (found generally on the plains) and availability of water (most abundant in the highlands). People and livestock congregate around limited water sources while other bush areas remain either heavily under-utilized or not used at all. In the wet season the distance traveled by Maasai to water sources varies from a few hundred meters to 3 km. In the dry season round-trip travel takes five hours, covering 10 km round-trip (Potkanski 1997). Energy loss from these treks as well as the consequent restrictions on feeding time "was the single most important constraint on milk production" for NCA livestock (Homewood and Rogers 1991:252).

3. PASTORALISTS AND PROTECTED AREAS

Ten percent of East Africa's land area is protected within national parks, game reserves, and other conservation areas (Reid et al. 2004). Although this percentage seems relatively low, these "important areas are the life-blood of rangelands when times are most difficult" (Reid et al. 2004:176). Landscapes within protected areas are important for pastoralists, their live-stock, and wildlife because these areas usually act as dry-season refuges. They maintain higher rainfall averages or include a year-round supply of water (e.g., rivers or swamps). Denying access to protected areas prevents pastoralists from obtaining key natural resources and ensuring the survival of both them and their livestock. So it is not surprising that in and surrounding most of the protected areas of East Africa are found pastoral populations as livestock and wildlife share the same resources. For example, for nearly two thousand years the area encompassing NCA has been inhabited by pastoralists and abundant wildlife (Homewood and Rogers 1991).

3.1 Tourism

Protected areas bring in outside revenues through tourism. In Tanzania, gross earnings from game-hunting tourism tripled between 1988 and 1993. Over a ten-year period (1985-1995) the number of tourists in the country increased by 1,441 percent (WRI 2000). During the mid-1990s, Tanzania's tourism industry was the country's top foreign exchange earner (Honey 1999).

The NCA Maasai were not permitted to benefit from tourism until the mid-1990s, and only then in a limited way. Cultural *bomas,* or tours of tradi-tional homes, are now permitted, though most profits benefit tour guides and little filters back to local residents. Other indirect benefits from the tourism industry include the sale of handicrafts, honey, fruits and vegetables, and ceremonial viewing and walking tours. Charnley (2005) suggests that most impacts of tourism are negative including: increased prostitution between Maasai girls/women and tour guides, as well as increased begging and social conflict among residents who compete for economic benefits. The Maasai Pastoral Council receives 10% of revenues though they demand more (Charnley 2005).

In addition to receiving little benefit from the tourism industry, the protected lands policy prevents increased job diversification opportunities for NCA Maasai. Only in the Endulen area are there opportunities for working at a hospital, trading center, police station, or veterinary center. In the Masai Mara Game Reserve, Kenya, where Maasai have a political voice, the majority of hotel and park staff are Maasai. In contrast, only 0.2% of NCA Maasai over the age of 15 are employed full-time in the tourist industry (Coast 2002). Though NCA Authority (NCAA) policy theoretically gives first priority to Maasai, very few are actually hired due to a lack of basic skills (English, driving, or primary education) necessary to perform tourism jobs (Charnley 2005).

While tourism, and therefore protected areas, are clearly an important source of income, some studies show that greater economic gain could be achieved through other forms of land use. If the protected areas within parks, reserves, and forests of Kenya were used for agricultural and livestock production, an annual net gain of $203 million could be earned compared to the $42 million annual net gain from the tourism and forestry industries (Norton-Griffiths and Southey 1995). The authors acknowledge that there are many indirect benefits for people, wildlife and biodiversity of protected areas, but they were unable to quantify them. "Our data have sufficient detail to address only the net benefits from direct uses" (Norton-Griffiths and Southey 1995:126). There is also not a discussion of the staggering loss of ecosystem services that would occur if protected areas were converted to agriculture.

4. LAND TENURE IN THE NCA

The NCAA, which manages the NCA, was established under the *Game Parks Laws Miscellaneous Amendments Act of 1975*. Under colonial rule the area was controlled by the governor, thereafter, the state. The functions of the NCAA include: conservation and development of natural resources, promotion of tourism, provision of the facilities necessary for the promotion of tourism, protection of the interests of Maasai citizens, promotion and regulation of the development of forestry, construction of infrastructure, and transportation to, from, and within the conservation area (Perkin 1997).

The NCAA and the resident pastoralists are not in clear agreement concerning pastoralist land tenure in the NCA. The Authority sees the land tenure status of the pastoralists as "vague" while the pastoralists regard their customary rights to land as unambiguous and sacrosanct (Tenga 1998).

The NCA Maasai pastoralists practice a system of transhumance, the rotating use of dry and wet season pastures from a permanent base. Maasai customary land tenure consists of all pastures belonging to all Maasai (Potkanski 1997). The system is not concerned with exclusive ownership of land but with access to natural resources. In practice, local communities have primary user rights to pastures within their customary area although, due to their system of reciprocity, these users are required to offer pasture to other groups in emergencies. Groups have secondary user rights to areas they visit seasonally (Potkanski 1997). Whether primary or secondary, all user rights are not passed on inter-generationally within families (except for the case of wells or springs), but rather are inherited through membership in a community. Decision-making regarding user-rights is negotiated by Maasai elders. For example, during the dry season, elders may collectively decide to prohibit access to certain pastures until there is no longer forage available elsewhere. There are exceptions for individual private rights to a small plot within the collective property, but this is only for protection or grazing of calves, sick, or aged animals (Seno and Shaw 2002).

In regions across Tanzania the 1975 *Villages and Ujamaa Village Act* and the *1999 Village Land Act* demarcate village land and empower village councils to serve as the local government with the responsibility of managing individual land titles. These rights refer back to the original *1923 Land Ordinance* giving 'deemed rights of occupancy' to individuals who used land according to customary law. Furthermore, under Article 24 of the constitution of Tanzania, 'customary' or 'deemed rights of occupancy' are equated with real property rights for specific pieces of land. This means that the rights of Maasai residents of the NCA to legally "own and enjoy" their specific customary lands are protected under the Tanzanian constitution (Shivji and Kapinga 1998).

Ngorongoro Division (which is the same as the NCA, one of three Divisions in Ngorongoro District) includes six wards and within those six wards (Shivji and Kapinga 1998) there are 16 villages (Charnley 2005). All are political units. Unlike the rest of Tanzania's villages, these sixteen do not have demarcated boundaries or the right to distribute legal land titles. The dispute over land tenure lies in the *1959 Ngorongoro Conservation Ordinance No. 14.* Under the 1959 Ordinance the NCA Authority has the right to prohibit, constrict, control, and manage cultivation, grazing, collection of firewood and residence in any part of the NCA except over freehold land, leasehold land, or land held under granted rights of occupancy. Essentially, the Authority was given power over the NCA Maasai's collective land. Yet of equal importance, the Ordinance of 1959 did *not* diminish Maasai 'deemed rights of occupancy' or grant NCA land titles to the Authority. Rather, under constitutional law, the land still legally belongs to the resident Maasai (Shivji and Kapinga 1998). So how can Maasai 'deemed rights of occupancy' and legal statutory powers of the Authority be simultaneously compatible?

The constitution of Tanzania does include limitation clauses which allow for the expropriation of property by the state. Under these circumstances, the carrying out of expropriation must: 1) be under the authority of the law and 2) provide compensation. Admittedly, the NCAA has the right to limit received benefits to the Maasai community from the NCA, but in order to remain under Constitutional law the Authority should also consult Maasai leaders before any decisions are made regarding land in the NCA and provide compensation to the community. Shivji and Kapinga (1998) argue that "it is a constitutional obligation of the NCAA, the breach of which results in the violation of the Maasai rights to property stipulated in Article 24" to include Maasai in decision-making and provide collective compensation in the form of water, health, schooling, etc. Unfortunately, since the *1975 Game Parks Laws Act No. 14*, all forms of democratic governance and rule of law in the NCA were taken away by giving the Authority ultimate control of all decision-making (Shivji and Kapinga 1998). The Authority is not acting according to Constitutional law unless it consults local leaders and compensates Maasai residents for their denied land rights. The Maasai emphatically state that this is not occurring (Poole 2006:28).

Land tenure policies in NCA continue to be hotly debated among development workers, conservationists, government officials, and NCA residents. Some argue, including many Maasai themselves, that residents should have legal titles to land in order to secure their permanent residency within NCA. Charnley (2005) maintains that legal village or private ownership would allow Maasai in the NCA to lease their land to tourist companies and benefit directly from the tourism industry. Other researchers have legitimate concerns about land privatization, despite the security of investment it would provide.

In most of East Africa land privatization leads to the building of fences, habitat loss and fragmentation, and restriction of wildlife and pastoralists from accessing key resources (Reid et al. 2004, Boone et al. 2005). Legalizing and privatizing individual land titles could allow the NCA Authority to restrict current Maasai "usufructuary rights" across the majority of the Area. If this were to become true and Maasai were confined to their village land, the result would "compartmentalize conservation and development functions over geographical and social space, thus disembodying the

Maasai not only as a cultural but also as an economically sustainable community" (Shivji and Kapinga 1998:36). Ultimately, giving the NCAA sole power to delineate village land (without Maasai input) and therefore dispense individual land titles could lead to further restrictions on the movements of people, livestock, and wildlife. Privatization would only increase habitat fragmentation, thereby limiting access to natural resources and reducing Maasai ability to adapt to climate variability or other drivers of change.

5. REGIONAL AND HISTORICAL SOURCES OF FRAGMENTATION OF THE GREATER SERENGETI ECOSYSTEM

5.1 Protected area policy

During colonialism, East Africa's protected areas were designed exclusively for white hunters, scientists, and tourists. Hundreds of thousands of rural poor were forced to relocate outside park boundaries. "The colonial philosophy, initially adopted by post-colonial governments, was that wildlife had to be protected from the local Africans with fences, fines, and firepower" (Honey 1999:29).

During the colonial era, Maasai across East Africa saw their most productive grazing land being taken over by white settlers and cultivators. As Arhem (1985:19) states, "The colonial land policies in Kenya and Tanganyika at the time favored settler agriculture and indigenous smallholder farming". Except for a brief period from 1926 to 1930 when the British Colonial administration tried to prohibit agricultural encroachment of Maasailand in Tanzania, the history of pastoral peoples in Tanzania since the beginning of the 1900s has been characterized by a loss of land and an attempt to industrialize the livestock economy. Today, Maasailand has shrunk to a fraction of its previous size (Arhem 1985).

Throughout the German (1885-1916) and British (1918-1961) administrations in East Africa the colonialists assumed that the indigenous occupants had no ownership rights over the land. All lands were considered to belong to the colonial government, whose goal was to exploit the land and labor of the colonized people. The British colonial government in Tanzania issued the *1923 Land Ordinance*. This Ordinance declared all lands as public lands, controlled by the governor. The ordinance was ambiguous about indigenous customary land rights, defining them without securing them by title or right. A 1928 Amendment expanded the meaning of "rights to occupancy" to include natives using land within native custom. This customary title was called "deemed rights of occupancy." These rights recognized customary law but, once again, did little to protect it. Another type of land right was called "granted rights of occupancy" (similar to a 99 year lease). This type of tenure allowed the colonial government to distribute land to foreign companies. Overall, the colonial system was feudalistic in nature and characterized by conflict between indigenous peoples and foreign settlers (Shivji 1998).

5.2 NCA policy

The NCA land tenure history is inextricably linked to the larger Greater Serengeti Ecosystem. Between 1836-1851 the Maasai gained control over the Ngorongoro highlands by driving out the Tatog (Barabaig), another pastoralist tribe. It was during this time, at the height of Maasai power in East Africa, that the entire Rift Valley stretching between central Kenya and central Tanzania was considered Maasailand (Arhem 1985). In the early 1900s, following a rinderpest epidemic, the Maasai Kisongo Section—the same group that remains in NCA today—replaced the weakened Loita Section for control of the Ngorongoro highlands (McCabe et al. 1992).

Under the *Game Ordinance of 1940*, which established Serengeti National Park (SNP), which included what is now the NCA, the NCA land tenure system diverged from the rest of Maasailand and even East Africa. The Serengeti and NCA Maasai residents relinquished claims to SNP in return for a government pledge to be permitted to continue to follow or modify their traditional way of life subject only to close control of hunting (Shivji and Kapinga 1998). Unfortunately, promises were not kept by the Colonial administration and Maasai residents reacted strongly against the prohibition of cultivation in 1954. Their protests resulted in the *Ngorongoro Conservation Ordinance No. 14 of 1959* separating NCA and SNP. All Maasai living in SNP were forced to move as well as abandon access to permanent water sources in the National Park. They were assured that permanent land rights and new water supplies would replace what they had lost (Arhem 1985).

After Tanzanian Independence in 1961, the official policy toward pastoralism changed somewhat resulting in the NCA's multiple land-use strategy allowing for wildlife and pastoralism to co-exist; however, with international conservation pressures, restrictions within the NCA began to increase starting in 1965.

6. PROXIMATE SOURCES OF RESOURCE (LAND AND WATER) FRAGMENTATION IN THE NCA

6.1 Grazing and settlement policy

Restrictions on livestock numbers within NCA do not exist, though there are regulations prohibiting grazing within Ngorongoro, Olmoti and Empakaai Craters, Olduvai Gorge, and the Northern Highland Forest Reserve, together a significant portion of the NCA.

Since 1974, The Ngorongoro Crater area within the NCA has been managed as a core protected area, with the remainder of the NCA being managed as a buffer zone to both the Ngorongoro Crater and the Serengeti National Park (Thompson 1997). The *Game Parks Laws of 1975 Act* declared that Maasai pastoralists would not be allowed to live in permanent *bomas* or graze livestock on the Crater floor after the 1974 dry season. According to former NCA conservator, Henry Fosbrooke, a paramilitary Field Force Unit carried out the eviction of Maasai and their cattle from the Crater floor. No explanations were given to Maasai for this change in policy (McCabe 2002).

In Olmoti Crater policymakers believed that livestock grazing contributed to the encroachment of a coarse tussock grass, *Eleusine jaegeri*, which is unpalatable to wildlife and livestock. Despite two decades without livestock in Olmoti, the grass still spreads there (Homewood and Rogers 1991). It may be the case that NCAA range management policies regarding fire restriction caused bush encroachment and the further spread of *Eleusine* (Kijazi 1997).

The Maasai are also restricted in their use of Olduvai Gorge, one of the most important prehistoric sites in the world, for grazing. The denied access to craters, gorge and plains' lands decreases the number of permanent water sources and nutritious grasses available for Maasai and their livestock's use. As a result, the highlands undergo more and more pressures from the now year-round, formerly only dry-season, grazing (Kijazi 1997, Perkin 1997).

6.2 Wildlife, livestock and disease

The Maasai living in the NCA have traditionally taken their livestock out onto the Serengeti Plains in the wet season when grasses are green and plentiful. As the grasses of the Serengeti plains diminish in the dry seasons the Maasai then return to the highlands where water is available and forage is more nutritious. This regular seasonal movement has been disrupted by the wildebeest movements. As the wildebeest move onto the Salei Plains and calve it is the young who carry a disease, malignant catarrhal fever, which is fatal to cattle (Boone et al. 2002). As the wildebeest numbers have grown and pushed into the NCA, the Maasai have to either take their livestock onto the plains for just a short while or not at all. This has caused a disruption of the movement patterns and access to vegetation for the livestock. Thus, NCA livestock are prevented from using much of the shortgrass plains during the wet season, due to the occupancy of nearly one million wildebeest. And to make matters worse, tick-borne diseases take their toll on the livestock in the highlands, reducing their productivity.

While wildebeest numbers have increased, cattle numbers have decreased due to tick-borne and other diseases. East Coast fever and *olmilo* (bovine cerebral theileriosis) incidences were 18% for adult cattle and 52% for calves in a recent survey (Rwambo et al. 2000). A study by Field et al. (1997) suggests that mortality rates of one-year old calves were between 30-90% in NCA due to East Coast fever. For adult livestock, individual owners lost 25-75% of their herds primarily because of *olmilo* (Field et al. 1997). Maasai have increased their reliance on small stock, which are less prone to disease. The trend towards small stock could be due to the number of diseases cattle face and an economic strategy, which emphasizes quick growth (Kijazi 1997).

6.3 Water

Water is a limiting resource within NCA. From a historical perspective the Tanzanian government initially recognized the need to address water issues for the Maasai and attempted to compensate Maasai for their loss of access to permanent water sources in SNP in 1959. With more humidity and less porous soils than NCA, the Serengeti surface waters were an important resource for the Maasai. The Serengeti Compensation Scheme (1958-1962) sought to provide water supply systems to NCA people and their livestock. In 1998 when the systems were inspected, only 10 out of 29 water scheme systems were functional and of these ten, four were designated for NCAA staff and tourists, four were for livestock and wildlife use, and only two provided water for Maasai domestic use. These systems are simply not enough for a population of around 300,000 livestock and 52,000 people (Aikman and Cobb 1997).

While it is clear that access to surface water is important to the NCA Maasai, the tourism industry and NCAA employees compete with local residents and wildlife for this resource. A hotel in the Irkeepus area was reported to have been diverting local spring water and causing water security issues for the local residents (Coast 2002). Most NCAA conservation policies have attempted to prevent infringement of water sources by outside commercial or large-scale cultivation projects. The exception to this principle is the tapping of the Lerai Spring (on the floor of Ngorongoro Crater) to supply water year-round to tourist lodges and the NCAA headquarters on the Crater rim. The Lerai Spring is also an important source of water for wildlife within the Crater itself (Aikman and Cobb 1997).

In 1991 the Serena Hotel group, owned by the Aga Khan (one of the richest men in the world), applied for approval to build on the Crater rim. They requested access to the Loloueru Spring for their main water supply. This spring happened to be the water source for the Oloirobi Maasai – a people who already faced a shortage of water for their livestock. Consultants realized that this system of joint-users would not be sustainable during the dry season and the Serena Hotel resorted to pumping water from the Lerai Spring (Potkanski 1997). Although the situation seemed to be resolved, a local Maasai organization still contends that the Serena Hotel was built on local village land and in an area important for their livestock. The group believes that residents were not consulted before the hotel was built (Honey 1999).

6.4 Security issues

Cattle-raiding by invading tribes prevents the Maasai of NCA from utilizing important rangelands. Incidents of WaSukuma and Tatog cattle-raiding along the southwestern boundaries of NCA were especially high in the 1980s. Even through 1992 the Ndutu area was avoided. From 1992-93 Somalis passing through Kenya and into Sukumaland (west of the NCA) were poaching and cattle-raiding in NCA (Potkanski 1997), as well as WaKuria from the north entering to cattle-raid (Kijazi 1997). These security fears place further grazing pressures on the already over-utilized, but more secure, NCA highlands.

6.5 Human population changes

Much has changed in the NCA over the last 40 years, including a large increase in the human population that is dependent on a relatively stable livestock population. Between 1966 and 1999 the pastoral population of NCA grew from an estimated 8,700 to 51,600 (POLEYC 2002). This increase is believed to be due to a combination of factors including natural increase and immigration. Population fluctuations are standard for transhumant populations because migrations are seasonal in nature. However, there appears to be a link between the rate of population growth and the prevailing management policy.

Human population rapidly increased in NCA during the mid-1970s as a result of policy change. From 1970-76, water catchment areas and dams were constructed and promises of land titles were made as part of the USAID's Maasai Land Range Development Project. Upon failure of the project, population numbers declined (Homewood and Rodgers 1991). Human population growth rates are also linked to management policies concerning cultivation. From 1975 until 1991, all cultivation was banned in the NCA. Two different studies (Arhem 1981, Makacha and Frame 1986) concur that the decrease in human population growth rates during this time was linked to the cultivation ban. Conversely, the lifting of the cultivation ban in 1991 by Tanzania's prime minister caused an influx of agriculturists to the area. Although actual figures are not known, there seemed to be an influx of cultivators, especially in the Naiyobi/Kapenjiro and Endulen areas following the ban's lifting (Kijazi 1997, Boone et al. 2006). The rising human population caused by migration of pastoral and cultivating populations into NCA could squeeze pastoral livestock and the area's wildlife onto smaller pieces of land. However, our research shows that under current conditions of less than 1% of the NCA under cultivation, it is not negatively affecting either wildlife or livestock populations at this time (Boone et al. 2006).

6.6 Cultivation

Pressures from outside the area's boundaries affected NCA in the past and presently. When Tanzania's socialistic *Ujamaa* (or Villagization) Program was intensifying during the late 1970s, agriculture was expanded in the areas surrounding NCA. This put land pressures on Maasai living outside the Area and resulted in many retreating to NCA (Kijazi et al. 1997). The well-watered and fertile agricultural communities to the east and south of the NCA have expanded tremendously over the last decades. This is creating a set of external pressures on the NCA which are distinct and separate from the pressures inherent in the growth of the resident human population. In particular the growth in these populations has led to large scale cultivation encroaching into the Northern Highland Forest Reserve from the southeast. These populations also engage in extensive illegal exploitation of fuel wood and building poles from the same forest (Perkin 1997). This may become detrimental to the reserve as a water catchment (Boone et al. 2006).

The ban on cultivation in the entire NCA remained in effect until 1991 when Tanzania's Prime Minister lifted it in order to improve food security (Perkin 1997). Previously, policymakers based their decisions to prohibit cultivation on a false stereotype that grain is not part of the Maasai diet or that Maasai do not practice agriculture (McCabe et al. 1992). In actuality, cultivation is an important component of the Maasai livelihood strategy within NCA (Galvin et al. 2002, McCabe 2003a). In 2001 the Tanzanian government considered banning cultivation once again but did not make a final decision (POLEYC 2002). Allowing cultivation resulted in a large population increase in the NCA whose purpose has been to cultivate more land. The use of NCA land by non-Maasai cultivators increased pressure on the natural resource base and reduced the resources available to Maasai residents and wildlife (Kijazi 1997). NCAA employees hired outside labor to perform their cultivation, thus contributing to the decrease in available resources for Maasai and wildlife (Potkanski 1997). Conservationists feared that uncontrolled cultivation would continue to encourage illegal immigration to the NCA and undermine the ecological integrity of the area (Kijazi 1997). Since then all non-Maasai cultivators were evicted from the NCA and NCAA personnel have been prohibited from further cultivation.

On the other hand, the re-introduction of agriculture in 1991 has had an immediate and positive effect on Maasai household food security and malnutrition levels (McCabe et al. 1997, Potkanski 1997, Galvin et al. 2000, McCabe 2002). A food security study conducted in 1995 estimated that 85% of Maasai households in the NCA participate in cultivation. A 1993 aerial survey revealed that small-scale plots of less than five acres in extent made up less than 30% of the area of cultivation recorded, while more than 50% of cultivation in the NCA was made up by only twenty-six plots, each averaging more than 20 acres in size. The food security study noted that plots of around four acres generally belonged to non-Maasai cultivators, while the average plot size among Maasai households was 0.86 acres, although there is variability in the size of the plots per family. The largest and wealthiest families had the largest plots (McCabe et al. 1997). It is probably the case that nearly all Maasai are cultivating to some extent. The issue of whether and how much cultivation should occur is a constant source of contention within the NCA in large part because it has the potential to affect access to resources for wildlife and livestock.

7. CONSEQUENCES OF DIMINISHED ACCESS TO RESOURCES

7.1 Livelihood strategies of pastoralists

Current constraints on pastoralist livelihood strategies have made people more vulnerable to natural and human-derived perturbations. Rising human populations, lack of water availability, intertribal conflict (cattle-raiding from WaSukuma and Tatog along the southwestern boundary), encroachment of unpalatable grass species (due to restrictions on burning), and many land-use restrictions have squeezed pastoral livestock onto smaller areas of land. This has led to a situation where the Maasai can no longer depend on their livestock for the sole basis of their livelihood while opportunities for livelihood diversification are few. Furthermore, livestock populations have tended to be stable rather than expand because of disease epidemics and low livestock condition. The result is a rising human population dependent on a stable livestock population. So while pastoralists have been able to track temporal and spatial variability very well in the past, their strategies, based on centuries of exposure to natural and human-derived perturbations are not working now due, in part, to an inability to implement them. Thus, while the current protected lands policies may result in less fragmentation from large-scale agriculture and privatization, the residents still have restrictions on their livelihood strategy options and their movements.

A symptom indicating the breadth of poverty in NCA includes the increasing disintegration of customary mutual assistance programs (Potkanski 1999, Johnson 2000). Potkanski (1999) explains that when poverty is too widespread these customary systems of livestock redistribution, which promote egalitarianism across economic strata, can no longer function sustainably. Many individuals no longer ask clan-members or neighbors for assistance because they are aware that the potential donors are as worse-off financially as themselves (Potanski 1999). Another symptom of widespread poverty in NCA is the rate of malnutrition among Maasai residents (McCabe et al. 1992), especially compared to neighboring Loliondo (Galvin et al. 2002, Charnely 2005).

7.2 Decreasing livestock numbers

The result of constraints on grazing and fire-burning is that disease-driven mortality rates are very high among NCA livestock and herd sizes are small and static. Annual mortality among cattle herds appears to be closely related to herd size. Small herds had much higher mortality rates than large herds in both NCA and Loliondo Game Controlled Area. This suggests that disease, the major mortality factor (of which East Coast fever, a tick-borne disease, is dominant), is density-dependent in occurrence, thus having a much greater effect proportionally on small herds than on large ones. Cattle herds in the NCA are small (modal herd size 0-49) and annual percent mortality is high at about 24% (Lynn 2000). Thus the NCA Maasai appear to be caught in a poverty trap from which escape is difficult under current circumstances.

The total biomass of livestock was approximately the same in the year 2000 as it was 40 years ago. In 1998, eighty-seven percent of Maasai households in NCA fell below the necessary minimum of 4.5–5 livestock units per capita needed to support subsistence pastoralism (Galvin et al. 2002). This has put serious food security stress on households, so that by 1991 people were selling their female livestock to purchase food, depleting their core reproductive herds. McCabe (2003a) suggests that the adoption of cultivation among NCA Maasai was initially caused by low livestock to human ratios, however, recent increases in the intensity of farming is not entirely caused by poverty or food security issues. Rather, cultivation is necessary to maintain the livestock-based economy and reproduce the pastoral identity of NCA Maasai. Cash demands (for health care, veterinary needs or school fees) place risks on livestock numbers. To avoid selling their cattle, wealthier Maasai invest in agriculture and thereby preserve their pastoral livelihood (McCabe 2003a).

7.3 Wildlife, migration corridors, disease and water

A dramatic expansion of wildlife populations due to veterinary control of the rinderpest virus has also led to increased land-use pressure. The most noticeable population increase has been in wildebeest whose population rose from 250,000 to 1.7 million between 1960 and 1990 (Perkin 1997). Expansion of the wildebeest population has increased land-use pressure because it has decreased the area available for livestock grazing and placed increasing pastoral pressure on the highlands. This is a reversal of traditional grazing patterns in which the plains were used in the wet season and the highlands in the dry season. Portions of the highlands are now used year round (Perkin 1997).

The Ngorongoro Crater is the only area in the NCA that has had repeated total censuses of the herbivore population. The crater within NCA is not a self-contained ecological unit, although it does have resident populations of herbivores and carnivores. There is movement of both herbivores and carnivores into and out of the caldera and it is dependent on the Ngorongoro highlands catchment area for water. From 1963 to 1992 there had been significant changes in the population sizes of individual species, but total herbivore biomass had not changed significantly. The removal of pastoralists, their livestock, and their range management practice has resulted in significant changes in the population trends of wildebeest and cape buffalo, as well as a significant change in the vegetation species composition (Moehlman et al. 1997).

Some issues facing the Ngorongoro Crater area in particular include pressure on migration corridors and water sources. Historically there have been migration corridors to and from the Ngorongoro Crater. The individual movement patterns of wildebeest are not known at this time and research is needed to determine the relative numbers moving in and out of the Crater and whether the corridors are remaining open. There are concerns that cultivation on the north-western slopes of the Crater Highlands are encroaching on an important wildlife route (Moehlman et al. 1997).

The NCA also provides corridors to and from SNP, Lake Manyara National Park, and even into Tarangire National Park (Kijazi 1997). When human population increases (as is happening by 3.5% each year), so do the rates of forest and woodland destruction. Through utilization of Landsat images, Misana (1997) calculated 12,380 ha of woodland were lost between 1979-87; this correlates to the 26% increase in human population during those same years.

8. CONCLUSIONS

8.1 Climate

Recent climate studies show that the amount of rainfall in southern Africa, a trend that is linked to the Indian Ocean's warming, is projected to decline by 10% by the year 2050. Scholars envisage that the February-April wet season will be 10-20% drier than the previous fifty years. Land areas of the Sahara and arid regions in southern Africa are set to increase in temperature by 1.6 degrees Celsius by the year 2050 (Simms and Reid 2005). Even the famous snows of Kilimanjaro decreased by 80% in the 20th century and are predicted to disappear within ten to fifteen years (Thompson et al. 2002). Compounding the state of failing rainfall and increasing temperatures in Africa, sea level is projected to rise about 25 cm by the year 2050 (Simms and Reid 2005). These combined factors create a forecast of unpredictability. On the other hand, some parts of East Africa may become wetter and experience longer growing seasons (Thornton et al. 2003). Allali et al. (2001) suggest that annual rainfall should increase by two to five percent (25-50 mm) with longer, wetter wet seasons. However, Serengeti National Park rainfall data do not agree with these predictions. Annual and wet-season rainfall over the first 50 years of the 20th century *significantly* increased in the Serengeti (Ritchie et al. in press). However, annual and wet-season rainfall has declined significantly since the 1960s despite the fact that dry-season (June-October) rainfall increased slowly through the 20th century

(Ritchie et al. in press). With such variable predictions for the region it is not clear how the NCA climate will change. However, it is likely that with climate change, an increase in disease transmission, constrained water availability, and conflict over access to resources is likely (Simms and Reid 2005).

Such variability in climate surely will make fundamental changes to ecosystem structure and function. These in turn will affect human land use and livelihoods and have the potential to make these populations more vulnerable. For NCA pastoralists, freedom to engage in transhumance in order to access various sources of water and patches of rangeland across the landscape is crucial for adaptability to and resilience within the projected climate changes.

8.2 Pastoral poverty

In the last few decades pastoralists across East Africa have become increasingly poor as livestock-to-human ratios decrease. As a result, subsistence systems based on nomadic livestock herding have been shifting towards diversification (Broch-Due and Anderson 1999, Zaal and Dietz 1999, Johnson 2000, Little et al. 2001, McCabe 2003a, b). For the pastoral poor, livelihood strategy diversification is a means for coping with risk as the pastoral economy declines. Another way is out-migration of young men (McCabe 2003b). The problems of the Maasai in NCA echo those of the rest of pastoralist East Africa, but their plight is exacerbated by NCAA policy limiting their ability to diversify and to prevent the spread of livestock disease.

Pastoralists as a whole are among the most impoverished group within the region of East Africa (Little et al. 2001). The traditional subsistence pastoral economy, one that is dependent entirely on livestock production, is under serious strain due to unequal increases in human and livestock populations (Little et al. 2001, Western and Nightingale 2002). Expanding human populations are trying to subsist on stable livestock numbers. Each subsequent generation of Maasai seems to be poorer than the one before it (McCabe 2003a). This is why in last 30-40 years there are increased levels of livelihood diversification in which pastoralists of East Africa seek income through non-livestock-keeping means (Talle 1999, Western and Nightingale 2002, McCabe 2003a).

Accurately quantifying poverty among Maasai, or any pastoralist group for that matter, is difficult because the absolutely destitute are pushed out of the pastoral system (Little et al. 2001). Marginalized pastoralists seek employment outside pastoral society in towns or cities or even exit the system completely by becoming agriculturalists (Broch-Due and

Anderson 1999, Little et al. 2001). In essence, poor pastoralists become non-pastoralists.

Major forces that drive poverty in pastoral systems include decreasing livestock-to-human ratios; sedentarization; landscape loss and fragmentation through expanding agriculture, rangeland privatization, and conservation schemes; and human-wildlife conflict (Western and Nightingale 2002). The loss of land, in particular dry season refuges, has intensified pastoral vulnerability to risk (Western and Nightingale 2002). "The loss to the Maasai of critical dry season grazing areas threatens to undermine an entire system of production and a way of life. Acute shortages of arable land persist and rural poverty is becoming more severe" (Parkipuny 1997:165). By restricting pastoral mobility, pastures become over-grazed and Maasai are forced to sustain their herds on shrinking and poorer-quality pastures (Johnson 2000).

8.3 Conservation policy

The protected lands policies restrict pastoral exploitation of resources across spatial scales within the greater NCA-Serengeti Ecosystem resulting in the bottleneck effect of livestock population dynamics as described in Hobbs et al. (Chapter 2). People and their livestock are forced to consistently use areas that were traditionally only used seasonally.

Conservation practice in the NCA has drawn heavily on the ban on cultivation and use of a core protected area, and a buffer zone approach. While restrictions on cultivation are likely to continue, restrictions on pastoralism persist despite a lack of evidence that pastoralism has conflicted with conservation values. The results in terms of achieving conservation and human development objectives have been mixed. For many years, the bulk of the NCAA's resources appear to have been devoted to achieving conservation objectives. Wildlife values have been fairly well preserved. Factors underlying this success have been the high priority accorded to natural resource conservation by the NCAA, the traditional compatibility of wildlife conservation and pastoralism in the area, and the maintenance of a core protected area in which wildlife conservation has been the overriding objective. However, there have been land use conflicts including a rapidly increasing human and wildlife population, changing human development needs, and the intensification of land use outside of the conservation area. Perhaps the greatest challenge now facing NCA management is to improve the socio-economic status of NCA residents in a fashion that reinstates an unfragmented landscape.

Maasai interaction with the Ngorongoro Conservation Area Authority has been generally adversarial. NCA residents believe policies favor wildlife to the detriment of local people (Arhem 1985, Honey 1999). Diversity and intensification of livelihood strategies (beyond limited cultivation), a process common among other East African pastoral populations, have been options generally not available to the Maasai of the NCA due to conservation policy (McCabe et al. 1992). Restricted movement, limited alternative livelihood strategies, and marginal benefit from the tourism industry have left the Maasai with few choices other than sedentarism and limited agriculture. Whether or not the Ngorongoro Conservation Area, a unique component of the Greater Serengeti Ecosystem, can maintain its rich heritage of wildlife, its human welfare and its tourism will, in large part, be dependent on the ability to insure access to habitat suitable for all its residents.

ACKNOWLEDGEMENTS

This work is supported by a grant from the U.S. National Science Foundation (DEB-0119618). Thanks to the reviewers for excellent comments.

REFERENCES

Aikman, D. I. and S. M. Cobb. 1997. Water Development. Pages 201-232 *In* D. M. Thompson, editor. Multiple land use: The experience of the Ngorongoro Conservation Area, Tanzania. IUCN (International Union for Conservation of Nature and Natural Resources), Gland, Switzerland and Cambridge, UK.

Allali, A., C. Basalirwa, M. Boko, G. Dieudonne, T. E. Downing, P. O. Dube, A. Githeko, T. Githendu, P. Gonzalez, D. Gwary, G. Jallow, J. Nwafor, R. Scholes. 2001. Africa. *In* J. J. McCarthy, O. F. Canziani, N. A. Leary, D. J. Dokken, and K. S. White, editors. Climate change 2001: Impacts, adaptation and vulnerability. Contributions of Working Group II of the Intergovernmental Panel on Climate Change. http://www.grida.no/climate/ipcc_tar/wg2/ index.htm.

Arhem, K. 1981. The ecology of pastoral land use in the Ngorongoro Conservation Area. A background paper for a new management plan for the NCAA. BRALUP Paper, University of Dar-es-Salaam.

Arhem, K. 1985. Pastoral man in the Garden of Eden: the Maasai of the Ngorongoro Conservation Area, Tanzania. University of Uppsala, Uppsala, Sweden.

Boone, R. B., M. B. Coughenour, K. A. Galvin, and J. E. Ellis. 2002. Addressing management questions for Ngorongoro Conservation Area, Tanzania, using the Savanna Modeling System. African Journal of Ecology 40:138-150.

Boone, R. B., S. B. BurnSilver, P. K. Thornton, J. S. Worden, and K. A. Galvin. 2005. Quantifying declines in livestock due to land subdivision. Rangeland Ecology & Management 58:523-532.

Boone, R. B., K. A. Galvin, P. K. Thornton, D. M. Swift, and M. B. Coughenour. 2006. Cultivation and conservation in Ngorongoro Conservation Area, Tanzania. Human Ecology 34:809-828.

Broch-Due, V. and D. M. Anderson. 1999. Poverty and the pastoralist: Deconstructing myths, reconstructing realities. Pages 3-19 *In* D.M. Anderson and V. Broch-Due, editors. The poor are not us: Poverty and pastoralism in Eastern Africa. Ohio University Press, Athens, Ohio.

Charnley, S. 2005. From nature tourism to ecotourism? The case of the Ngorongoro Conservation Area, Tanzania. Human Organization 64:75-88.

Coast, E. 2002. Maasai socioeconomic conditions: A cross-border comparison. Human Ecology 30:79-105.

Ellis, J. and K. A. Galvin. 1994. Climate patterns and land use practices in the dry zones of east and west Africa. BioScience 44:340-349.

Field, C. R., G. Moll, and C. ole Sonkoi. 1997. Livestock development. Pages 181-199 *In* D. M. Thompson, editor. Multiple land use: The experience of the Ngorongoro Conservation Area, Tanzania. IUCN, Gland, Switzerland and Cambridge, UK.

Galvin, K. A., P. Thornton, and S. Mbogoh. 2000. Integrated modeling and assessment for balancing food security, conservation and ecosystem integrity in East Africa. Final Report. Socio-economic modeling component, 1997-2000 to GL-CRSP, U.S.AID. University of California, Davis, California USA.

Galvin, K. A., R. B. Boone, N. M. Smith, and S. J. Lynn. 2001. Impacts of climate variability on East African pastoralists: Linking social science and remote sensing. Climate Research 19:161-172.

Galvin, K. A., J. E. Ellis, R. B. Boone, A. L. Magennis, N. M. Smith, S. J. Lynn, and P. Thornton. 2002. Compatibility of pastoralism and conservation? A test case using integrated assessment in the Ngorongoro Conservation Area, Tanzania. Pages 36-60 *In* D. Chatty and M. Colchester, editors. Conservation and mobile indigenous peoples: displacement, forced settlement, and sustainable development. Berghahn Books, New York.

Galvin, K. A., S. Polasky, C. Costello, and M. Loibooki. In press. Human responses to change: modeling household decision-making in Western Serengeti, Tanzania. *In* A. R. E. Sinclair, C. Packer, S. A. R. Mduma and J. M. Fryxell, editors. Serengeti III. Human impacts on ecosystem dynamics. University of Chicago Press, Chicago.

Homewood, K. M. and W. A. Rogers. 1991. Maasailand ecology: Pastoral development and wildlife conservation in Ngorongoro, Tanzania. Cambridge University Press, Cambridge, England.

Honey, M. 1999. Treading lightly? Environment 41:4-9, 28-33.

Johnson, N. 2000. Placemaking, pastoralism, and poverty in the Ngorongoro Conservation Area, Tanzania. Pages 148-172 *In* V. Borch-Due and R. Schroeder, editors. Producing nature and poverty in Africa. Nordiska Afrikainstitutet, Stockholm.

Kijazi, A. 1997. Principal management issues in the Ngorongoro Conservation Area. Pages 33-44 *In* D. M. Thompson, editor. Multiple land use: The experience of the Ngorongoro Conservation Area, Tanzania. IUCN, Gland, Switzerland and Cambridge, UK.

Kijazi, A., S. Mkumbo, and D. M. Thompson. 1997. Human and livestock population trends. Pages 169-180 *In* D. M. Thompson, editor. Multiple land use: The experience of the Ngorongoro Conservation Area, Tanzania. IUCN, Gland, Switzerland and Cambridge, UK.

Little, P. D., K. Smith, B. A. Cellarius, D. L. Coppock, and C. Barrett. 2001. Avoiding disaster: diversification and risk management among East African herders. Development and Change 32:401-433.

Lynn, S. J. 2000. Conservation policy and local ecology: Effects on Maasai land use patterns and human welfare in northern Tanzania. M.S. Thesis, Colorado State University.

Makacha, S. and G. W. Frame. 1986. Population trends and ecology of Maasai pastoralists and livestock in Ngorongoro Conservation Area, Tanzania. IUCN, Gland, Switzerland.

McCabe, J. T. 2002. Giving conservation a human face? Pages 61-76 *In* D. Chatty and M. Colchester, editors. Conservation and mobile indigenous peoples: displacement, forced settlement and sustainable development. Berghahn Books, New York.

McCabe, J. T. 2003a. Sustainability and livelihood diversification among the Maasai of Northern Tanzania. Human Organization 62:100-111.

McCabe, J. T. 2003b. Disequilibrial ecosystems and livelihood diversification among the Maasai of northern Tanzania. Implications for conservation policy in eastern Africa. Nomadic Peoples 7:74-91.

McCabe, J. T., S. Perkin, and C. Schofield. 1992. Can conservation and development be coupled among pastoral people? An examination of the Maasai of the Ngorongoro Conservation Area, Tanzania. Human Organization 51:353-366.

McCabe, J. T., N. Mollel, and A. Tumani. 1997. Food security and the role of cultivation. Pages 397-416 In D. M. Thompson, editor. Multiple land use: The experience of the Ngorongoro Conservation Area, Tanzania. IUCN, Gland, Switzerland and Cambridge, UK.

Misana, S. B. 1997. Vegetation change. Pages 97-110 *In* D. M. Thompson, editor. Multiple land use: The experience of the Ngorongoro Conservation Area, Tanzania. IUCN, Gland, Switzerland and Cambridge, UK.

Moehlman, P. D., V. A. Runyoro, and H. Hofer. 1997. Wildlife population trends in the Ngorongoro Crater. Pages 59-70 *In* D. M. Thompson, editor. Multiple land use: The experience of the Ngorongoro Conservation Area, Tanzania. IUCN, Gland, Switzerland and Cambridge, UK.

Norton-Griffiths, M. and C. Southey. 1995. The opportunity costs of biodiversity conservation in Kenya. Ecological Economics 12:125-139.

Parkipuny, M. S. 1997. Pastoralism, conservation and development in the Greater Serengeti Region. Pages 143-168 *In* D. M. Thompson, editor. Multiple land use: The experience of the Ngorongoro Conservation Area, Tanzania. IUCN, Gland, Switzerland and Cambridge, UK.

Perkin, S. L. 1997. The Ngorongoro Conservation Area: Values, history and land-use conflicts. Pages 7-18 *In* D. M. Thompson, editor. Multiple land use: The experience of the Ngorongoro Conservation Area, Tanzania. IUCN, Gland, Switzerland and Cambridge, UK.

Polasky, S., J. Schmitt, C. Costello and L. Tajibaeva. In press. Larger-scale influences on the Serengeti ecosystem: national and international policy, economics and human demography. *In* A.R.E. Sinclair and C. Packer, editors. Serengeti III. University of Chicago Press, Chicago.

POLEYC (Policy Options for Livestock-based livelihoods, and Ecosystem Conservation). 2002. Annual Report to the Global Livestock Collaborative Research Support Program. University of California, Davis, California.

Poole, R. M. 2006. Heartbreak on the Serengeti. National Geographic 209:2-29.

Potkanski, T. 1997. Pastoral economy, property rights and traditional mutual assistance mechanisms among the Ngorongoro and Salei Maasai of Tanzania. International Institute for Environment and Development (IIED) Drylands Programme, London.

Potkanski, T. 1999. Mutual assistance among the Ngorongoro Maasai. Pages 199-217 *In* D. M. Anderson and V. Broch-Due, editors. The poor are not us: Poverty and pastoralism in Eastern Africa. Ohio University Press, Athens, Ohio.

Reid, R. S., P. K. Thornton, and R. L. Kruska. 2004. Loss and fragmentation of habitat for pastoral people and wildlife in East Africa: concepts and issues. African Journal of Range and Forage Science 21:171-181.

Ritchie M. E., M. Coughenour and C. Costello. In press. Global environmental changes and their impact on the Serengeti. *In* A. R. E. Sinclair, C. Packer S. A. R. Mduma and J. M. Fryxell, editors. Serengeti III. Human impacts on ecosystem dynamics. University of Chicago Press, Chicago.

Rwambo, P., J. Grootenhuis, J. De Martini, and S. Mkumbo. 2000. Assessment of wildlife and livestock disease interactions in the Ngorongoro Conservation Area of Tanzania. Report to the Global Livestock CRSP, University of California, Davis, California.

Seno, S. K. and W. W. Shaw. 2002. Land tenure policies, Maasai traditions, and wildlife conservation in Kenya. Society and Natural Resources 15:79-88.

Shivji, I. G. 1998. Not yet democracy: reforming land tenure in Tanzania. International Institute for Environment and Development (IIED), London.

Shivji, I. G. and W. B. Kapinga. 1998. Maasai rights in Ngorongoro, Tanzania. International Institute for Environment and Development (IIED) and Land Rights Research and Resources Institute, London.

Simms, A. and H. Reid. 2005. Africa: up in smoke? International Institute for Environment and Development (IIED), London.

Sinclair, A. R. E. 1995. Serengeti past and present. Pages 3-30 *In* A. R. E. Sinclair and P. Arcese, editors. Serengeti II: Dynamics, management, and conservation of an ecosystem. University of Chicago Press, Chicago.

Talle, A. 1999. Pastoralists at the border: Maasai poverty and the development discourse in Tanzania. Pages 106-124 *In* D. M. Anderson and V. Broch-Due, editors. The poor are not us: poverty and pastoralism in Eastern Africa. Ohio University Press, Athens, Ohio.

Tenga, R. 1998. Legislating for pastoral land tenure in Tanzania: The draft land bill. Internet document. http://www.whoseland.com/paper8.html.

Thompson, D. M. 1997. Integrating conservation and development. Pages 7-18 *In* D. M. Thompson, editor. Multiple land use: The experience of the Ngorongoro Conservation Area, Tanzania. IUCN, Gland, Switzerland and Cambridge, UK.

Thompson, L. G., E. Mosley-Thompson, M. E. Davis, K. A. Henderson, H. H. Brecher, V. S. Zagorodnov, T. A. Mashiotta, P. N. Lin, V. N. Mikhalenko, D. R. Hard, and J. Beer. 2002. Kilimanjaro ice core records: Evidence of Holocene climate change in tropical Africa. Science 298:589-593.

Thornton, P. K., K. A. Galvin, and R. B. Boone. 2003. An agro-pastoral household model for the rangelands of East Africa. Agricultural Systems 76:601-622.

Western, D. and D. L. M. Nightingale. 2002. Environmental change and the vulnerability of pastoralists to drought: A case study of the Maasai in Amboseli, Kenya. United Nations Environmental Program UNEP.Net, Africa.

World Food Program (WFP). 2000. Kenya's drought: No sign of any let up. WFP, Rome, Italy www.wfp.org/newsroom/in_depth/Kenya.html.

World Resource Institute (WRI). 2000. World resources. Oxford University Press, New York.

Zaal, F. and Dietz, T. 1999. Of markets, meat, maize and milk: pastoral commoditization in Kenya. Pages 163-198 *In* D. M. Anderson and V. Broch-Due, editors. The poor are not us: poverty and pastoralism in Eastern Africa. Ohio University Press, Athens, Ohio.

Chapter 12

NORTH-WEST PROVINCE, SOUTH AFRICA: COMMUNAL AND COMMERCIAL LIVESTOCK SYSTEMS IN TRANSITION

Kathleen A. Galvin[1,2], Randall B. Boone[2], Philip K. Thornton[3,4], and Linda M. Knapp[1,2]

[1]*Department of Anthropology, Colorado State University, Fort Collins, CO 80523, USA;* [2]*Natural Resource Ecology Laboratory, Colorado State University, Fort Collins, CO 80523, USA;* [3]*International Livestock Research Institute, Nairobi, Kenya;* [4]*Institute of Atmospheric and Environmental Sciences, School of Geosciences, University of Edinburgh, Edinburgh, UK*

1. INTRODUCTION

South Africa is a nation rich in natural resources, from diamonds and gold, to the world-famous fauna in Kruger and other national parks, to areas of extreme diversity in flora (the southern coast of South Africa is rich in plant species) (Kemper et al. 1999). South Africa is also home to 43 million hectares of savanna, contributing to Africa possessing the world's highest proportion of these savannas (Hudak 1999). But what is the state of South Africa's savanna ecosystems? Multiple studies show that land degradation in the country as a whole is extremely high (Dean and Macdonald 1994, Snyman 1998, Hoffman and Todd 2000, Dube and Pickup 2001) and that livestock carrying capacities on rangelands are decreasing (Dean and Macdonald 1994). What are the causes for this decline in rangeland sustainability and productivity? A long process of land fragmentation is certainly one reason.

Fragmentation of landscapes across South Africa occurs for various reasons. In the Cape the main causes of recent fragmentation of an indigenous

K. A. Galvin et al. (eds.), Fragmentation in Semi-Arid and Arid Landscapes: Consequences for Human and Natural Systems, 281–304.

shrubland (renosterveld) are agricultural expansion and urbanization (Kemper et al. 1999). In contrast, in the southern Kalahari savanna ecosystem of the North-West Province, a region too dry for extensive agriculture, drivers of habitat fragmentation are linked to processes of land tenure, climate and livelihood strategies. However, the ultimate cause of fragmentation is historical and political, that is, apartheid. Drought, common in the region, exacerbates the problem of an already vulnerable system that was ultimately fragmented through politically-induced racial discrimination leading to an uneven distribution of and privatization of land (Vogel and O'Brien 2004).

The Republic of South Africa is divided into nine provinces of geo-political and administrative units, which are then sub-divided into districts. The North-West Province, located in north-central South Africa, had 28 districts. Since 2005 administrative boundaries and their names have changed (Statistics South Africa 2001); however we report here information from the districts as they were in 2000 when we conducted research to assess drought coping strategies and use of climate forecasts of commercial and communal farmers (Hudson 2002, Boone et al. 2004, Thornton et al. 2004). The five western-most districts, Vryburg 1, Vryburg 2, Ganyesa, Kudamane, and Taung, which encompass our study area, comprise nearly half of the province's land area (Hudson and Vogel 2003) (Figure 12-1). These five districts are known as the western region of the North-West Province and can be classified as a savanna environment (Schultze 1997, Hudson 2002). Also considered part of the southern Kalahari ecosystem, the western region does not have enough rainfall, surface water, or fertile soils to sustain large-scale crop production (Tyson 1986, Vogel 1994, Cowling et al. 1997, SADA 1999). Thus, the most prevalent livelihood strategy for people in these five districts is livestock production (mainly cattle, sheep and goats) with very limited amounts of irrigated agriculture (Hudson 2002, Boone et al. 2004). The area provides a large proportion of South Africa's livestock production (SADA 1999). There are two major types of farms in this region, commercial farms and communal farms, the first usually occupied by white farmers, the latter operated by small-scale, black farmers. In this chapter we look at land fragmentation in the North-West Province within the context of apartheid, both for communal and commercial farmers.

In the following sections we first establish the historical setting of land use and land tenure in South Africa due to apartheid and second, we determine the consequences of apartheid on policy, labor and population. This is followed by an in-depth look at our study site, to first, look at the history of land use in the North-West Province through the era of apartheid; second, to examine natural resources and climate, especially drought, in the

Figure 12-1. North-West Province, South Africa, with the five commercial and communal districts studied identified. The location of the province within South Africa and the country within the continent shown in insets.

region; and third, to examine current land use, both commercial and communal in our study sites in the province. Finally, we explore the effects of apartheid on the environment and on how people have responded to fragmentation of their environment.

2. A RACIALLY FRAGMENTED SYSTEM OF LAND TENURE AND LAND USE: THE ERA OF APARTHEID

The ultimate agent of change of land use in South Africa within the last 300 years has been policy that has racially segregated land (Christopher 1986). Apartheid is the term used to describe a system of land use planning in which political power, as well as most of the country's valuable natural or economic resources, was left in a small (white) minority of the South African population. Another goal of apartheid was to protect European culture and identity (Jones 1999). In 1864 the British Colonial government was already beginning a system of racial segregation by forcing Africans onto reserves in 42 different locations on 12.5 million acres. Africans who continued to live on "Crown Land" were forced to rent land from whites. The British intent was to create space and land availability for whites within the colony (Thompson 2001). In 1913 the separation of land fragments for different nationalities became legal under the new *Union of South Africa's Land Act*. These allotted fragments of land, called homelands, were for ten different linguistic groups of black South Africans (Christopher 1986).

Apartheid was officially institutionalized as the Afrikaner National Party gained control of South Africa's government in the 1948 election (Deegan 2001). One of the new legislature's early reforms was to impose the *Group Areas Act*, a measure designed to geographically separate ethnic groups. Mass removals forced blacks and colored peoples onto fragments of land exclusively reserved for people of their same language and ethnic group. These parcels of land were referred to as 'Bantustans' or 'homelands' (Deegan 2001). People themselves were officially classified into racial categories (i.e., 'white', 'colored', 'Asian', and 'native') under the *Population Registration Act of 1950* (Deegan 2001). In 1960, black and colored Africans lost their rights to representation in parliament (Thompson 2001) and, under the guise of gaining citizenship in their specific homelands, black Africans lost their citizenship to the Republic of South Africa with the *Bantu Homelands Citizenship Act in 1970*. Many other stipulations, such as restricttions on how long they could visit urban areas (72 hours) or where they could find labor, prevented non-white South Africans from enjoying the bounty of their natural-resource-rich country (Deegan 2001).

The Tswana homeland of Bophuthatswana (in today's North-West Province) was granted 'independence' in 1977 and consisted of seven different fragments of land totaling 40,000 sq. km (Bophuthatswana no date). Between 1968 and 1971 alone 79,000 Tswana were unwillingly forced to move into Bophuthatswana (Jones 1999). The South African government's motives for granting so-called self-governance to each homeland were ambiguous. They cloaked their rhetoric in altruistic terms of preserving tradition and cultural identity as well as bringing about modernization and reform. Realistically, all they accomplished was the furthering of white control and the entrenchment of black dependency upon outside white work (Hudak 1999).

The apartheid system began to crumble in the 1980s as blacks were given back their citizenship and internal migration laws were loosened, due to international and domestic pressures, allowing blacks the right to own property in urban areas once more (Christopher 1986). Apartheid was officially ended in 1994 with the first nonracial election resulting in the African National Congress' victory and Nelson Mandela's appointment as president of the Republic of South Africa (Thompson 2001). The new government then faced the enormous task of rebuilding and reforming a fragmented society and landscape. Among many other problems, the new government had a country with high economic disparity between ethnic groups. The 1996 census revealed that most white South Africans were "well-to-do, well educated, and well housed" (Thompson 2001:266). Most black South Africans, on the other hand, were "poor, badly educated, and ill housed" (Thompson 2001:266). The legacy of apartheid is being felt today though the country is in a process of transition.

3. CONSEQUENCES OF APARTHEID

3.1 Government policy

During the latter half of the 20th century there were 80 Acts of South Africa's Parliament that economically helped commercial farmers (Hudson 2002). These policies have impacted market accessibility, government supported land/soil conservation schemes (such as subsidies), and access to labor for farmers - all variables that affect the livelihood strategies adapted by communal and commercial farmers. While communal farmers are still feeling the impacts of a government that historically helped commercial farmers alone, the change of government in 1994 has tried to some degree to reverse the trend so that communal farmers can also benefit from soil conservation, fair market prices, private land tenure, and climate forecasts, among other transitional opportunities (Hudson 2002).

For decades government policies regarding market access generally benefited commercial farmers and hindered communal farmers. The *Marketing Act of 1931* restricted communal farmers from accessing markets outside their homelands (Simbi 1998). The rationale behind this Act was to diminish the competition from black farmers and to create a large labor force for the mining industry and white agricultural sector (Piesse et al. 2005). Since the demise of apartheid, the deregulation of agricultural markets has occurred and communal farmers' livelihood strategies are becoming more heavily dependent on the market economy (Hudson 2002).

Government concern over soil erosion began in the 1920s and its involvement in preventing it intensified in the 1930s (Delius and Schirmer 2000). Hoffman and Todd (2000) believe that the country's approach to dealing with land degradation has been hindered by two aspects of its policy: 1) commercial areas have gotten more attention than communal areas and 2) the overarching paradigm links political and ecological perspectives. The government's underlying approach was based on racist ideals (Delius and Schirmer 2000).

> Racism rather than research was central to the way in which conservation was being appropriated. Numerous speakers in Parliament argued either that white farmers were being forced by economic circumstances to denude the soil or that their land was being destroyed by black tenants. In contrast to whites...[black] culture or their innate backwardness led them into harmful farming methods. The outcome of this racist logic was clear. If white farmers could be helped economically, they would improve their farming practices. Black farmers would not respond to such inducements. They would either have to be forced into the correct farming methods or forced out of farming altogether. In this way, politics, racism and populism combined to turn soil conservation policies into something that would benefit all white farmers (Delius and Schirmer 2000:735).

The underpinnings of racism led policymakers to reorganize black rural society under the following reforms: stock reductions, fencing of lands, concentrated settlements, and increased agricultural education. It was acknowledged by the government that black farmers needed more land, but blacks were disappointed when their reserve enlargements mostly encompassed land that tenant farmers were already using. Furthermore, there was an increase in government-enforced cattle culling and agricultural plot reductions (Delius and Schirmer 2000). In sharp contrast, white soil conservation schemes were less coercive and centered on providing financial assistance. Subsidies were granted in the 1930s and 1940s for helping white farmers build fences, houses or dams (Hudson 2002). In addition, weed and cacti eradication, cattle breeding improvements, and increased fodder production were implemented (Delius and Schirmer 2000).

Government policies were such that white farmers, regardless of their efforts at conservation, were compensated for loss of stock during droughts (Vogel 1994). This led to more farmers wanting to live in drought-prone environments and over-stock their pastures so that their losses during drought could result in greater allotments of funds. “A curious anomaly is that land prices in South Africa, in relation to productive capacity, are relatively higher in areas where droughts occur frequently, suggesting that the potential receipt of financial aid, rather than agricultural potential, is an important factor in farming of such areas” (Dean and Macdonald 1994:291). Before the end of apartheid, commercial farmers received far greater drought relief subsidies than communal farmers. In 1992-93, sixty-four percent of drought relief funds were allocated to white commercial farmers and only eight percent was given to black communal farmers (Hudson 2002). Following the 1994 election there have been no more drought relief programs. The emphasis is shifting to more long-term sustainable policies.

3.2 Access to labor

Communal farmers of South Africa have had limited access to outside labor over the last 150 years. In 1860, over 83% of white-owned farms were worked by indigenous tenants (Piesse et al. 2005). Tenant farmers soon became too competitive for the white farmers because their costs of production were lower (Shillington 1995). White farmers' complaints about a lack of cheap labor resulted in the *Masters and Servants Acts of 1911* and *1932*, prohibiting blacks from breaking contracts or changing employers (Simbi 1998). A further detriment to the black labor cause was the *Native Land Act of 1913,* preventing labor tenancy or sharecropping and requiring blacks to only own land on reserves (Thompson 2001). Access to labor in urban areas became more difficult as well. The *1964 Bantu Labor Act* prevented blacks from seeking work in towns and prevented white employers from hiring them. This policy was aimed to keep black South Africans from living in "white cities" or "white towns" (Deegan 2001). The main source of income for blacks living in the homelands continued to be the remittances sent back from migrant workers or government subsidies (Christopher 1986).

Today, in the five districts of our area of study, most communal farmers receive income from outside the livestock industry; however, commercial farmers receive (on average) ten times more income than communal farmers from these outside sources (Hudson 2002). With inconsistent access to outside labor throughout history, communal farmers have had limited means for generating other forms or adequate amounts of income. Thus, as we will see, the scope of their farming operations is significantly smaller than commercial farmers' (Hudson 2002).

3.3 Population

Human population growth rates exploded on rural black homelands as a result of the apartheid system. Blacks were not welcome in urban areas but were forced to move to fragments of rural land. Colonel Stallard's 1923 dictum expressed the white opinion that "the town is a European area in which there is no place for the redundant native" (Christopher 1986:332). It was this politically-enforced racial fragmentation that contributed to over-crowding and over-grazing on communal lands. By 1941 there was a four times greater human population density in homelands than there was in white-controlled areas (Hudak 1999).

Today, on the other hand, South Africa has a rapidly increasing urban population and the rural-to-urban migration is expected to continue to grow. Compared to other countries in the Southern African Development

Community (SADC), the Republic of South Africa has a much greater percentage of its population (50% versus 25-30%) living in urban areas (Vogel and O'Brien 2003). However, disease, particularly the HIV/AIDS pandemic along with other health concerns (such as malaria and cholera) is creating rapid changes to the age and sex structure of the population, resulting in serious costs to society in both the rural and urban areas. South Africa had 4.2 million cases of AIDS or about 25% of the black population was infected. This number has been higher than any other country in the world (Vogel and O'Brien 2003). By 2005 South Africa continued to have the largest number of people living with HIV/AIDS in the world (Shisana et al. 2005).

In the five western districts of the North-West Province, population densities in communal and commercial districts are relatively low and are 1 person/4 ha. Communal farms support 75% more family members than commercial farms. When permanent workers as well as the farm family are considered, however, commercial farms are shown to support four times as many people as the communal farms (Hudson 2002). As can be seen from the above description, commercial and communal farmer livelihoods were affected very differently by the processes of apartheid. We now look specifically at farmers in the North-West Province.

4. HISTORY OF LAND USE IN THE NORTH-WEST PROVINCE

4.1 Early inhabitants of the north-west region

South Africa has had nearly every type of land use from hunting and gathering, to livestock grazing, extensive and intensive farming, mining, commercial forestry, urbanization, and land set aside for conservation (Hoffman and Todd 2000, Thompson 2001). Before the 16th century, South Africa had little outside human influence due to the flow of ocean currents and the lack of natural harbors along the country's coastline. Pastoralism spread from eastern Africa to southern Africa sometime before the first century A.D. It began as an extension of a hunting-gathering way of life but eventually led to a new classification of people known as the Khoikhoi (or in Afrikaans, the Hottentots). Khoikhoi culture diverged from their Khoisan (Bushmen) neighbors in that they began to acquire wealth in the form of livestock, which they first obtained through trading in the area of present-day Botswana (Thompson 2001). Today, southern Africa's pastoralists include the Nama people (Khoisan speakers) as well as the

Bantu-agropastoralist tribes of the Tswana, Zulu, Herero, and Swazi (Fratkin et. al. 1994).

The movement of the Bantu-speaking peoples into southern Africa began in the 4th century A.D. and by the late 18th century these agriculturalists occupied nearly the entire mesic eastern region of southern Africa. The Bantu migrated in small groups, not massive waves, originating from West Africa, then spreading to the Great Lakes region of East Africa and finally reaching South Africa. Each new pocket of immigrants shared similar language, metallurgy, pottery, settled or semi-permanent village life, complex political organization based on chiefdoms, mixed farming (which included livestock), a varied diet, and a system of trade. South of the Drakensburg Escarpment the Bantu tribes of the Xhosa and Zulu settled, while north of the escarpment were the Tswana (the ancestors of the indigenous inhabitants of the North-West Province today), Sotho, and Pedi. It is believed that the Tswana arrived during two separate migrations sometime between 1300-1400 A.D. (Wilson and Thompson 1969).

4.2 Colonial settlers

In the mid-1600s settlers of European descent began to arrive in southern Africa (at that time generally occupied by hunter-gatherers and pastoralists in the west, and Bantu agro-pastoralists in the east). Colonists of Dutch descent (Afrikaaners or Boers) settled along the Cape of Good Hope but then were followed and eventually conquered (in 1795 and again in 1806) by the British. As European settlers were pouring in along the south-western coast, trouble began between Bantu tribes. Up until the early 19th century, South Africa's indigenous peoples had not engaged in warfare. The expansion of the Zulu kingdom during the *Mfekane* Wars (1816-1828), however, caused widespread chaos and people escaped from their homes, hid in the bush, and eventually fled to the white-dominated Cape region in search of subsistence from white employment. This period of intra-Bantu warfare not only began black dependency on white populations, but also led white settlers to believe the eastern region of South Africa was unclaimed and uninhabited (since large populations had fled their homes). Shortly thereafter, the famous "Great Trek" (1835-1840) brought thousands of Afrikaaners in search of freedom from British control, to the supposedly newly available land north of the Orange River and along the Vaal River (in the present-day Free State Province). The Afrikaaners' defeat of the Zulu kingdom in 1843 led to their expansion into the "Natal" area and into the north-central grasslands of the Tswana, an area known today as the North-West Province (Thompson 2001).

The Tswana people had only just begun to come out of hiding after the tragedy of the Mfekane Wars when the Afrikaaners began to settle on their land. The British government vacillated greatly in its policy toward protecting Boer (Afrikaaner) versus black African (Tswana) rights. Ultimately, it gave the Boers political power against the indigenous Africans in 1854 (Thompson 2001). With a history of their own intra-tribal skirmishes, the Tswana were unable to unite against the invading Europeans. Instead, they were forced to pay rent for living on "white land" and many of their children were stolen in order to become "apprentices" (that is, slaves) to colonial settlers (Thompson 2001). The last attempts to rebel against European control by the southernmost Tswana came in 1896-97; their attempts were unsuccessful and the subsequent rinderpest epidemic that killed their livestock left many dependent on white farmers. Among the northern Tswana, the British colonial government was more helpful in preventing Afrikaaner control over traditional chiefdoms, but like their southern neighbors, the northern group was hard-hit by rinderpest and was forced to send young men off to earn wages in the cities.

4.3 North-West Province land tenure policy

Land tenure policy has been the dominant form of land fragmentation in the North-West province. Farmers of European descent embraced land tenure policies and livelihood strategies that differed from the systems and strategies of Tswana farmers. Apartheid led to an uneven distribution of *quantity* and *quality* of land. Thus, exclusivity of resource use was based on ethnic classifications and resulted in the oppressed ethnicities having little control over what land they were given access to. Decision-making rights and access to resources over space were bequeathed to farmers of European descent alone.

In 1854, following the Great Trek, Afrikaaner settlers in the *Highveld* of the Tswana land began to hold their wealth in the form of cattle. This was similar to the economic system of the Bantu agro-pastoralists except that Afrikaaners owned their land individually (Thompson 2001). The early European government in what is now the North-West Province initially had little monetary capital and reverted to paying their government officials with large chunks of land. These vast lands were often so large that individual white farmers were not able to use all the land they were given. As a result, white farmers hired black labor or rented out portions of their land to black tenant farmers (Piesse et al. 2005).

In Bantu cultures of pre-colonial South Africa (including the Tswana of the North-West Province), there was no concept of individual ownership of land. All land was shared by a community and used for hunting, grazing livestock, and gathering plants. Outside the village, land was granted for individual women to plant and harvest crops; however, during the winter months these fields became open-access for all community members' livestock to be grazed (Thompson 2001). Tswana polity was originally broken down into small individual tribes connected by a common chief.

Today, two districts in the North-West Province, Vryburg 1 and Vryburg 2, are solely populated by commercial farmers of European descent. There are no black land owners in these two districts. The only blacks living in Vryburg 1 and 2 hold labor or service jobs in towns or work on white farms (Hudson 2002). Commercial farms are dominated by a land tenure system of private ownership or 'freehold tenure' where farmers have documented proof of ownership. This entitles farmers to sell their land (Hoffman and Todd 2000). Commercial farms are demarcated by fences which prevent the traditional movements of livestock and wildlife across the rangeland (Hudak 1999).

Three other districts in the western region, Taung, Ganyesa, and Kudumane, are communal farming districts and each formerly belonged in the Bopthuthatswana homeland. The majority of inhabitants in these districts are farmers of Tswana descent. Communal farmers do not own the land that they farm, nor do they have the legal rights to sell or buy this land. Technically the land of communal farms is owned by the state but administered by local authorities. The forms of management on communal farms usually fall under one of the following systems: traditional (chiefs), democratically elected officials, a combination of the two, or an open access system (Hoffman and Todd 2000). Under apartheid, communal tenure and chieftanship were viewed by blacks as important strategies for buffering the total authority of whites over their land (Delius and Schirmer 2000). Today each village governs its communal grazing land differently—from autocratic leaders to complex social frameworks. Within some of these systems there is no limitation on the amount of livestock each farmer can have (Hudson 2002).

South African Development Trust (SADT) farms are one of the few examples of communal farmers privately owning their farms. SADT farms were started in 1992 under a government initiative of economic development for black farmers. The farms, roughly 2,500 ha in size, were able to be purchased by individual farmers after seven years of leasing and environmentally-conscientious management of the land (Hudson 2002). Another anomaly within communal lands is the BOP 4-40 plan which leased 40 ha of pasture to groups of four farmers. The goal of the BOP 4-40 plan was to increase economic capabilities of communal farmers (Hudson 2002).

5. NORTH-WEST PROVINCE NATURAL RESOURCES AND CLIMATE

The five districts of the North-West Province in this study are part of the southern Kalahari ecosystem and specifically classified as Kalahari thornveld and shrub bushveld (Acocks 1975). The most prevalent soils in this region are red-yellow apedal soils with glenrosa and Mispah soils in southern Vryburg 2, Taung, and northern Ganyesa (SADA 1999). The Kalahari is usually labeled as a 'desert,' though it is really semi-arid or arid savanna as numerous shrubs and grasses are the norm. Marked by a hot climate with a low annual rainfall, flat topography, and high evaporation rates, the savanna of the North-West Province lacks sufficient amounts of surface water suitable for crop or livestock production (Cowling et al. 1997). Across all five districts of the western region, the most common source of water for farms is from boreholes, or wells, that access the relatively high water table (Hudson 2002).

The greatest determinant of productivity within a savanna ecosystem is rainfall levels (Hudak 1999). Rainfall acts as the limiting factor during dry periods but Snyman (1998) also cites soil nitrogen as a limiting factor for southern Africa rangeland productivity during periods of above-average rainfall. Savannas are amazingly resilient to disturbance such as fire, grazing, and drought, but after decades of over-grazing and fire suppression, some areas of the southern Kalahari are showing signs of vulnerability (Hudak 1999).

Across the North-West Province, inter-annual and intra-annual rainfall is highly unpredictable and varies both spatially and temporally (Hudson and Vogel 2003, Vogel and O'Brien 2003, Boone et al. 2004). There is usually one rainy season between the months of November and April (Vogel and O'Brien 2003). The more mesic eastern section (Vryburg 1 and 2) of the five districts in our study area receives approximately 500 mm of annual rainfall while the western-most area (along the Kalahari) receives about 300 mm annually (Boone et al. 2004). Drought is a common phenomenon across southern Africa with some droughts occurring only locally while others affect vast areas (Vogel 2002). The causes for droughts and/or rainfall variation are often, although not always, correlated with the El Niño Southern Oscillation (ENSO) phenomena, rising sea surface temperatures, and convectional activity between tropical and mid-latitude weather systems (Vogel and O'Brien 2003). ENSO-induced droughts often result in severe impacts. The 1992/93 El Niño brought devastating drought effects to South Africa while the 1997/98 El Niño affected the country far less than in other areas of the continent or the world, where for instance, floods occurred in East Africa (Hudson 2002, Galvin et al. 2004).

Effects of drought in the southern Kalahari historically include reduced forage production, crop yields, and livestock and wildlife populations (Boone et al. 2004). Effects upon vegetation include changes in species composition and an increase in poisonous plants (Cowling et al. 1997). Drought often occurs undetected by humans unless its implications impact socio-economics such as decreased employment, income, nutrition and health (Glantz 1994). Vogel et al. (2000) list multiple ripple effects that resulted from South Africa's droughts between 1980 and 1994 including migration, poverty, unemployment, malnutrition, loss of biodiversity, permanent loss of biological productivity of the landscape, reduced market prices for livestock, and increased financial debt.

Though drought has caused serious socio-economic and environmental disasters in South Africa, it is acknowledged that the intensity of drought effects are greatly enhanced by human actions and are, therefore, not just an outcome of decreasing rainfall (Beinart and Coates 1995, SADA 1999, Hudson 2002). Vogel and O'Brien (1994:153) explain that, "...the severity of drought impacts [in South Africa] has been more a consequence of the mishandling of drought situations, farm management, and agricultural systems in the country than a consequence of a reduction in rainfall." Generally it is people living on marginal land in underdeveloped regions that suffer the worst consequences of drought (Hudak 1999).

6. THE NORTH-WEST PROVINCE STUDY AREA

The political history of South Africa has led to a system of land privatization in which a minority of the population held control of vast sums of land. White commercial farmers, who comprise 28% of the rural population, control 88% of the agricultural land (Hudson 2002). Our sites in the North-West Province show similar patterns. The commercial and communal farmers have contrasting economic goals for their farms and therefore also different livestock production and range management schemes. This impacts the way they respond to recommended stocking rates (determined by SADA) and how they cope with drought. Grazing capacity is the number of livestock that can be supported on a farm without causing long-term degradation to the rangeland (Boone et al. 2004). In 1999, the eastern Vryburg districts' grazing capacity was set at 7 ha/TLU (large stock units), while northern Ganyesa was 25 ha/TLU and central Taung was 30 ha/TLU (SADA 1999). This variation in carrying capacities is thought to be appropriate based on the rainfall gradient across the province and/or levels of bush encroachment and rangeland degradation over time (Figure 12-2).

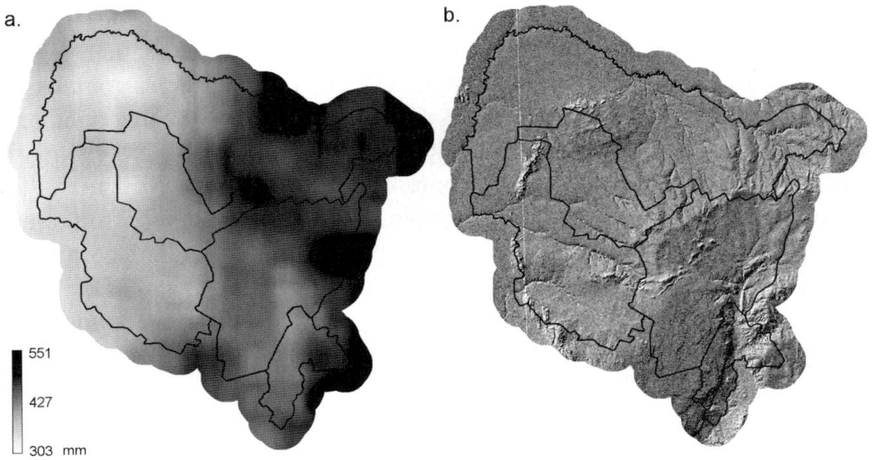

Figure 12-2. Mean (a) annual precipitation from 2001 to 2005 in the western portion of the North-West Province, South Africa, and (b) topographic variation. Precipitation is from estimates of rainfall based on satellite images (Xie and Arkin 1997), and topography from space-borne radar (SRTM 2004).

The livestock carrying capacity of a farm is an indicator of the nature and condition of the rangeland (Snyman 1998). Stocking rates affect how vulnerable the landscape is to environmental disturbance, the long-term vegetation composition, animal performance, and the economic advantages for the farmer. The stocking rates on commercial farms are about 9 ha/TLU and commercial farmers are thought to be conservative about stocking levels on their land (Table 12-1).

A paradox for communal lands is that only a segment of the land is used, so large parts of it are left under-grazed. It is being under-utilized because some pastures are too far from the boreholes. Grazing on communal land is generally restricted by access to fodder within a certain range of water points. Stocking rates on communal lands in the study area is almost 16 ha/TLU and communal farmers tend to overstock their range (Table 12-1).

Table 12-1. Farm statistics from Hudson (2002).

	Commercial Farm	Communal Farm
Average size of farm	3,000 ha mode and median (while mean equals 4,100 ha)	476 ha – 714 ha (based on estimates)
Stocking rates Grazing capacity is linked to rangeland condition (Snyman 1998)	8.6 ha/TLU (indicates higher quality land)	15.8 ha/TLU
Farmers' perception of pasture condition	64% said pastures are in "good condition"	51% said pastures are in "good condition"

6.1 Commercial farmers' management goals

Commercial farmers' primary economic goal is to sell livestock on the commercial market (Hoffman and Todd 2000). Objectives include profit maximization or maintaining particular income levels, to utility maximization (i.e., accounting for risk attitudes), and quite possibly for tax minimizing purposes too (Thornton et al. 2004). To attain these goals, commercial farmers, within the constraints of a drought-prone environment, will generally maintain a constant herd size to prevent too much over-grazing of the landscape regardless of levels of rainfall (Boone et al. 2004). The SADA holds that commercial farmers will usually under-stock their farms and this is a good strategy, in terms of maximizing profits over the long term. Thornton et al. (2004) used economic modeling linked to an ecosystem model to estimate the economic value of climate forecasts and showed long-term average income to increase with conservative stocking. A smaller herd size ensures minimal loss (in numbers and in condition of livestock) during drought so that fewer animals have to be sold during drought when prices are low, and better conditions mean more rapid herd recovery when grass starts to grow again. Ninety-two percent of commercial farmers' livestock is cattle and the other 8% is sheep. With a goal of receiving a profit in commercial markets, these farmers invest mostly in cattle because they receive the highest market price of any commercial animal (Hudson 2002). Commercial farmers carry out their production goals through a system of rotational grazing. Eighty-eight percent of commercial farmers claimed they embraced a rangeland management system in which grazing is rotated between three or more pastures. In contrast, 51% of communal farmers rotate livestock between two or more pastures with less than 10% allowing a full growing season rest in between use (Hudson 2002).

The average (mean) commercial farms size equals 4,100 ha with the mode and median of 3,000 ha (Hudson 2002) (Table 12-1). The scale of operation for commercial farms is larger than communal farms, defined as the number of people supported by the farm, the number of animals produced, the market value of those animals, and the number of boreholes on the farm. Commercial farmers have about six times more pasture land than do communal farmers.

Some studies (Hudak 1999, Hoffman and Todd 2000, Hudson 2002) suggest that commercial farmland is not as good as farmers may perceive, although the fact that many commercial farms in our study site are somewhat under-stocked (and this being an economically efficient strategy) suggests that their perceptions, when translated into management action, are actually pretty accurate. Vogel (1994) explored how the government encouraged the spread of commercial farming onto marginal lands—areas that should never have been used for livestock. Thus, 21 million hectares of commercial grazing land have been degraded by wind and water erosion, though this would not have happened if white farmers were not engaged in over-production of the landscape (Vogel 1994). Poorly planned government schemes have led to many commercial farmers on marginal lands selling their farms to more successful commercial farmers. Therefore, there are fewer commercial farmers than before, but these remaining successful farmers have ever-increasing farm sizes, a process similar to what is occurring in the U.S. Great Plains (Lackett and Galvin, Chapter 5).

6.2 Communal farmers' management goals

Communal farmers' primary economic goal is to retain wealth in the form of livestock (Schmidt 1992). This system is focused on minimizing risk (but risk of loss, rather than production risk per se), rather than maximizing production (Thompson 2001). These farmers operate under a system that is similar (in some regards) to pastoralist peoples in East Africa. Under this system, communal farmers strive for risk-avoidance from disease or climate variability by building the herd during wet and normal years (Galvin et al. 1994, Hudson 2002, Boone et al. 2004). This provides "insurance" that at least some animals will survive in the face of drought or disease. "Communal farmers manage their herds to build security and meet short-term needs such as to pay school fees and for health care" (Boone et al. 2004:324).

Communal farmers' herd sizes are determined then by the amounts of grass production, which is ultimately determined by amounts of rainfall (Hudson 2002). Thus, during drought, herd sizes tend to decrease. Following the 1991-92 El Niño, 243,000 cattle and 101,000 small stock died on communal lands throughout South Africa due to lack of forage or from poisonous plants (Hudson 2002). Goats comprise 40% of communal farmers' livestock holdings while only 33% of communal farms' livestock are cattle and 27% are sheep (Hudson 2002). This reflects the differing economic or production goals of the two farming groups as well as farm sizes. Goats are resilient in drought and therefore can act as a buffer from the devastating decrease in overall herd size as communal farmers seek to retain wealth in their herds. Another drought-coping strategy utilized by about a third of communal farmers is the purchase of fodder. Only 3% of farmers decreased their herd size as a drought management strategy (Boone et al. 2004).

No one, not even South Africa's Department of Agriculture, is certain of the size of communal farms at our sites. Estimates suggest the average communal farmers' pastureland is between 476 ha and 714 ha (Hudson 2002) (Table 12-1). Few communal farmers wish to know how much land their livestock use for fear that if their communal lands are surveyed, measured, and fenced, then grazing restrictions could be enforced and further grazing land would be lost.

Land and soil degradation levels are higher in communal than commercial areas of South Africa but this is not necessarily correlated with land tenure (Hoffman and Todd 2000). It is suggested that socio-economic variables, such as number of household dependents or unemployment levels, and bio-physical variables (i.e., steeper slopes or higher annual temperatures) are the underlying causes of land degradation. In South Africa, the tendency is to blame communal farmers for over-grazing and over-stocking their land. Rather, it may be that decades of racial land fragmentation is the cause for an overall decrease in the quality of communal land.

6.3 Water

In commercial lands, farmers have had the financial flexibility and political clout to have many boreholes constructed, which are water sources for their livestock. In contrast, the communal areas have very few working boreholes. Multiple boreholes allow farmers to rotate animals among pastures to distribute forage off-take more evenly. The dense patterning of boreholes provide commercial farmers some buffer against fragmentation as farms are subdivided, which appears to be happening in the past decade based on satellite imagery (Boone et al. 2007). Communal farmers graze their animals in areas near boreholes and village centers. Those areas are heavily grazed, while areas more distant from available water have vegetation that goes unused by livestock.

6.4 Wildlife conservation areas

Within the North-West Province there are two areas set aside for the dual purpose of wildlife conservation and tourism. These protected areas include the Madikwe Game Reserve and Pilanesberg National Park. Madikwe is famous for being the site of the largest project for the relocation of game, including leopard, wild dog, rhino, and buffalo. Madikwe, comprised of 60,000 ha of veld, was previously over-grazed white farmland that was given to the Bophuthatswana (BOP) Homeland Government. The BOP government decided to restore this area to its original natural state while also opening it up for tourism (Hudak 1999).

7. CONCLUSIONS

We have seen thus far that the policies of apartheid have had major consequences for land tenure, land use, and thus livelihood strategies of both commercial and communal livestock owners in South Africa. It also has had consequences for ecosystem structure and function and thus resources. The different constraints imposed by historical apartheid policies, and the resulting land fragmentation impacts, are further exacerbated by the impacts of climate variability. Similarly, urban population growth and the effects of the HIV/AIDS pandemic within the context of land fragmentation and marginalization have created a fragile population. These issues are discussed below.

7.1 Ecological consequences of fragmentation

In the semi-arid rangelands of the southern Kalahari surface water is limited. Commercial farmers build boreholes to spread out grazing over a wide area (Hudak 1999). In general, more boreholes result in over-grazing of pastures but in the case of communal farmers, the lack of boreholes means that large parts of the region go unused. Over-grazing of pastures causes bush encroachment that leads to land degradation. Dean and Macdonald (1994) suggest that the building of stock-water points leads to changes in species composition and decreased carrying capacities. Both are results of the greater livestock numbers around fixed water points depleting the availability of forage and increased hoof action degrading the rangeland.

Africa as a whole contributes 36% of the world's total land degraded by over-grazing and 49-90% of the continent's rangelands are believed to be already desertified (that is, in the process of long-term damage) (Hudak 1999). Multiple studies show that degradation of South Africa's rangelands is occurring through bush encroachment, the replacement of palatable grasses by unpalatable trees and shrubs, and changes to plant species composition (Dean and Macdonald 1994, Snyman 1998, Hoffman and Todd 2000, Dube and Pickup 2001). Essentially, bush encroachment forms an impenetrable, thorny thicket which usurps access to light, water, and nutrients from palatable species. Bush encroachment not only leads to decreased grazing capacities, but also increased transpiration (which lowers soil moisture available for grass growth), decreased calf production, and overall economic hardship for farmers (Hudak 1999). "Changes in the absolute and relative abundance of fodder plants, together with irreversible losses of topsoil and changes in infiltration rates in semi-arid and arid rangelands has reduced the biomass of domestic livestock that can be carried on...[South Africa's] rangelands" (Dean and Macdonald 1994:293). Though South Africa still maintains

43 million ha of savanna, rangeland scientists estimate that 66% of its rangelands are in "a moderate to serious phase of degradation" (Snyman 1998).

Some scientists (cf Hudak 1999) link rangeland degradation with global climate change, while others (Dean and Macdonald 1994, Vogel 1994, Hudak 1999) could not find evidence for climate change in South Africa during the past 150 years. It is not that average rainfall has decreased over time but that the efficacy of rainfall is decreasing (Dean and Macdonald 1994) leading to human-induced drought (Vogel 1994). On the other hand, the paper on climate change in Africa over the period 1900 to 2100 by Hulme et al. (2001) shows that the continent is warmer than it was 100 years ago (they mention 0.5°C); they also note that southeast Africa has had a relatively stable rainfall regime since 1900, but with marked inter-decadal variability.

Our research certainly suggests that livestock owners in both the commercial and communal districts must cope with a climate that is extremely variable and harsh. In the last 40 years, the study area has averaged about 400 mm of rainfall annually, but the year-to-year variation in rainfall is extreme (e.g., the coefficient of variation in rainfall from 1960 to 2002 was 37% while variation in rainfall from 1900 to 1960 was just 25%). Droughts occur every three to six years (Dilley 2000, NOAA 2002), and often reduce livestock conditions and populations (Ellis and Galvin 1994). Even in years with normal precipitation, patterns of rainfall can be very patchy. These changes accompany a long-term decline in the quality of ungulate forage and stocking (Dean and Macdonald 1994).

7.2 Human responses to fragmentation

The southern Kalahari ecosystem of the North-West Province was once self-sustaining. Yet whenever complex ecosystems are simplified through fragmentation, they become dependent on outside investments of monetary capital to be sustainable (Ellis and Peel 1995). As Hobbs et al. (Chapter 2) explain, habitat fragmentation in the form of land privatization is often justified by the economic gain it should provide. And yet, as natural capital is lost, other costs accrue. Economic inputs, such as the building of boreholes or buying of fodder, are necessary in climatically variable systems when access to heterogeneity of resources is lost. In the North-West Province, commercial and communal farmers alike are dealing with the consequences of habitat fragmentation across their rangelands. Communal farmers are the least equipped to deal with the consequences of habitat fragmentation because of financial constraints limiting their ability to make economic inputs into the system. For commercial farmers it is easier to obtain capital to invest in their fragments of rangeland and make them economically sustainable. Commercial farmers have a greater scope and scale of operation which provides the means to make these investments possible though the ecological consequences mentioned above still remain.

Privatization of land and its exclusive use forces farmers to rely on boreholes for the main source of water. Commercial farmers average six more boreholes per farm than communal farmers and they are able to sustain more than twice the amount of livestock per borehole because of their capacity to pipe water from a borehole in one pasture to a different pasture (Hudson 2002). This prevents over-grazing around boreholes on commercial farms. Communal farmers, however, do not have the monetary capital to make such developments possible. Therefore, it is more difficult for them to embrace rotational grazing patterns. Many of their pastures are under-utilized because of a lack of water supply in those areas. This leads to over-grazing around the few (average two boreholes per communal farm) water sources that they do have (Hudson 2002).

Significant numbers of both communal and commercial farmers seek income outside their farms. Hudson (2002) showed that 49% of communal farmers receive supplemental income. In contrast, 36% of commercial farmers receive supplemental income but their monthly average from these outside sources was more than ten times greater than the average communal farmer's income from such sources.

Finally, when asked how they could cope with one year of severe drought, only 4% of commercial farmers thought this would be difficult while 37% of communal farmers predicted it would be hard to cope (Hudson 2002). During the early 1990s, poor rural communities, as opposed to commercial farmers, shouldered the worst effects of major drought (Vogel 1994). Thus, it is argued that communal farmers—who lack the monetary capital to compensate for lost access across the landscape—feel the consequences of habitat fragmentation in the North-West Province more acutely.

High variability in precipitation demands flexibility among livestock producers in the study area. Many commercial farmers have sufficient flexibility and resources to reduce the stocking rates on their lands to levels below those recommended by the Department of Agriculture (Hudson 2002, Boone et al. 2004). They receive less overall profit from their livestock production systems, but the profit is more predictable from year to year (Thornton et al. 2004). Communal farmers do not have this flexibility. Areas grazed by communal farmers are not well defined but Hudson (2002) estimated that those areas were stocked at 125 to 150 percent above the level recommended. Whereas commercial farmers may sell stock in the face of drought, communal farmers, like many pastoral people, will retain as much stock as possible, in hopes of having the greatest number survive the drought

(Galvin et al. 1994). In general, though, communal farmers have had to diversify their livelihoods in other ways, such as seeking more outside labor. Finding work has been more difficult, however, as commercial farms have become more mechanized, and some commercial farmers have changed to less labor-intensive production schemes (Francis 2002).

South Africa is undergoing rapid change climatically, politically, and demographically. Transition to a changing world is dependent on choices people have under change (Galvin et al. 2005). Climatically, the region is becoming warmer with a net drying of soils in certain regions (IPCC 2001, Millennium Ecosystem Assessment 2005). This will affect food production, water supplies and ecosystems. HIV/AIDS has affected about 25% of South Africa's population (DOH 2003) and it has been suggested that a huge loss of the African adult population will affect human well-being, among other things, in the loss of human capital, diversion of government resources, and increasing dependency of the young on government (Harvey 2003). Politically, the country has dropped many subsidies such as market controls and drought relief. Changing land tenure has been a much slower process; it still reflects the policies of apartheid. Whether or not the livestock sector and particularly the communal livestock sector can cope with these changes is dependent on the ability and will of government and policymakers to enable them to access the resources needed to maintain their livelihoods.

REFERENCES

Acocks, J. P. H. 1975. Veld types of South Africa. Memoirs of the Botanical Survey of South Africa 40:1-128. Botanical Research Institute, Pretoria.

Beinart, W. and P. Coates. 1995. Environment and history: The taming of nature in the USA and South Africa. Routledge, London.

Boone, R. B., K. A. Galvin, M. B. Coughenour, J. W. Hudson, P. J. Weisberg, C H. Vogel, and J. E. Ellis. 2004. Ecosystem modeling adds value to a South African climate forecast. Climatic Change 64:317-340.

Boone, R. B., J. M. Lackett, K. A. Galvin, D. S. Ojima, and C. J. Tucker, III. 2007. Links and broken chains: evidence of human-caused changes in land cover in remotely sensed images. Environment, Science and Policy 10:135-149.

Bophuthatswana (n.d.). Wikipedia Encyclopedia. Retrieved on November 18, 2005 from http://www.reference.com/browse/wiki/Bophuthatswana.

Christopher, A. J. 1986. The inheritance of apartheid planning in South Africa. Land Use Policy 3:330-335.

Cowling, R. M., D. M. Richardson, and S. M. Pierce. 1997. Vegetation of Southern Africa. Cambridge University Press, Cambridge, UK.

Dean, W. R. J. and I. A. W. Macdonald. 1994. Historical changes in stocking rates of domestic livestock as a measure of semi-arid and arid rangeland degradation in the Cape Province, South Africa. Journal of Arid Environments 26:281-298.

Deegan, H. 2001. The politics of the New South Africa: Apartheid and after. Pearson Education Limited, London.

Delius, P. and S. Schirmer. 2000. Soil conservation in a racially ordered society: South Africa 1930-1970. Journal of Southern African Studies 26:719-742.

Department of Health (DOH). 2003. National HIV and syphilis antenatal sero-prevalence survey in South Africa: 2002. Republic of South Africa, Pretoria.

Dilley, M. 2000. Reducing vulnerability to climate variability in Southern Africa: the growing role of climate information. Climatic Change 45:63-73.

Dube, O. P. and G. Pickup. 2001. Effects of rainfall variability and communal and semi-commercial grazing on land cover in southern African rangelands. Climate Research 17:195-208.

Ellis, J. E. and K. A. Galvin. 1994. Climate patterns and land-use practices in the dry zones of Africa. Bioscience 44:340-349.

Ellis, J. E. and M. Peel. 1995. Economies of spatial scale in dryland ecosystems. Arid Zone Ecology Forum. Kimberly, South Africa.

Francis, E. 2002. Rural livelihoods, institutions and vulnerability in North-West Province, South Africa. Journal of Southern African Studies 28:531-550.

Fratkin, E., K. A. Galvin, and E. A. Roth. 1994. Introduction. Pages 1-15 *In* E. Fratkin, K. A. Galvin, and E. A. Roth, editors. African pastoralist systems: An integrated approach. Lynne Rienner Publishers, Boulder, CO.

Galvin, K. A., D. L. Coppock, and P. W. Leslie. 1994. Diet, nutrition, and the pastoral strategy. Pages 113-131 *In* E. Fratkin, K. A. Galvin, and E. A. Roth, editors. African pastoralist systems: An integrated approach. Lynne Rienner Publishers, Boulder, CO.

Galvin, K. A., P. K. Thornton, R. B. Boone, and J. Sunderland. 2004. Climate variability and impacts on East African livestock herders. African Journal of Range and Forage Science 21:183-189.

Galvin, K. A., D. S. Ojima, R. B. Boone, M. Betsill and P. K. Thornton. 2005. DRU: Decision-making in Rangeland systems: an integrated Ecosystem-Agent-based Modeling Approach to Resilience and change (DREAMAR). An NSF funded research project: http://www.nrel.colostate.edu/projects/dru.

Glantz, M. H. 1994. Drought, desertification and food production. Pages 7-30 *In* M. H. Glantz, editor. Drought follows the plow: cultivating marginal areas. University Press, Cambridge, UK.

Harvey, P. 2003. HIV/AIDS: What are the implications for humanitarian action? A literature review. Humanitarian Policy Group. Overseas Development Institute, London.

Hoffman, M. R. and S. Todd. 2000. A national review of land degradation in South Africa: the influence of biophysical and socio-economic factors. Journal of Southern African Studies 26:743-758.

Hudak, A. T. 1999. Rangeland mismanagement in South Africa: failure to apply ecological knowledge. Human Ecology 27:55-78.

Hudson, J. W. 2002. Responses to climate variability of the livestock sector in the North-West Province, South Africa. Master's Thesis: Colorado State University.

Hudson, J. and C. Vogel. 2003. Use of seasonal forecasts by livestock farmers in South Africa. Pages 75-96 *In* K. O'Brien and C. Vogel, editors. Coping with climate variability: the use of seasonal climate forecasts in Southern Africa. Ashgate Publishing Company, Burlington, VT and Aldershot, UK.

Hulme, M., R. Doherty, T. Ngara, M. New and D. Lister. 2001. African climate change: 1900-2100. Climate Research 17: 145-168.

IPCC. 2001. Climate change 2001: The scientific basis. Contribution of Working Group I to the Third Assessment Report of the Intergovernmental Panel on Climate Change. J. T.

Houghton, Y. Ding, D. J. Griggs, M. Noguer, P. J. van der Linden, S. Dai, K. Maskell and C. A. Johnson, editors. Cambridge University Press, Cambridge, UK and New York, NY.

Jones, P. S. 1999. 'To come together for progress': modernization and nation-building in South Africa's Bantustan periphery – the case of Bophuthatswana. Journal of Southern African Studies 25:579-605.

Kemper, J., R. M. Cowling, and D. M. Richardson. 1999. Fragmentation of South African renosterveld shrublands: effects on plant community structure and conservation implications. Biological Conservation 90:103-111.

MEA (Millennium Ecosystem Assessment). 2005. Ecosystems and human well-being: synthesis. Island Press, Washington, DC.

NOAA (National Oceanic and Atmospheric Administration). 2002. Climate prediction center: El Niño/La Niña home. http:www.cpc.ncep.noaa.gov/products/analysis_monitoring/ensostuff/.

Piesse, J., T. Doyer, C. Thirtle, and N. Vink. 2005. The changing roles of grain cooperatives in the transition to competitive markets in South Africa. Journal of Comparative Economics 33:197-218.

SADA (South African Department of Agriculture). 1999. North-West Province Database. Vryburg District Office.

Schmidt, I. 1992. The relationship between cattle and savings, a cattle owner perspective. Development and Southern Africa 9:433-434.

Schulze, R. E. 1997. Climate. Pages 21-41 *In* R. M. Cowling, D. M. Richardson and S. M. Pierce, editors. Vegetation of Southern Africa. Cambridge University Press, Cambridge, UK

Shillington, K. 1995. History of Africa. St. Martin's Press, New York.

Shisana, O, T. Rehle, L. C. Simbayi, W. Parker, K. Zuma, A. Bhana, C. Connolly, S. Jooste and V. Pillay. 2005. South African National HIV Prevalence, HIV Incidence, Behaviour and Communication Survey. HSRC Press, Cape Town, S.A.

Simbi, T. 1998. Agricultural policy in South Africa – a discussion document. Ministry for Agriculture and Land Affairs, RSA. Government Printing Office, Pretoria.

Snyman, H. A. 1998. Dynamics and sustainable utilization of rangeland ecosystems in arid and semi-arid climates of Southern Africa. Journal of Arid Environments 39:645-666.

SRTM. 2004. Shuttle Radar Topography Mission: mapping the world in 3 dimensions. US Geological Survey, Sioux Falls, SD. http://srtm.usgs.gov/site_map.html.

Statistics South Africa. Census 2001 with New Demarcation boundaries as of 9 December 2005. http://www.statssa.gov.za/census01/html/C2001Interactive.asp.

Thompson, L. M. 2001. A History of South Africa. Yale University Press, New Haven, CT.

Thornton, P. K., R. H. Fawcett, K. A. Galvin, R. B. Boone, J. W. Hudson, and C H. Vogel. 2004. Evaluating management options that use climate forecasts: modeling livestock production systems in the semi-arid zone of South Africa. Climate Research 26:33-42.

Tyson, P. D. 1986. Climatic change and variability in Southern Africa. Oxford University Press, Cape Town, S.A.

Vogel, C. 1994. South Africa. Pages 151-170 *In* M. H. Glantz, editor. Drought follows the plow: cultivating marginal areas. University Press, Cambridge, UK.

Vogel, C. 2002. Living in a changing earth system. Newsletter of the International Human Dimensions Programme on Global Environmental Change. Issue 4/02.

Vogel, C., M. Laing, and K. Monnik. 2000. Drought in South Africa, with special reference to the 1980-94 period. Pages 348-365 *In* D. A. Wilhite, editor. Drought: A global assessment. Routledge, London.

Vogel, C. and K. O'Brien. 2003. Climate forecasts in Southern Africa. Pages 3-33 *In* K. O'Brien and C. Vogel, editors. Coping with climate variability: the use of seasonal climate forecasts in Southern Africa. Ashgate Publishing Company, Burlington, VT.

Vogel, C. and K. O'Brien. 2004. Vulnerability and global environmental change: rhetoric and reality. Aviso 13, March, http://www.gechs.org.

Wilson, M. and L. Thompson, editors. 1969. The Oxford history of South Africa: South Africa to 1870. Oxford University Press, New York.

Xie, P. and P. A. Arkin. 1997. A 17-year monthly analysis based on gauge observations, satellite estimates, and numerical model outputs. Bulletin of the American Meteorological Society 78:2539-2558.

PART III. ISSUES OF FRAGMENTATION AND COMPLEXITY: A SYNTHETIC PERSPECTIVE

Chapter 13

THE DRIVERS OF FRAGMENTATION IN ARID AND SEMI-ARID LANDSCAPES

Roy H. Behnke
Macaulay Institute, Craigiebuckler Aberdeen, AB15 8QH, U.K.

> We have good Statutes made for the Commonwealth, as touching commoners and inclosers, many meetings and sessions; but in the end of the matter there cometh nothing forth. Hugh Latimer, First Sermon preached before King Edward VI, 1549 (Tawney 1912:311)

> Communal-property regimes do not work well under stress from colonialism, population pressure, technology change, and transformation of subsistence economies to cash economies (Berkes 1996:100).

1. INTRODUCTION

The tone of Latimer's preaching was both weary and impatient – "many meetings and sessions; but…there cometh nothing forth." It might therefore amaze him to discover, four hundred and fifty years later and still counting, that he was observing a global process that continues to this day – the displacement of feudal or tribal systems of land holding to make way for exclusive tenure and commercial agriculture. In the 16th century, the Tudor English state was ambivalent about this novel process, afraid that enclosure would depopulate the countryside and sap the fighting strength of English armies relative to their continental rivals. But the authorities were either powerless or unwilling to halt commercial developments that they had helped to initiate and from which they profited (Tawney 1912).

K. A. Galvin et al. (eds.), Fragmentation in Semi-Arid and Arid Landscapes: Consequences for Human and Natural Systems, 305–340.

This chapter picks up the story in the 20th century on the semi-arid rangelands of Africa and Asia. In these areas rangeland enclosure and landscape fragmentation have been promoted by the spread of exclusive systems of private ownership that legitimate the subdivision of communal rangelands. Frequently, the most valuable rangeland resources have been converted to more intensive non-pastoral forms of land use by indigenous pastoralists, encroaching agriculturalists or colonial settlers. For three pastoral regions in East Africa, southern Africa, and Central and Inner Asia, this chapter supplements the evidence provided by the case studies in this book (see Alimaev and Behnke, Chapter 7; Ojima and Chuluun, Chapter 8; Reid et al., Chapter 9; BurnSilver et al., Chapter 10; Galvin et al., Chapter 11; Galvin et al., Chapter 12).

The objective of this analysis is to identify the causes of fragmentation in semi-arid and arid landscapes in developing regions. The proximate causes of rangeland fragmentation are myriad, ranging from technical innovations (barbed wire, boreholes, maize and ploughs, to name a few) to population pressure, legal changes, government policies, and a wide variety of commercial incentives. Often these factors work in combination, or affect different segments of a population, producing enclosure movements that operate at multiple geographical scales and embody the resources and interests of different economic classes. Despite these complexities, this review will show that rangeland fragmentation has been made possible by two enabling conditions – the growing power of centralized, bureaucratic states and the spread of capitalism. It is this combination of factors that allows the political dimensions of resource control to recede and permits indigenous land users to view natural resources in modern economic terms as a commodity and as a source of personal profit. Once this shift in perspective occurs, enclosure and landscape fragmentation can be set in motion by any number of locally specific considerations that encourage the exclusive appropriation and exploitation of natural resources.

Emergent capitalism and the bureaucratic state came together with global implications in early modern England, and with results that Latimer inveighed against. Four centuries later, the power of this combination is now felt even in the remote, relatively unpopulated and economically marginal rangeland areas considered in this analysis. This chapter therefore has many instances of rangeland fragmentation to report, and it concentrates on doing so. This is not to suggest that rangeland fragmentation is an inevitable process. To substantiate this conclusion we would need to examine why some rangelands remain open and collectively managed while others become privatized and fragmented, a task that is beyond the scope of this chapter. For several regions of the developing world, the following analysis demonstrates how rangeland fragmentation occurs and why the rate of fragmentation has accelerated in the last century, not that the process is unavoidable. This material nonetheless documents a long-term trend towards fragmentation in the semi-arid grazing lands of African societies undergoing industrialization. During much of the 20th century, rangeland fragmentation was arrested in socialist Inner and Central Asia with the imposition of state ownership, but has re-emerged suddenly with the demise of the Soviet Union and market reforms within Chinese communism. In different ways in both Africa and Asia, fragmentation has been codified and reinforced by increasingly exclusive systems of property ownership.

The industrial pastoral societies created in the 1800s by European settlers present a different picture (Figure 13-1). In commercial ranching areas there has been a long-term trend towards the aggregation or engrossment of pastoral properties. This process has occurred through the buying and selling of land and leases with only marginal adjustments in the legal conditions under which land is owned (Galvin et al., Chapter 12; Stokes et al., Chapter 4; Lackett and Galvin, Chapter 5). Whether the fragmentation occurring in the developing world will eventually evolve into landscape consolidation analogous to that occurring in commercial ranching settings remains unclear. Figure 13-1 suggests that these opposing trends are part of a single hypothetical development trajectory, with the fragmentation process eventually reversing itself after the most productive parts of the landscape have been excised and re-consolidation has occurred in the residual grazing lands (Stafford Smith 2002). This chapter examines only half of this larger question – the movement from open range pastoral systems to increasingly privatized and settled systems of production on fragmented landscapes. Whether we are dealing with a single syndrome of land consolidation/

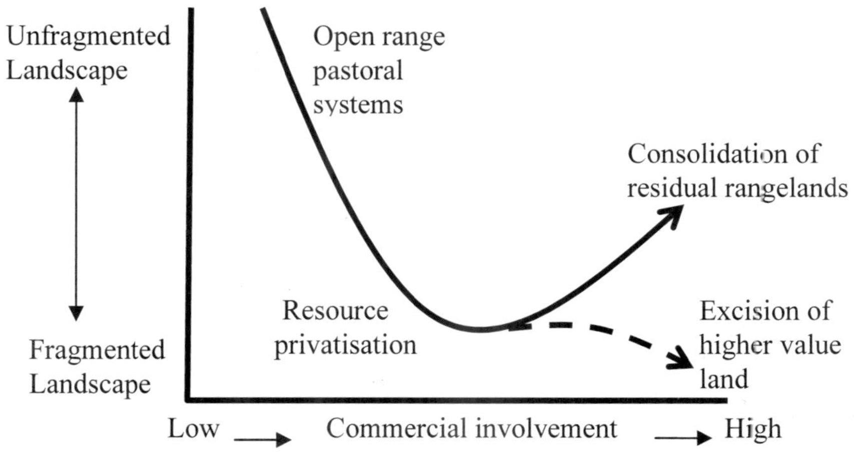

Figure 13-1. Hypothetical patterns of rangeland consolidation and fragmentation.

fragmentation or with several isolated processes is a matter for future research.

2. THE CAUSES OF RANGELAND FRAGMENTATION IN EAST AFRICA

The pastoralists of East Africa, and especially the Maasai who are examined in three chapters of this book, occupy some of the most intensively studied rangelands in Africa or Asia. East Africa also provides numerous examples of rangeland enclosure through "fragmentation of the resource base, from communally-managed rangelands down to subdivided individual parcels…" (BurnSilver et al., Chapter 10). Table 13-1 summarizes available field studies. Given the wealth of research in the region, Table 13-1 is certainly incomplete, but it is not intentionally selective and should give a balanced impression. There are two relatively distinct types of rangeland enclosure documented for East Africa in Table 13-1 – fragmentation caused by increased cultivation (cases 1-9) and fragmentation undertaken to promote the shift from subsistence/nomadic to commercial/settled livestock production (cases 10-14).

2.1 The displacement of pastoralism by cultivation

Early recorded examples of cultivation-induced fragmentation – among the Gusii, Sukuma, Kamba and Kipsigis (cases 1-4, Table 13-1) – occurred during the colonial period in agro-pastoral areas subject to increasing land pressure due either to population growth or to the spread of commercial agriculture. More recently, increased cultivation and resource privatisation has "spread down the ecological gradient" (Campbell 1993) and contributed to landscape fragmentation in areas that were once almost exclusively pastoral – among the Narok and Kajiado Maasai, the Borana of Ethiopia, and in Somalia (cases 5-8, Table 13-1).

Changing land use among the Kipsigis of western Kenya exemplifies the spread of cultivation during the colonial period in areas of mixed arable and livestock husbandry. The Kipsigis traditionally depended upon their livestock for subsistence, but they also farmed, recognized individual use rights to arable plots during the cropping seasons, and fenced these plots to exclude livestock and minimize crop damage. Grazing areas remained unfenced and were used and defended collectively as pasture and as a source of new fields. Cultivated fields were small because households produced for their own consumption, and the fields were annually relocated because the

Table 13-1. Rangeland privatisation and fragmentation in East Africa.

	Site	Date	Legal basis	Proximate causes	Source
1	Sukuma-land, Tanzania	1920 - late 1930s	Endogenous	Population pressure and resource scarcity	Smith 1938, Malcolm 1953
2	Kipsigis, Kenya	1920s to 1950s	Endogenous	Plough-based commercial maize cultivation	Manners 1964
3	Gusii, Kenya	1925-1950	Endogenous	Pacification and land pressure	Mayer and Mayer 1965
4	Machakos Distict, Kenya	1930 - 1980	Dual	Population pressure and intensification	Tiffen et al. 1994
5	Borana Plateau, Ethiopia	1980s ongoing	Endogenous	Population pressure and cultivation	Kamara 2000
6	Maasai, Narok District, Kenya	1980s ongoing	Official	Commercial maize cultivation	Thompson and Homewood 2002
7	Kajiado District, Kenya	1980s ongoing	Dual	Group ranch subdivision for cultivation, settlement	Grandin 1986, Rutten 1992
8	Central Somalia	1970s - mid-1980s	Endogenous	Water development and cultivation	Behnke 1988
9	Il Chamus, Baringo District, Kenya	Began in 1970s	Dual	Cultivation and tenure insecurity	Little 1992
10	Ankole, Uganda	1960s and early 1970s	Dual	Allocation of individual ranches; land pressure	Doornbos and Lofchie 1971, Muwonge 1978
11	Kajiado District, Kenya	Late 1950s to 1960s	Official	Local government allocation of individual ranches	Hedlund 1971, Evangelu 1984
12	Samburu District, Kenya	1980s	Dual	Sale of individual ranches; group ranch subdivision	Sperling 1987, Schlee 1991, Lesorogol 2003, 2005
13	Orma, Tana River District, Kenya	1970s and 1980s	Dual	Land loss, ethnic conflict, commercialization	Ensminger 1992
14	Boran Isiolo District, Kenya	1960 - 1985	Dual	Ethnic conflict and population pressure	Hogg 1987

primary crop, millet, required freshly opened ground. Uncultivated land was readily available and the use of as much land as they could cultivate was deemed by the Kipsigis to be a fundamental right (Manners 1964).

Arable land rights took on a new meaning in the 1930s with the adoption of maize and the shift from hoes to ploughs and from subsistence household provisioning to surplus production for sale. Field sizes increased dramatically to produce commercial surpluses and continuous occupation was achieved by shifting from millet to maize, which extended five-fold the period of time a field could be cropped before fallowing. In these changed circumstances fencing, which had previously been employed to protect crops, became a claim to the permanent ownership of cultivated fields and to any fallow land that might be ringed by cultivation and reserved as private pasturage. In the course of several decades the Kipsigis moved from small, shifting arable fields and unenclosed pasture to engrossed, enclosed, geographically-stable mixed crop and livestock farms:

> Now, in 1964, the appearance of the countryside is changed considerably. From the air or from a high hill, the Kipsigis Reserve resembles the rolling hedgerow lands of rural England….The common pasturelands are gone. And during the past 25 years the Kipsigis have built fences which not only protect their crops from destruction but separate paddock and pasture from garden, and, on the style of the prototypical fences of the whites, mark clearly the boundaries between Kipsigis and Kipsigis. Live fences of Mauritius thorn, euphorbia, or aloes, and, with increasing frequency, non-nonsense fences of barbed-wire, now dissect most of the Kipsigis countryside into a pattern of rectangular fields (Manners 1964:270).

Kipsigis enclosure is typical of the colonial process; it occurred spontaneously and in response to the increased security, commercial opportunities, and population growth that accompanied colonial rule, and it was sanctioned by endogenous changes in African customary law. By the time of Manners' analysis in 1964, individual land ownership was tacitly condoned by the colonial authorities but there had been no official registration of individual titles.

Land use around the Mara National Reserve provides a later and more contentious case of cultivation-induced rangeland fragmentation in an area that remains predominately pastoral. Land-use change since independence among Mara Maasai is an officially sanctioned process involving the formation of Group Ranches and their subsequent legal subdivision, the registration of titles to individual ranches, and the creation of wildlife management associations. Pastoralism, cultivation, and income from game viewing and tourism are all possibilities. Small-scale subsistence cultivation (typically occupying less than two acres of land next to the homestead) is a widespread and established practice among the Mara Maasai, and "is easily accommodated in a landscape which is dominated by grazing, and remains open to use by wildlife as well as livestock" (Thompson and Homewood 2002:121). As among the Kipsigis, however, the effects of commercial farming by Maasai households (using 10-50 acres) and especially by entrepreneurs (on farms of 2,000-4,000 acres) are much different:

> Statistical analyses of long-term data sets suggest that wildlife decline is strongly linked to the spread of commercial cultivation removing key sites and resources, but hitherto not to smallholder agriculture, which is more dispersed (Thompson and Homewood 2002:130).

Revenue from tourism in Maasai Mara is captured by a small elite of large Maasai landowners who control tourist concessions and own properties in prime locations adjacent to the Masai Mara National Reserve. Deprived of any substantial income from wildlife, ordinary Maasai herders have increasingly turned to commercial agriculture, especially in areas back from the reserve where crops are less prone to damage from wildlife. By depressing wildlife numbers, expanding commercial cultivation renders ever larger areas open to profitable cultivation, a positive feedback process that encourages additional cultivation and furthers landscape fragmentation (Thompson and Homewood 2002).

Around the national reserve, wildlife conservancy and tourism are potentially profitable but are not a realistic option for the majority of land owners since they derive little benefit from these activities. Pastoralism, on the other hand, remains the basic livelihood option and is likely to remain so for some time. But pastoralism, in orders of magnitude, is less profitable per hectare than either cultivation or tourism in this environment (Thompson and Homewood 2002:124). There is also little evidence to suggest that pastoralism is likely to become more profitable in the immediate future. Homewood (1992) examined indices of livestock productivity and Maasai diet and nutritional status over two decades in Kenya and Tanzania. Despite divergent state policies and massive development interventions, Homewood's results suggest that there was little intensification. There is, on the other hand, substantial evidence for human population growth in Ngorongoro, Kajiado and among the Mukogodo Maasai (Grandin 1987, Herren 1991, McCabe 1997). In line with the data for the Athi-Kaputiei plains presented in this volume (Reid et al., Chapter 9), these increases in human numbers have been accompanied by either stable or declining livestock populations. Maasai *per capita* livestock wealth is declining, a trend documented elsewhere in east Africa (Hogg 1986, Desta and Coppock 2004, Bollig 2006, Sandford 2006).

The Akamba of Machakos, situated immediately to the north of the Kajiado Maasai, illustrate one response to the problem of increasing land pressure and declining livestock wealth as a result of growing human populations. From 1930 to the 1980s the human population of the district increased about five-fold, livestock populations increased marginally, and livestock output per person declined by two-thirds. While livestock production may have improved during this period due to more productive breeds and the use of cultivated forages, these improvements were more than offset by human population growth. *Per capita* agricultural output nonetheless grew overall due to spectacular increases in the volume of horticultural and non-food cash crops (Tiffen et al. 1994:92-96). While it was economically important, the pastoral sector was stagnant and incapable of producing the increases in output that would support a growing human population. Economic growth was achieved, instead, by privatisation of natural resources and the expansion and intensification of commercial crop production – a process that promoted landscape fragmentation but halted degradation (Tiffen et al. 1994).

The extent to which these processes are being replicated in Kajiado District, which is marginally more arid than Machakos, remains unclear. In Kajiado human populations have grown four-fold in thirty years (Reid et al., Chapter 9). Attempts to intensify livestock production through the institution of Group Ranches failed, and the introduction of exotic breeds of sheep and hybrid cattle, which are more productive but drought intolerant, may increase output at the expense of increasing risk from drought, unless other aspects of the farming system are modified to accommodate this change (BurnSilver et al., Chapter 10). In these circumstances, diversification into non-pastoral activities would seem to offer land owners an opportunity for increasing economic output. Changing patterns of land use and tenure suggest that land owners are pursuing this option, with subdivision of group ranches occurring first under conditions suitable for cropping – adjacent to Nairobi city, on small plots, and in areas of relatively high rainfall (Kimani and Pickard 1998):

> A closer look at the seven group ranches which have already implemented subdivision shows that proximity to urban centres, availability of arable and irrigable land and a long experience with group ranching seem to have added influence. A common feature of those which have resolved not to subdivide seem to be lack of arable land. All of these are located in the drier part of Western, Southern and Southeastern parts of the districts' (Bekure and Ole Pasha 1986).

These changes have been accompanied by large increases in the area devoted to both rainfed and irrigated agriculture and by substantial declines in the populations of both wild and domesticated large herbivores (Reid et al., Chapter 9; BurnSilver et al., Chapter 10).

Mean annual rainfall on the Athi-Kaputiei Plains ranges from 500-800 mm per year, and from 300-600 mm per year in the Amboseli area; in contrast, median rainfall in Machakos ranges from less than 600 mm to over 1000 mm (Reid et al., Chapter 9; BurnSilver et al., Chapter 10; Tiffen et al. 1994). At some point increasing aridity must render impractical the intensification of rainfed arable agriculture on the model of Machakos. Some observers argue that this threshold has already been crossed in Kajiado, with enclosure and landscape fragmentation rendering households more drought prone and impoverished (Rutten 1992, Kimani and Pickard 1998, Thornton et al. 2006, BurnSilver et al., Chapter 10).

However, another and possibly more fundamental difference between Machakos and the Maasai Districts of Kenya is the definition of resource conservation employed by outside observers. Rhinoceros and elephant, though once plentiful, are long gone from Machakos. Conservation in this environment entails the prevention of soil erosion and improved soil fertility – the maintenance, in short, of a sustainable and productive *agricultural* environment. The working definition of conservation in Kenya Maasailand, where large wildlife are still plentiful, is more exacting and requires the successful integration of agriculture with wildlife conservation and tourism. It is now widely appreciated that this integration can only come about when ordinary Maasai landowners can profit financially from preserving wildlife, which impose both direct and opportunity costs on agricultural production. Reid et al. (Chapter 9) describe attempts to create new institutions that will pass some of the profits from wildlife conservation back to landowners in Kajiado District. Thompson and Homewood (2002) describe an earlier, failed attempt around the Mara National Reserve. Both these initiatives assume that wildlife can 'pay their own way' if there exist institutions that equitably distribute the proceeds of tourism. Norton-Griffiths and Southey (1995) question this optimistic assumption with respect to biodiversity conservation in Kenya as a whole. They show that wildlife conservation imposes huge opportunity costs on Kenya's predominantly rural and agricultural economy with its growing population, and that the proceeds from tourism are unlikely to offset these costs. From this perspective, declining livestock populations and landscape fragmentation in Kenya's Maasailand may be nothing more than an unattractive externality generated by the rational and self-interested behaviour of local land owners.

The preceding material documents rangeland fragmentation caused by agricultural intensification. Two final cases – the central rangelands of Somalia and the Il Chamus of northern Kenya - complicate the analysis by raising questions about the role of law in accelerating or retarding this process.

Range enclosure in central Somalia was precipitated by the drilling of a limited number of deep boreholes in an area where water was exceptionally scarce, which made the land around the boreholes unusually valuable. In a manner reminiscent of the Kipsigis, Somali agro-pastoralists sought to control these areas by amending their arable practices and reinterpreting their arable land rights. Small, shifting farms that had previously been abandoned to fallow were enlarged and made permanent. Fences that had once been constructed solely to protect crops were used to claim both arable land and enclosed pastures. But it would be an oversimplification to say that the spread of arable agriculture caused rangeland enclosure in this case. Because water supplies were limited, property adjacent to water was valuable for a host of reasons that included cultivation, grazing, and tactical considerations regarding control of water points by competing political groups. Cultivation was the quasi-legal indigenous rationalization for land- use changes that had multiple causes, only one of which was the desire to extend cultivation. Enclosure was driven by the intrinsic scarcity and high value of certain limited types of land, which encouraged individuals to explore various culturally acceptable justifications for expropriating this resource (Behnke 1988). Once underway, enclosure spread rapidly as individuals competed to privatize their own portion of the shrinking commonage.

The Il Chamus of Lake Baringo in northern Kenya provide a final example of enclosure associated with expanding cultivation and complicated by the interplay between legal and economic factors. The Il Chamus occupy a swamp bordering a savannah, and are therefore in a position to switch from irrigated agriculture to pastoralism, something they have done more than once in the last century and a half. At present the trend is towards increasing cultivation, both by Il Chamus themselves and by neighbouring agricultural groups that have encroached on and appropriated prime grazing areas for cultivation. Expanding agriculture is clearly causing rangeland fragmentation and the retreat of pastoralism, but for economic reasons that are not immediately clear. According to Peter Little, the returns to labour from farming are low and most larger farms are uneconomic, in contrast to large pastoral operations which are generally profitable and provide better returns to labour. Continued interest in agriculture is explained, in part, by households directly provisioning themselves in order to minimize their exposure to erratic markets for food. Also, while large farming operations may be uneconomic when analyzed in isolation, these farms complement large pastoral operations and are profitable in conjunction with them. Finally, there is the issue of tenure insecurity. "Intermittent pastoral use of land and water is not a viable means of acquiring secure tenure in Kenya" notes Little, and he adds:

> Uncertainties surrounding tenure also motivate herders to cultivate irrigated, and in some cases dryland, farms that are larger than can be profitably worked....Some have demarcated their farms with expensive metal fencing that highlights the permanency of their land claim, while adding to the general unprofitability of the farm.... [M]ost irrigated maize farms of over two hectares are uneconomical, but by claiming large plots of irrigated land, the wealthy herder can fend off competing claims to the land (either from within or outside of the community) and gain official recognition of 'ownership' by the state (Little 1992:102, 103).
>
> Among wealthy homesteads, agricultural investment, therefore, can be seen more as a method of securing access to land (land speculation) than as a strategy for increasing short-term profits. It is symptomatic of the insecurity over land rights in the area (Little 1992:104).

Among the Il Chamus, securing land rights through cultivation went beyond rationalizing changes that would have occurred for other reasons. By distorting decisions about how to use the land, legal considerations actively promoted a fragmented landscape.

Summing up the case material reviewed thus far, in East Africa cultivation erodes pastoral forms of collective land ownership for both technical and legal reasons. On the technical side, cultivation lends itself to intensification and is likely to be preferred whenever land is the scarce factor in production, as a result either of population growth or commercial agricultural expansion. In East Africa, expanding rural populations and markets will thereby create pressure for rangeland fragmentation, which may be triggered for different reasons, at different times, in different localities. On the legal side, there is a persistent bias to confer security of tenure on cultivated in preference to grazing land, and for permanent in preference to seasonal use. This bias arises in African customary tenure, which often recognizes increased tenure rights in proportion to the labour that owners invest in their land. It is reinforced by national land policies in countries like Kenya which are numerically, politically, and economically dominated by agriculturalists, many of whom were preyed upon militarily by pastoralists prior to and during the early colonial period (Waller 1976). Finally, as Little (1992) noted in a review of colonial and post-colonial perceptions of the Il Chamus, there is the presumption in development circles that cultivators, and irrigated agriculturalists in particular, are modern and progressive, with pastoralists serving as the logical foil, traditional and conservative.

Agricultural intensification and concerns about tenure security are likely to develop in tandem. Land users will undertake agricultural intensification when land becomes scarce and valuable (Boserup 1965). It is precisely at this point that land users are also likely to become concerned about their security of tenure. Agricultural encroachment, motivated by a combination of technical and legal considerations, therefore serves as a powerful and persistent inducement to rangeland fragmentation in East Africa.

2.2 Fragmentation and the transition to settled pastoralism

There are climatic limits to the spread of rainfed cultivation in arid and semi-arid environments. Beyond these limits in pastoral areas of East Africa, rangeland fragmentation has been brought about by attempts to improve livestock production.

The Ankole ranching scheme in south-western Uganda was funded by USAID in the 1960s, and individual ranches in Kenya's Maasailand were an outgrowth of colonial grazing schemes (Doornbos and Lofchie 1971, Rutten 1992). Both cases (rows 10 and 11 of Table 13-1) exemplify the attempt – popular among the colonial authorities and with development agencies from the late 1940s through the 1970s – to transform African pastoralists into settled, commercial ranchers (see Goldschmidt 1981 for a review and assessment of these policies).

Despite the foreign origin of the ranching concept, in both Kenya and Uganda individual ranching acquired an indigenous following. In Uganda the donor-financed ranching scheme was supported by the local elite who exploited it to their own profit. Of the first forty ranches allocated, fifteen were owned by absentee landlords who were members of an ethnically distinct cattle-owning upper class, with four of the six Ankole members of the Uganda parliament receiving ranches (Doornbos and Lofchie 1971). The scheme also helped precipitate a spontaneous enclosure movement among smallholders outside the scheme where land pressure was increasing due to population growth, increasing cattle numbers, and the alienation of land to the ranches and wildlife conservation. In a situation where typical agricultural holdings barely exceeded two hectares, by 1973 enclosures (excluding the ranching scheme area) ranged in average size from 30 to about 190 hectares in different parts of Ankole, and were held for a mixture of pastoral, arable, and legal reasons:

> While it is true that there are individuals in Ankole whose motive to fence is associated with the need to institute a system of modern animal husbandry, to many the fencing movement is nothing more than a device to protect land. The need for protecting land has been heightened by the

> marked growth of the human and livestock populations of the area (Muwonge 1978:184).

Local politics also played a role in the creation of individual ranches in Kajiado District, Kenya, where 24 ranches in 1963 had grown to 300 individual ranches by 1981 (White and Meadows 1981:3, Rutten 1992:212). Individual ranches received both a disproportionate amount of land and the most productive land in the district (White and Meadows 1981:14-15, Rutten 1992:212, 273). While they received development loans, individual ranches were not allocated by the Kenya Livestock Development Project (KLDP phase I, II and III) which represented national and international (World Bank) interests, but by County Councils that were dominated by local elites.

> The procedure of establishing an individual ranch as been as follows: An individual, usually with some education, some political influence, and some capital, convinces the county council that it is of advantage for the economic development of the area to allot him a piece of land, usually the land where he has been grazing (Jahnke et al. 1974:26).

The technical benefits anticipated from the creation of individual ranches never materialized, but their effect on rangeland fragmentation was profound:

> Among the Maasai, rich households are almost invariably those which through unequal land registration at sub-division or by use of production surpluses have acquired individual ranches and agricultural lands. The original aims of the ranching policy, seen as increasing proper management of range resources and reduction in stocking levels, is a notable failure. However, its operation as the basis for unequal accumulation and resource exploitation has indeed been a notable success (Kituyi 1990:178-9).

As early as 1980, three-quarters of the Maasai individual ranchers surveyed in Kajiado lived in modern houses and 42% owned vehicles (White and Meadows 1981:69). The individual ranchers were better educated, wealthier, politically more prominent, and had greater access to non-pastoral sources of cash and food than the rest of the Maasai population (Hedlund 1971, Jahnke et al. 1974, White and Meadows 1981, Rutten 1992). Members of this 'newly created distinct social and economic class' withdrew from cattle exchanges with friends and relatives, declined traditional tribal offices that were not financially remunerative, but became involved in government development schemes and national political activities that were profitable (Hedlund 1971:16). These social changes were accompanied by the development of a form of commercial pastoralism that deemphasized milk production in favour of calf growth, employed increased levels of hired labour, and specialized in cattle trading and finishing for resale (White and Meadows 1981).

Economic and institutional change among the Orma (or Galla) of Tana River District, Kenya provides a final instance of rangeland fragmentation caused by the spread of commercial pastoralism in an area where rainfed cultivation was not an option (Ensminger 1992). After droughts in the 1970s and 1980s, Orma pastoral households settled in villages. When post-drought cattle populations began to recover, grazing pressure increased in the vicinity of settlements, and local cattle owners looked for ways to keep others from using local pastures that were now in short supply. The settled pastoralists appealed to different arguments to justify their exclusive access to grazing resources that had once been common property open to use by all Orma herd owners:

- Initially settled herd owners invoked Orma custom that reserved grazing around a household for the household's milking animals and excluded other herds and non-milking cattle. As settled cattle numbers increased, villagers expanded the exclusion zones around settlements, lengthened the period of closure until it became year-round, and began pasturing all village-owned cattle in the reserved area, which violated the reasons they had initially given for creating the exclosure.
- Having undermined their appeal to custom, village pastoralists argued that they needed a grazing reserve so that they could simultaneously pasture their cattle and send their children to school.
- When nomadic Orma started sending their own children to school, villagers promoted grazing restrictions as a way of suppressing banditry or of protecting Orma pastures from incursions by outside ethnic groups.
- Finally, by the late 1980s some wealthy, settled herd owners were considering an explicit appeal to the government to set up a private ranch on the border of the restricted territory.

The justifications for village grazing exclusions therefore rested increasingly on government authority – initially by appealing to tribal custom, then to programs that promoted education or public order, and finally on a formal request for land reform. Paralleling this shift in rhetoric, the involvement of government officials in enforcing grazing restrictions also increased over time (Ensminger 1992).

The Orma progressively fragmented their rangelands for a variety of reasons, but the overriding cause was their involvement in settled, commercial pastoralism. Settlement had initially occurred due to drought and herd loss, but once it was established it promoted commercial beef production at the expense of subsistence-oriented dairy production. Commercial beef production was attractive to settled herd owners because it eliminated milking and the need for families to move with their herds, which were instead left to the care of herders. These changes in husbandry practices were encouraged by rising prices for cattle that allowed the profitable substitution of purchased foods for home-produced milk products, a trend that was reinforced by new livestock breeds that maximized weight gain at the expense of milk output and drought resistance. Settlement and commercial pastoralism were further encouraged by restrictions on movement due to encroachments on Orma territory from irrigation schemes, game reserves, company ranches, and incursions by Somali pastoralists. Under pressure from the Somali, there were strong incentives to settle to establish more recognized rights to land (Ensminger 1992:132-134).

Commercial pastoralism produced rangeland fragmentation through its ramifying impact on Orma institutions. By providing new options for expenditure and investment, commercial involvement provided an alternative to traditional forms of surplus redistribution, increased economic inequality, and undermined collective solidarity. In a multitude of ways, as Ensminger (1992:150) expresses it, economic diversification increased the 'costs of engineering a consensus'. When land scarcity and commercialism drove up the value of pastures near villages, some individuals had a vested interest in the private appropriation of a newly valuable resource and there were diminished communal sanctions against such behaviour. There was also an alternative source of social control in the form of the state:

> As the Orma were fighting over property rights, the central government of Kenya was getting closer and closer. Those who had most contact with the government were in a favored position to use the new institutions of the nation-state to their advantage. The costs of enforcing more exclusive control of pastoral land would previously not have paid the gains from such change in property rights. Once the costs of enforcement could be at least partially shifted to the central government, however, it was certainly in the interests of some Orma to do so.... Thus by 1985 most Orma agreed that it was legitimate for the state to use force against their own members. It was common practice for the chief to use his police to arrest encroachers on the restricted grazing, Orma and Somali alike, and this policy was considered legitimate by the vast majority of settled Orma (63 percent of the population) (Ensminger 1992:141-142).

Commercial livestock husbandry encourages the spread of fencing, but among the Maasai and Orma, commercialization also indirectly fostered fragmentation by promoting institutional change. It is this latter, less tangible process that is both more fundamental and difficult to understand. With the increasing market value of livestock products and rangeland resources, politically well-connected Maasai and Orma abandoned or co-opted collective institutions so that they could privatize communal land. These tribal elites could afford to isolate themselves because they had government backing, from local County Councils, national administrators and international development agencies. In the words of Kituyi's book title, the Maasai were "becoming Kenyans" (Kituyi 1990), and so were the Orma.[1] The combination of commercial opportunity with external government control produced fragmentation.

2.3 Fragmentation in East Africa: general trends

Where rainfall is sufficient to sustain cultivation or where irrigation is possible, rangeland fragmentation has been caused by the intensification of arable agriculture. Beyond the environmental limits of cultivation, fragmentation has been promoted by the adoption of settled, commercial pastoralism, with a complex mixture of both arable and pastoral enclosure in transitional environments like Kenya's Maasailand or in Ankole.

Fences are the tangible marker of enclosure, but the causes run deeper. There is a persistent legal bias in African customary law, national legislation, and international development programmes to favour property claims established by continuous agricultural use rather than those based on intermittent pastoral occupation. When ownership is uncertain, claimants to land may adopt cultivation, or the appearance of cultivation, to strengthen their land rights. This legalistic bias reverses an earlier and opposite trend that favoured pastoral land rights in the period prior to colonial rule, when their mobility, military culture, and superior political organization gave pastoralists control over areas that were also suitable for cultivation (Waller 1976). With central government control and mounting land pressure, many of these transitional areas are now being converted to cultivation accompanied by increased landscape fragmentation.

When combined with effective government control, settled pastoralism also promotes rangeland fragmentation. It offers wealthy pastoralists the opportunity to distance themselves from redistributive claims on their livestock resources, to engage in new forms of market-oriented livestock production and trading, and to invest in land, a less volatile form of wealth than livestock. Stripped of many of the technical characteristics of industrial ranching, ranch ownership has proved popular among East African pastoral elites.

It frequently is difficult to draw a firm distinction between spontaneous enclosure based on the evolution of indigenous law and driven by endogenous causes versus fragmentation imposed by government policy. East African customary law traditionally recognized a degree of individual control that has provided a basis for creative reinterpretation and resource privatisation as conditions changed. Conversely, government programs such as the promotion of individual ranches in Ankole or Kajiado or group ranches among the Isiolo Boran (case 14, Table 13-1) or Orma often developed local constituencies that appropriated external policies for their own personal or local purposes. Even when the coercive power of government is formidable, the authorities cannot dictate how ordinary producers will respond to their directives. Arable encroachment after the lifting of the ban on cultivation in the Ngorogoro Conservation Area illustrates the limits, and unforeseen consequences, of government action (Galvin et al., Chapter 11). If rangeland fragmentation proceeds apace in East Africa, it is because it suits the interests of at least some producers.

3. THE CAUSES OF RANGELAND FRAGMENTATION IN SOUTHERN AFRICA

The southern African case study in this book (Galvin et al., Chapter 12), examines land use among the Tswana and among the white commercial ranchers who displaced the Tswana from parts of North-West Province in the Republic of South Africa. Coercive state policy in the form of *apartheid* overwhelmingly caused rangeland fragmentation in this instance (Galvin et al., Chapter 12), making it difficult to identify other contributing factors or to imagine the evolution of land use in this environment under less oppressive political conditions. In this regard, the neighbouring republics of Botswana and Namibia provide an illuminating contrast. In Namibia, an arid country in which agriculture is oriented to pastoralism, the enclosure of communal rangelands began in the 1980s. Namibia was ruled by South Africa for over half a century and one of the principal causes of rangeland fragmentation was the imposition of apartheid policies in the last decades of South African rule. On the other hand, before independence in 1966 Botswana was the Bechuanaland Protectorate, created in the late 19th century to advance British imperial interests and shield the northern Tswana from the land expropriation and political domination occurring across the border in South Africa. The Tswana are agro-pastoralists, but the political elites who indirectly ruled Bechuanaland and still run Botswana are culturally and economically committed to cattle husbandry. Botswana is an African state controlled by indigenous pastoral interests. It is also a state with a long history of rangeland fragmentation.

3.1 The fencing of Botswana's ranges

In Botswana water is scarce and the evolution of water rights has provided the catalyst for land-use change. Restricted supplies of permanent water occur in eastern Botswana, where the bulk of the population is settled, but Kalahari sands cover the western two-thirds of the country which is almost devoid of surface water aside from ephemeral seasonal sources. In this waterless environment, anyone who controlled a supply of stock water effectively owned the rangelands around that water.

In common with traditional East African legal systems, Tswana custom-mary law endorsed the importance of labour in establishing land and water rights. This legal principle meant that natural sources of water tended to be freely available, while man-made sources (such as hand-dug wells in the beds of dry rivers) were owned by those who made them and by their descendants. These rules took on new significance with the introduction of borehole technology, which permitted the construction of water points on an unprecedented scale. Borehole drilling began in eastern Botswana and spread west into the Kalahari over the course of seven to eight decades. Early wells in the 1930s were financed by the Protectorate government but were privatized through the creation of borehole syndicates, which assigned private property rights over boreholes to groups of cattle owners. This transitional form of collective private tenure was about as far as the Tswana elite could go towards privatizing natural resources in the first half of the 20th century, for although they privately owned capital goods like ploughs and vehicles, "to seek exclusive individual control of scarce water sources in the grazing area was not politically possible at this period" (Peters 1994:67). Wealthy cattle owners instead defused opposition to exclusive water owner-ship by exploiting what Peters calls the syndicates' "double image". Before their tribal subjects, the syndicate could be depicted in conservative terms as a *kgotla*, a traditional assembly of resource users and managers. Before their colonial masters, on the other hand, the syndicate was presented as a modern form of private enterprise "counterpoised against the irresponsible, retrogressive aspects of communal organization and yet also … evoking hardy traditions of collective, even democratic, effort" (Peters 1994:69). By expelling poorer syndicate members, restricting the purchase or hire of water by non-members, or by simply chasing away prior occupants, syndicate ownership became more exclusive over time. Restrictive water ownership also eventually undermined the legal status of collective grazing rights. Originally, boreholes conferred *de facto* exclusive grazing rights by giving their owners control over pastures that were otherwise inaccessible. This was changing by the late 1970s, with some rural authorities maintaining that borehole owners had legal rights to communal rangelands around their boreholes, even if natural sources of water rendered these pastures open to others (Peters 1994:132-37).

Introduced in 1975, the *Tribal Grazing Lands Policy* (TGLP) unequivocally legalized rangeland privatisation in designated parts of the country, granting several hundred individuals private leases to large tracts of land (typically about 6,400 ha per ranch) in the Kalahari. Like the Ankole and Maasai schemes, TGLP was justified to donor agencies in economic and environmental terms as an attempt to transform African pastoralists into resource conserving commercial ranchers. Like the East African ranching projects, TGLP achieved almost none of its technical goals, but it did cause a major shift in land rights, which in retrospect appears to have been the implicit purpose of the policy.

The spread of fencing has been a gradual process. Surveys conducted in the mid-1970s indicated that many borehole owners viewed fences as expensive impediments to herd mobility, which was required in the event of drought, range fires, or borehole breakdown (Hitchcock 1978:388). Ranch development was also uneconomic and conferred no advantage in terms of herd performance (CARL BRO 1982, McGowan 1988). By 1991, however, just under half of all TGLP ranches were perimeter fenced (Tsimako 1991:18), and by 1996 fencing was seen as a good way to exclude the neighbours' cattle and was spreading rapidly in western portions of Botswana (Perkins 1996:508). Finally, in 1991 the government's *New Agricultural Policy* extended the legal right to fence to all communal rangelands (GOB 1991), though by 2002 little fencing had actually taken place in the communal areas outside of pilot schemes (Darkoh and Mbaiwa 2002).

TGLP was designed by those who most benefited from it – a small group of senior civil servants and politicians who were also large, absentee cattle owners (Picard 1980, Parson 1981, Holm 1985). To explain how TGLP caused rangeland fragmentation we must understand the economic interests and political power of this group. Even before colonial incorporation, the Tswana were semi-urbanized, living around their chiefs in capitol towns organized like huge villages that moved periodically when arable resources were exhausted or political power shifted. Political control and cattle interests were projected outwards from these capitols, with grazing rights linked to the control of subject peoples who were incorporated into the state as serfs (Schapera 1970, Hitchcock 1985). With independence, the old tribal elites became civil servants and politicians in the new state, but retained their interest in absentee pastoralism and in the control of rural natural resources and labour. In this hierarchical social order, fences and private property provided suitably modern justifications for maintaining established privilege. In a survey of Kalahari water points conducted in 1975, Hitchcock found:

> Of those water source owners who said they liked the idea of TGLP, virtually every one said that they wanted exclusive rights. Although few favoured fencing, and none wanted resident managers, they did want rights to the land. The main reason, according to every single water source owner I spoke with, was so that they could force the 'squatters' off the land (Hitchcock 1978:379).

A string of surveys, the most recent conducted in 1991, confirm that rural pastoral employees (squatters and their descendants) live at subsistence level, often receiving little more than a basic food ration (Perkins 1996). Cheap labour, government subsidies and soft loans to large commercial cattle owners (Isaksen 1984:92-93, McGowan 1988, Harvey and Lewis 1990:89), and sub-economic ranch rents that have remained unchanged for three decades (Daily News 2006) all contribute to low production costs. Income for large commercial herd owners initially came from the marketing of cheap meat to the mines at Kimberly and on the Witwatersrand in South Africa, then from selling beef to Europe at subsidized EC prices, and finally has been derived from domestic meat sales to Botswana consumers whose purchasing power has grown with expanding diamond and copper exports (Hubbard 1986, Stevens and Kennan 2005). The privatisation of Botswana's rangelands has therefore long been an attractive financial prospect for large herd owners, despite the social costs associated with growing economic inequality and the negative impact of fencing on wildlife and tourism.

3.2 Enclosure in Namibia

Rangeland enclosure in Namibia has occurred among two of the country's largest and politically most important tribes – the Herero, who led the unsuccessful resistance to colonial domination at the beginning of the 20th century, and the Ovambo, who were at the forefront of the struggle that led to Namibian independence in 1990. On the pattern exemplified by Ankole and elsewhere in East Africa, enclosure has occurred at two distinct spatial and economic scales, with the elite (often as absentee owners) enclosing large tracts of remote rangelands (known as *ofarama* among the Ovambo and *outemba* among Herero) while ordinary stock owners fence small areas near their dwellings (*ekove* among Ovambo and *ozokamba* among Herero) (Christian 1998, Kerven 1998, Stahl 2000).

Large-scale enclosure began in the late colonial period. In Hereroland, the colonial government assigned responsibility over land affairs to local government bodies that by 1989 had authorized a total of 100 fenced farms on communal land (Adams and Werner 1990:163, Werner 2000). In parts of Ovamboland, the customary Tribal Authorities authorized the allocation of fenced ranches to create a barrier to further white encroachment (Fuller et al. 1996, Kerven 1998, Werner 1998). By 1996 the Tribal Authorities in one Ovambo region held records on more than one hundred approved farms (Werner 1998:39). However, the rate at which new farms were registered dropped off markedly after independence, while the fenced area continued to expand (Holme and Kooiman 1994, Cox 1998). It would appear that much of the enclosure that took place after independence was unregistered and unregulated by any authority and was illegal. Loss of communal land, water points and migratory routes damaged the welfare of local residents in the areas being enclosed. The absentee businessmen, politicians and senior civil servants who were erecting fences defended them as an essential component of modern commercial ranching, which was needed to control stocking rates, introduce improved breeds, and segregate herds to prevent the spread of disease (Adams and Werner 1990, Fuller et al. 1996, Kerven 1998).

Fencing was also taken up by ordinary Ovambo and Herero pastoralists who enclosed small areas adjacent to their homesteads. In Herero areas small-scale fencing was commonly used to promote commercial methods of livestock keeping, reduce labour costs and prevent straying and livestock theft (Stahl 2000). In Ovambo areas fencing was primarily a response to land pressure. In peripheral grazing areas encroached by private ranches, village headmen fenced to control access to the remaining communal boreholes, which were overused (Kerven 1998:73). In the densely settled areas of central Ovamboland where pastures were in short supply, ordinary households were fenced to create private grazing reserves (Christian 1998).

In sum, enclosure was caused by two sets of factors that came together in the late colonial period – the commercialisation of livestock keeping and the growing power of local government authorities backed by central government. The convergence of these factors reflected *apartheid* policy, which attempted to divide the African opposition along economic and ethnic lines by supporting the emergence of a black middle class and creating separate, tribally demarcated Bantustans (Werner 2000). In Hereroland legislation expanded the power of local authorities over land; in Ovamboland the same result was obtained by allowing an ethnically-based regional government to develop standard procedures for processing and registering requests for private ranches (Werner 1998, 2000). These innovations in customary tenure were supported by a black, urban, commercial and salaried elite that viewed fenced ranching both as a contribution to community development and as a source of personal wealth (Kerven 1998, Werner 1998). Villagers opposed large-scale fencing by absentee owners but engaged in small-scale fencing themselves. They fenced to reinforce the authority of village headmen or create private grazing reserves, and doubted their own ability to collectively manage resources without recourse to individual fencing (Stahl 2000).

3.3 Fragmentation in southern Africa: general trends

In neither Botswana nor Namibia has the expansion of cultivation been a major cause of rangeland fragmentation. A cursory examination of evidence from Zimbabwe and Lesotho (cited below) supports the same conclusion for these countries. Southern African agriculturalists mix herding with farming at the household level. Shifting between these activities appears to be a domestic decision that does not excite inter-community tensions comparable to those in East Africa where identification with pastoralism or cultivation is often part of ethnic identity. Small-scale farmers in southern Africa also sell their output in markets dominated by high volumes of cheaply produced grain from mechanized, freehold farms. This situation may have discouraged many smallholders from expanding their cultivated area to produce surpluses for sale (Low 1986), and thereby reduced the competition between pastoralism and cultivation for land. In any case, the material reviewed here suggests that there are regional differences between the causes of rangeland fragmentation in East and southern Africa, with southern Africans being more likely to restrict pasture access in order to improve livestock production and less inclined to convert pasture into arable land.

But there are also fundamental similarities that cut across the two regions. The most important of these are the prevalence of inter-community resource competition and intra-community institutional fragmentation. In East Africa, Ensminger (1992) attributed the breakdown of common property institutions to increased economic differentiation that undermined traditional social controls. Virtually identical processes occurred in Namibia (see the previous discussion) and are reported for Lesotho and Zimbabwe, where diverse economic opportunities and interests encouraged both rich and poor to opt out of collective resource management arrangements (Lawry 1988, Cousins 1989, 1992). Enclosure in Botswana, on the other hand, followed the Maasai pattern with elites pursuing personal gain and detaching themselves from their traditional social and economic responsibilities.

Like East Africans, southern African livestock keepers do not want to fence themselves in but to fence others out. In Botswana ranch owners retained the right to pasture their herds both on common rangelands and exclusively on their private land, a system of 'dual' grazing rights that eliminated any restraints that private rangeland ownership might theoretically have placed on herd growth (Tsimako 1991). In Lesotho in the early 1980s, a USAID sponsored range management scheme paid range riders to police the scheme's boundaries and exclude outsiders, which led to a 30% decrease in grazing pressure 'creating an ideal management environment for a privileged community, while imposing much of the costs onto those excluded' (Lawry 1988:143). Excluding outsiders was enthusiastically supported by livestock owners inside this scheme who were nonetheless unwilling or incapable of imposing any additional controls on their own herding activities. In five of the six Zimbabwe grazing schemes reviewed by Cousins, one of the major benefits of fencing cited by herd owners was the exclusion of the neighbours' animals, and exclusion was one of the most consistently enforced aspects of scheme operation (Cousins 1992). In a second survey of 30 grazing schemes, half reported major boundary disputes with their neighbours, another quarter of all schemes reported minor disputes, and fence-cutting by neighbours had taken place on a third of all schemes with fencing (Cousins 1989:359). As among the Borana and Orma of East Africa, government-sponsored grazing schemes in Lesotho and Zimbabwe were being manipulated by communities or individuals to control resources by excluding competing users rather than by disciplining their own resource use or that of their community.

The causes of rangeland fragmentation in East and southern Africa are therefore remarkably similar – the internal disintegration of collective resource management institutions combined with the appropriation of commercially valuable resources by modernizing elites. In southern Africa, issues of economic interests and political power stand out more clearly than is sometimes the case in East Africa. Instead of ethnic rivalries, in southern Africa the 'outsiders' who are excluded may simply be an adjoining community or the poorer members of one's own community.

4. THE CAUSES OF RANGELAND FRAGMENTATION IN CENTRAL AND INNER ASIA

In the 20th century most of Asia's pastoralists lived at one time or another in centralized, socialist economies that explicitly challenged the hegemony of capitalism – the USSR, Mongolia, or mainland China. If there are alternatives to the market-driven regularities that characterize resource fragmentation in Africa, then we might reasonably expect to find these alternatives on the rangelands of post-socialist Asia.

Table 13-2 provides an overview of the extent of Central/Inner Asian rangelands and the size of the human populations that they support. The

Table 13-2. The importance of rangelands and rangeland populations in Central and Inner Asia.

Country	Sq km pasture[1]	Grasslands as % of total land area[3]	Grasslands as % of agricultural land[3]	Human pop. in grasslands[2]	% rural pop. in grass-lands[4]
Kazakhstan	1,851,000	69%	74%	4,700,000	68%
Mongolia	1,293,000	83%	94%	2,051,000	84%
Turkmenistan	307,000	65%	81%	1,537,000	43%
Uzbekistan	222,190	52%	60%	1,478,000	6%
Kyrgyzstan	93,650	49%	53%	256,000	7%
Tajikistan	31,980	23%	61%	205,200	4%
China	4,000,000	41%	72%	19,500,000	1.5%

[1]'FAOstats 2000 – 'permanent pasture'
[2]Thornton et al. 2002
[3]Grasslands as defined by Thornton et al. 2002 and total or agricultural land area from FAO 2000b
[4]Grassland population as defined by Thornton et al. 2002 and rural population from FAO 2000a

region contains the world's largest contiguous grassland and the largest rangeland population of any country in the developing world, 19.5 million people in China. In terms of both area and population size, Kazakhstan, Mongolia, and China are the region's primary pastoral countries, and are the focus of this brief review.

Mongolia and the five Central Asian Republics all began decollectivization at roughly the same time in the early 1990s following the dissolution of the Soviet Union. Two of these counties, Turkmenistan and Uzbekistan, still retain centralized agricultural economies, with pastoralists working to fulfill state production targets within what amount to reformed and renamed soviet farms. In these two countries there has been no subdivision, privatesation, or individual leasing of rangelands or water points and at least in Turkmenistan levels of flock mobility appear to be comparable to those in the Soviet period. Rangeland fragmentation has not increased, but neither has comprehensive decollectivization taken place (Behnke et al. 2005).

Market reforms in the other four counties have been accompanied by increases in *de facto* rangeland fragmentation caused by pastoral settlement and declining levels of herd and household mobility. The extent of these changes varies by country. In Kazakhstan and Kyrgyzstan, the privatisation of land and livestock was completed by the end of the 1990s, but was accompanied by dramatic declines in fodder production, the loss of about three-quarters of the national flocks, and emigration from pastoral areas into towns and larger rural settlements. In Kyrgyzstan pastoral settlements became smaller, poorer, and less geographically dispersed (Farrington 2005).

In some parts of eastern Kyrgyzstan, flock mobility has now returned to Soviet levels, but there are also large expanses of remote, high altitude pastures that are largely depopulated (Farrington 2005). In Kazakhstan, rural farmsteads and wells were destroyed and many remote seasonal pastures were abandoned in the mid-1990s, as flock owners retreated to the larger rural settlements and flocks concentrated around these settlements (Robinson 2000, Behnke 2003, Alimaev and Behnke, Chapter 7). Around 2000, these downward trends were reversed as flocks sizes expanded for the first time in a decade and larger flock owners began to re-colonize isolated farmsteads, wells, and seasonal pastures (Kerven et al. 2003). Whether fragmentation following decollectivization will become permanent in Kazakhstan or is a passing phase in a recovery process therefore remains unclear (Alimaev and Behnke, Chapter 7). In Tajikistan following decollectivization, most high elevation summer pastures in the eastern Pamirs are now occupied in the summer months (Robinson 2005), though stocking rates are higher closer to villages, and only the largest herds still use the most remote pastures in the western Pamirs (Hangartner 2002).

In Mongolia, decollectivization left the pastoral sector 'atomized and demechanized' as the old collective institutions that regulated movement and land use were stripped away and the costs of movement escalated with the rising cost of fuel, which had to be obtained at international market rather than subsidized Soviet prices (Sneath 1993, Sneath 2004:163). In 1994-95 immobility was also correlated with poverty and small herd sizes, since poorer households routinely lacked pack animals and could not afford to pay for mechanized transport (Fernandez-Gimenez 2000:1323). By 1999, however, poorer households were equally as mobile as wealthier ones, apparently as a result of new land tenure legislation that strengthened private use rights at winter camp sites. Poor herders were forced to move because the official contracts to these sites and access to adjacent pastures had been won predominately by long established, better connected and wealthier pastoral households (Fernandez-Gimenez and Batbuyan 2000:25).

A strong positive correlation between herd size and mobility has nonetheless been repeatedly demonstrated in studies across Central Asia. The explanation for this recurrent phenomenon rests on the economics of herd and household movement in temperate climates. Portable shelters must be substantial and the costs of transporting household equipment is high in cold climates relative to the tropics. Movement costs – everything from the opportunity costs of the lost services and economic opportunities associated with settlements, herding labour, and the costs of transporting household equipment – tend to be uniform irrespective of herd size. Large herd owners can offset these movement costs against the increased productivity of many mobile animals. For small herd owners, however, there comes a point when the improved output from a few animals is insufficient to pay for the costs of movement, and immobility is cost-effective despite being less productive (Kerven et al. 2003). More exclusive systems of rangeland tenure (in Mongolia, Fernandez-Gimenez and Batbuyan 2000) or inappropriate tenure models based on Russian peasant farms (in Kazakhstan, Alimaev and Behnke, Chapter 7) may therefore have contributed to rangeland fragmentation. But it would appear that the bulk of the sedentarization that has occurred in post-soviet Central Asia has been caused by the declining size of herds and the fragmentation of livestock holdings and service provision, which has rendered movement uneconomical for small herd owners.

A different picture emerges from the decollectivized rangelands of western China. Here fragmentation is being driven overwhelmingly by state policies and colonization, as it was in *apartheid* South Africa or in Kazakhstan under Stalin (Galvin et al., Chapter 12; Alimaev and Behnke, Chapter 7).

In contrast to the Soviet Union, the brief imposition of rangeland collectivization in China (roughly from 1960 to 1980) left rural community-level institutions intact, while decollectivization was accompanied by an urban economic boom that created a demand for meat and fostered increased rural prosperity and herd sizes (Benson and Svanberg 1998). Many decollectivized pastoralists therefore had large enough flocks to maintain extensive systems of mobile production, and their communities were institutionally capable of regulating access to natural resources. In many Chinese rangeland areas a revival of indigenous systems of pastoral land use did occur following decollectivization, but it was tolerated by administrators as a temporary expedient and was never officially sanctioned (Banks 1999, 2001, Banks et al. 2003). Augmented and amended, the legal basis for rangeland tenure is the *1985 Grassland Law* which provides for state ownership of grasslands with use rights contractually assigned to individual households, typically for fifty years. According to official statistics, individual contracting of usable grasslands is virtually complete in the major pastoral provinces of western China (Banks et al. 2003). However, the actual implementation of the programme has varied between and within provinces and is more complex than official figures suggest (Zhaoli et al. 2005). Some of this complexity is revealed by a brief comparison of the situation in two ecologically and historically distinct regions: the Inner Mongolian Autonomous Region and the Tibetan plateau (consisting of the Tibetan Autonomous Region and portions of Qinghai, Gansu, Sichuan, and Yunnan provinces).

The Han Chinese colonization of Inner Mongolia began in earnest in the 1800s, intensified after 1911 when the Chinese Republic annexed Mongol lands and declared existing land titles invalid, and expanded yet again with improved rail links in the 1920s that brought in millions of drought and famine-stricken Chinese peasants. In 1912 the Han Chinese barely outnumbered the Mongol population; by 1990 the total population of the region had grown tenfold and there were six Han for every Mongol. Following the Great Leap Forward Movement in the 1950s, Marxist/Maoist ideology reinforced the antagonism of the Chinese settlers towards indigenous Mongolian forms of land use. Communist doctrine held that unimproved rangelands had no intrinsic value because they embodied no human labour, that pastoralism represented a primitive stage of social evolution, and that under socialism human effort could conquer nature (Williams 1997, Jiang 2002:189). Finally, with decollectivization in the 1980s, nomadic settlement, private pasture enclosure, irrigated agriculture and tree planting expanded rapidly, accompanied by rangeland degradation, a falling water table, landscape homogenization, and reduced ecosystem resilience (Williams 1996b, Ellis et al. 2002, Jiang 2002). Despite the dominant role of government policy, this process was not driven wholly from above. Agricultural intensification in the 1980s improved livestock output and farm incomes and was generally supported by livestock owners (Jiang 2002), and some well-placed pastoralists profited from and supported rangeland enclosure (Williams 1996a, b). Like ranching in East Africa, Inner Mongolian enclosure and intensification had its origins in national policy, but it also had a local constituency.

In contrast to Inner Mongolia, enclosure came late to the Tibetan plateau. Government programmes to settle nomads and divide rangeland among individual households began in the mid-1990s in eastern parts of the plateau, which were at lower elevation, relatively more productive, possessed better communication links and market outlets, and were more exposed to Han Chinese influence and settlement. Fencing subsequently spread westward into higher, less productive and more remote pasture areas, and became more intensive. Initially limited to hay fields and small spring pastures, enclosure has gradually been extended to include summer pastures, especially in districts where government subsidies are available (Wu and Richard 1999, Zhaoli et al. 2005). Areas that were once collectively managed by whole villages or groups of kin-related households have increasingly been subject to subdivision (Banks et al. 2003, Zhaoli et al. 2005).

The policy rationale for government intervention has also shifted dramatically, abandoning the communist struggle to subdue nature and instead promoting conservation and 'ecological construction', which may include rangeland clearances, the complete removal of people and animals, as well as enclosure and privatisation (Yeh 2005). Complete bans on grazing have been introduced on rangelands at the headwaters of large river systems such as the Yangtze and Yellow Rivers, in the interests of downstream environmental conservation and agricultural production. Despite the top-down nature of these developments, researchers nonetheless report support for fencing by some Tibetan pastoralists, both for practical reasons of improved animal husbandry and to capture government subsidies (Bauer 2005).

In sum, in China as in the former Soviet Union and Mongolia, land-use practices are changing rapidly in directions that are difficult to predict. During much of the 20th century, fragmentation was arrested in socialist Inner and Central Asia by the imposition of state ownership, but has re-emerged suddenly with the demise of the Soviet Union and following market reforms in communist China. State socialism certainly altered the timing of rangeland fragmentation in much of Asia, but there is little evidence that it has permanently deflected the course of development.

5. CONCLUSION: THE GLOBAL DRIVERS OF RANGELAND FRAGMENTATION

For the pastoral societies examined here, colonial rule promoted landscape fragmentation in multiple ways. In Kazakhstan, Namibia, Kenya, and, most famously, along China's Great Wall, colonial authorities used military settlements and fortified lines to disrupt pastoral movements that threatened imperial control. In secure areas behind the fortifications, pastoral movements were curtailed by the alienation of land to incoming settlers. As colonialism became entrenched and administrative regulation replaced military coercion, specific native populations were tied to demarcated territories – tribal reserves, state farms, communes, homelands, etc. – and routine movement remained possible only within these enclaves. In southern Africa and Central Asia, the colonial period lasted for over three centuries and formally ended only in the 1990s; Chinese colonization of its western pastoral provinces continues up to the present.

Landscape fragmentation was not an incidental by-product of colonial rule, but was a consciously fashioned instrument of external control. The impact of colonialism also went beyond simple coercion. In her study of imperial Russian and tribal Kazakh law, Martin argued that Russian land law became influential not because it was imposed but because it was appropriated by the colonized for their own purposes – "Kazakhs incurporated colonial laws and legal practices into their land use strategies in order to prevent and resolve land disputes among Kazakhs" (Martin 2001:114). An exogenous cause of rangeland fragmentation had become endogenous, and in so doing had become an even more powerful agent of change, a process that is occurring today on the Tibetan plateau and in Inner Mongolia (Williams 1996a, b, Jiang 2002, Bauer 2005).

There is widespread evidence of similar processes in pastoral Africa, with ordinary pastoralists manipulating their legal traditions independently of official government or donor programmes to accommodate novel forms of land use. Spontaneous enclosure in Kajiado District in the 1980s provides an example. These enclosures exploited the traditional Maasai practice of creating *olopololi*, small fenced pastures adjacent to homesteads that were reserved for the use of calves. By the mid-1980s in some parts of Kajiado district, a fifth of all pastures were fenced. The size of these enclosures bore little relationship to the grazing needs of the household, adult cattle were pastured in them (contrary to traditional practice), and some enclosures were leased to neighbouring individual ranchers (Peacock et al. 1982, de Souze 1984, Grandin 1986). Government policy only indirectly influenced this spontaneous movement, as herders jockeyed to occupy the best sites in the run-up to formal group ranch subdivision.

Sperling reports a similar evolution of the Samburu use of *lokere* – reserved grazing areas for calves, sick or milking stock – which had in some cases expanded to fifteen to twenty times their original size (Sperling 1987:81), while wealthy Samburu were at the same time buying individual ranches through official channels (Schlee 1991). Already in the 1930s, the Sukuma of Tanzania had modified the indigenous practice of *ngitiri* – private areas set aside for fallows and the collection of thatching grass – to construct grazing reserves, which colonial officials were trying to transform into rotational grazing schemes (Smith 1938). The Boran and Orma both employed the customary distinction between *hawicha* and *fora* (milking versus dry herds) to exclude Somali pastoralists and garner local support for government-sponsored group ranches. For the Herero and Ovambo of Namibia, the presence of a homestead was used to legitimate the individual enclosure of village grazing areas and wells (Kerven 1998, Stahl 2000). Cultivation served the same purpose among Somali, Kipsigis, Il Chamus and numerous other agro-pastoralists, as did the reinterpretation of water rights among the Tswana.

Exclusive ownership of rangeland is uncommon in dry Africa and Asia because the erratic and low productivity of this resource generally renders individual control unattractive, not because indigenous pastoral tenure systems were incapable of recognizing individual entitlements. The above material nonetheless documents a spontaneous shift towards increased exclusivity under some circumstances. The factors that enable this transformation have been cited repeatedly in this review and are two-fold: the establishment of central government authority and the differentiation of economic interests within pastoral communities. In addition to the proximate causes of enclosure – factors as diverse as inter-ethnic tension, the availability of barbed wire, donor policies, the economic interests of local elites, or borehole technology – two characteristic features of modern life, comercialization and centralized administration, have promoted the decline and fragmentation of communal systems of rangeland use.

This particular kind of rangeland fragmentation has been characterized by the displacement of territoriality by land tenure. In the absence of central government control, individual African and Asian pastoralists did not own land in the sense of holding legal titles to particular plots. Rather, they secured rights of access through their membership in political groups that appropriated land in competition against other similar groups. The sovereignty and survival of the territorial group substituted for legal title, and military prowess rather than administrative authority established possession. In these fissile political systems, groups often mobilized in response to external threats and dissolved into their constituent parts when the threat had passed. Even if colonial or independent governments incorporated pastoral institutions into their administrative systems, externally imposed security undermined the political incentives that had previously sustained group cohesion. Individuals might continue to have access to much the same natural resources, but they defended their claims in novel ways. Provided central government control was effective, a political concept of land as collectively defended territory could give way to a legal concept of land as property secured by the state, which could be owned either collectively or individually. The advent of commercialization, by increasing the economic value of land, provided an incentive for individuals to see land not as a patrimony but as a commodity. External markets also absorbed livestock surpluses that previously had circulated internally and had mitigated the economic inequality within pastoral communities. It now became possible for pastoral elites, their authority protected or enhanced by indirect rule, to profitably detach themselves from the systems of local patronage that had previously sustained their power but placed demands upon their wealth. If the indigenous groups that had once managed natural resources could no longer act in a coordinated or disciplined way, pastoralists of all economic backgrounds had little option but to appropriate land and pursue their private interests by private means. Finally, as the commonage shrank, pre-emptive enclosure became the only safe course of action, so that enclosure begat enclosure in a self-reinforcing process.

Commercialization and a central administration may not cause rangeland fragmentation in the sense of triggering the process, but they enable it. Price fluctuations, population pressure, or technical changes might explain why a Maasai pastoralist would choose to convert open rangeland into a greenhouse for cut flowers or into a wheat field, but they do not reveal how the individual came to view the land as an economic good in the first place, or how he acquired the ability to appropriate it privately. This is explained by the decline of territoriality and the spread of land tenure into rangeland areas, the combined effects of markets and government control.

ENDNOTES

[1] Lesorogol makes the same point with respect to the Samburu (Lesorogol 2003, 2005), in articles that came to my attention too late to be referenced in the text.

REFERENCES

Adams, F. and W. Werner. 1990. The land issue in Namibia: An enquiry. NISER, University of Namibia, Windhoek.

Banks, T. 1999. State, community and common property in Xinjiang: synergy or strife? Development Policy Review 17:293-313.

Banks, T. 2001. Property rights and the environment in pastoral China: evidence from the field. Development and Change 32:717-740.

Banks, T., C. Richard, L. Ping and Y. Zhaoli. 2003. Community-based grassland management in western China. Mountain Research and Development 23:132-140.

Bauer, K. 2005. Development and the enclosure movement in pastoral Tibet since the 1980s. Nomadic Peoples 9:53-82.

Behnke, R.H. 1988. Range enclosure in Central Somalia. Pastoral Development Network, Paper 25b. Overseas Development Institute, London.

Behnke, R. 2003. Reconfiguring property rights and land use. Pages 75-107 *In* C. Kerven, editor. Prospects for pastoralism in Kazakstan and Turkmenistan: From state farms to private flocks. RoutledgeCurzon, London.

Behnke, R., A. Jabbar, A. Budanov and G. Davidson. 2005. The administration and practice of leasehold pastoralism in Turkmenistan. Nomadic Peoples 9:147-170.

Bekure, S. and I. Ole Pasha. 1986. Response of the Kenyan Maasai to changing land policies. Paper presented at the Plains Indian Cultures: Past and Present Meanings, a symposium sponsored by the Center for Great Plains Studies, University of Nebraska, Lincoln. (ILCA KE 631.585:333.013 SOL x. 3).

Benson, L. and I. Svanberg. 1998. China's last nomads: The history and culture of China's Kazaks. M.E. Skarpe, Inc., New York.

Berkes, F. 1996. Social systems, ecological systems, and property rights. Pages 87-107 *In* S. Hanna, C. Folke, and K-G Maler, editors. Rights to nature: Ecological, economic, cultural, and political principles of institutions for the environment. Island Press, Washington, D.C.

Bollig, M. 2006. Risk management in a hazardous environment: A comparative study of two pastoral societies. Springer, New York.

Boserup, E. 1965. The conditions of agricultural growth: The economics of agrarian change under population pressure. Allen and Unwin, London.

Campbell, D.J. 1993. Land as ours, land as mine: economic, political and ecological marginalization in Kajiado District. Pages 258-272 *In* T. Spear and R. Waller, editors. Being Maasai: Ethnicity and identity in East Africa. James Currey, London.

CARL BRO International A/S. 1982. Livestock management and production in Botswana. Ministry of Agriculture, Gaborone, Botswana.

Christian, I. 1998. Grazing management study. Northern Regions Livestock Development Project, Windhoek, Namibia.

Cousins, B. 1989. Community, class and grazing management in Zimbabwe's communal lands. *In* B. Cousins, editor. People, land and livestock: Proceedings of a workshop on the socio-economic dimensions of livestock production in the communal lands of Zimbabwe. Belmont Press, Masvingo, Zimbabwe.

Cousins, B. 1992. Managing communal rangeland in Zimbabwe: Experiences and lessons. Commonwealth Secretariat, London.

Cox, J. 1998. An assessment of fencing activity in east Oshikoto. *In* J. Cox, C. Kerven, W. Werner, and R. Behnke, editors. The privatisation of rangeland resources in Namibia: Enclosure in Eastern Oshikoto. Overseas Development Institute, London.

Daily News. 30 January, 2006. TGLP ranches rent up. Gaborone, Botswana. http://www.gov.bw/cgi-bin/news.cgi?d=20060130.

Darkoh, M.B.K. and J.E. Mbaiwa. 2002. Globalisation and the livestock industry in Botswana. Singapore Journal of Tropical Geography 23:149-166.

De Souza, M. and P. de Leeuw. 1984. Smallstock use of reserved grazing areas on Merueshi Group Ranch. ILCA, Nairobi.

Desta, S. and D. Coppock. 2004. Pastoralism under pressure: tracking system change in southern Ethiopia. Human Ecology 32:465-86.

Doornbos, M. and M. Lofchie. 1971. Ranching and scheming: a case study of the Ankole Ranching Scheme. Pages 165-187 *In* M. Lofchie, editor. The state of the nations: Constraints on development in independent Africa. University of California Press, Berkeley.

Ellis, J., K. Price, F. Yu, L. Christensen and M. Yu. 2002. Dimensions of desertification in the drylands of northern China. Pages 167-190 *In* J. Reynolds and D. Stafford Smith, editors. Global desertification: Do humans cause deserts? Dahlem University Press, Berlin.

Ensminger, J. 1992. Making a market: The institutional transformation of an African society. CUP, Cambridge.

Evangelou, P. 1984. Livestock development in Kenya's Maasailand. Westview Press, Boulder, CO.

FAO. 2000a. FAOSTAT PopSTAT module. http://faostat.fao.org/site/452/default.aspx.

FAO. 2000b. FAOSTAT ResourceSTAT module. http://faostat.fao.org/site/405/default.aspx

Farringon, J. 2005. De-development in eastern Kyrgyzstan and persistence of semi-nomadic livestock herding. Nomadic Peoples 9:171-198.

Fernandez-Gimenez, M. 2000. The role of Mongolian nomadic pastoralists' ecological knowledge in rangeland management. Ecological Applications 10:1318-1326.

Fernandez-Gimenez, M. and B. Batbuyan. 2000. Law and disorder in Mongolia: implementation of Mongolia's land law. Paper presented at the IASCP Biennial conference 'Constituting the Commons: Crafting Sustainable Commons in the New Millennium', Indiana University, May 31-June 4, 2000.

Fuller, B., S. Nghikembua, and T.F. Irving. 1996. The enclosure of range lands in the Eastern Oshikoto Region of Namibia. Research Report No. 24, Social Sciences Division, University of Namibia, Windhoek.

GOB (Government of Botswana). 1991. New agricultural development policy. Government Printer, Gaborone, Botswana.

Goldschmidt, W. 1981. The failure of pastoral economic development programs in Africa. Pages 101-118 *In* J.G. Galaty, D. Aronson, P.C. Salzman and A. Chouinard, editors. The future of pastoral peoples: Proceedings of a conference held in Nairobi, Kenya, 4-8 August 1980. IDRC, Ottawa, Canada.

Grandin, B.E. 1986. Land tenure, subdivision, and residential change on a Maasai group ranch. Development Anthropology Network 4:9-13. Institute of Development Anthropology, Binghamton, NY.

Grandin, B. 1987. Pastoral culture and range management: recent lessons from Maasailand. ILCA Bulletin 28:7-13.

Hangartner, J. 2002. Dependant on snow and flour: Organisation of herding life and socio-economic strategies of Kyrgyz mobile pastoralists in Murghab, Eastern Pamir, Tajikistan. M.A. Thesis, University of Berne.

Harvey, C. and S.R. Lewis. 1990. Policy choice and development performance in Botswana. Macmillan Press, London.

Hedlund, H.G.B. 1971. The impact of group ranches on a pastoral society. University of Nairobi, Institute for Development Studies Staff Paper No. 100, Nairobi.

Herren, U. 1991. Socioeconomic strategies of pastoral Maasai households in Mukogodo Kenya. University of Bern, Bern.

Hitchcock, R.K. 1978. Kalahari cattle posts: A regional study of hunter-gatherers, pastoralists, and agriculturalists in the Western Sandveld Region, Central District, Botswana, Volumes I and II. Ministry of Local Government and Lands, Gaborone, Botswana.

Hitchcock, R.K. 1985. Water, land, and livestock: the evolution of tenure and administration patterns in the grazing areas of Botswana. *In* L.A. Picard, editor. The evolution of modern Botswana: Politics and rural development in Southern Africa. Collins, London.

Hogg, R. 1986. The new pastoralism: poverty and dependency in Northern Kenya. Africa 56:519-555,

Hogg, R. 1987. The politics of changing property rights among Isiolo Boran patoralists in northern Kenya. Pages 20-31 *In* P.T.W. Baxter and R. Hogg, editors. Property, poverty and people: Changing rights in property and problems of pastoral development. Department of Anthropology, University of Manchester, Manchester, U.K.

Holm, J.D. 1985. The state, social class, and rural development in Botswana. *In* L.A. Picard, editor. The evolution of modern Botswana: Politics and rural development in southern Africa. Collins, London.

Holme, D. and A. Kooiman. 1994. Mapping fences in Oshikoto Region, Namibia. Technical Notes 2, National Remote Sensing Centre, NRSC/Directorate of Forestry/Ministry of Environment and Tourism, Windhoek, Namibia.

Homewood, K.M. 1992. Development and the ecology of Maasai pastoralist food and nutrition. Ecology of Food and Nutrition 29:61-80.

Hubbard, M. 1986. Agricultural exports and economic growth: A study of Botswana's beef industry. KPI, London.

Isaksen, J. 1984. Some aspects of the cattle industry in the economy of Botswana. Norsk Geografisk Tidsskrift 38:85-94.

Jahnke, H., H. Ruthenberg and H. Thimm. 1974. Range development in Kenya: A review of commercial, company, individual and group ranches. IBRD, Africa Rural Development Study Background Paper, Washington, D.C.

Jiang, H. 2002. Culture, ecology, and nature's changing balance: sandification on Mu Us Sandy Land, Inner Mongolia, China. Pages 181-196 *In* J. Reynolds and D. Stafford Smith, editors. Global desertification: Do humans cause deserts? Dahlem University Press, Berlin.

Kamara, A.B. 2000. Ethiopian case study. Pages 396-426 *In* N. McCarthy, B. Swallow, M. Kirk and P. Hazell, editors. Property rights, risk, and livestock development in Africa. IFPRI, Washington, D.C.

Kerven, C. 1998. The knife cuts on both blades: redefining property rights in Eastern Oshikoto. *In* J. Cox, C. Kerven, W. Werner and R. Behnke, editors. The privatisation of rangeland resources in Namibia: Enclosure in Eastern Oshikoto. Overseas Development Institute, London.

Kerven, C., I.I. Alimaev, R. Behnke, G. Davidson, L. Franchois, N. Malmakov, E. Mathijs, A. Smailov, S. Temirbekov and I. Wright. 2003. Retraction and expansion of flock mobility in Central Asia: costs and consequences. Proceedings of the VII International Rangelands Congress, Durban, South Africa.

Kimani, K. and J. Pickard. 1998. Recent trends and implications of group ranch sub-division and fragmentation in Kajiado District, Kenya. The Geographical Journal 164:202-213.

Kituyi, M. 1990. Becoming Kenyans: Socio-economic transformation of the pastoral Maasai. Acts Press, African Centre for Technology Studies, Nairobi.

Lawry, S.W. 1988. Private herds and common land: Issues in the management of communal grazing land in Lesotho, Southern Africa. Doctoral dissertation at the University of Wisconsin-Madison.

Lesorogol, C. 2003. Transforming institutions among pastoralists: inequality and land privatization. American Anthropologist 105:531-542.

Lesorogol, C. 2005. Privatizing pastoral lands: economic and normative outcomes in Kenya. World Development 33:1959-1978.

Little, P.D. 1992. The elusive granary: Herder, farmer, and state in northern Kenya. CUP, Cambridge.

Low, A. 1986. Agricultural development in Southern Africa: Farm household-economics and the food crisis. James Currey, London.

Malcolm, D.W. 1953. Sukumaland: An African people and their country. OUP, London.

Manners, R.A. 1964. Colonialism and native land tenure: a case study in ordained accommodation. *In* R.A. Manners, editor. Process and pattern in culture: Essays in honor of Julian H. Steward. Aldine Publishing Company, Chicago.

Martin, V. 2001. Law and custom in the steppe: the Kazakhs of the Middle Horde and Russian colonialism in the nineteenth century. Curzon Press, Richmond, Surrey, U.K.

Mayer, P. and I. Mayer. 1965. Land law in the making. Pages 51-78 *In* H. Kuper and L. Kuper, editors. African law: Adaptation and development. University of California Press, Berkeley and Los Angeles.

McCabe, T. 1997. Risk and uncertainty among the Maasai of the Ngorongoro Conservation Area in Tanzania: a case study in economic change. Nomadic Peoples 1:54-65.

McGowan International & Coopers Lybrand. 1988. National Land Management and Livestock Project: Incentives/Disincentives Study, Volumes I, II, and II. Gaborone, Botswana.

Muwonge, J.W. 1978. Population growth and the enclosure movement in Ankole, Uganda. East African Journal of Rural Development 11:168-184.

Norton-Griffiths, M. and C. Southey. 1995. The opportunity costs of biodiversity conservation in Kenya. Ecological Economics 12:125-139.

Parson, J. 1981. Cattle, class and the state in rural Botswana. Journal of Southern African Studies 7:236-255.

Peacock, C.P., P. de Leeuw and J.M. King. 1982. Herd movement in the Mbirikani area. ILCA, Nairobi.

Perkins, J.S. 1996. Botswana: fencing out the equity issue. Cattleposts and cattle ranching in the Kalahari Desert. Journal of Arid Environments 33:503-517.

Peters, P.E. 1994. Dividing the commons: Politics, policy, and culture in Botswana. University Press of Virginia, Charlottesville and London.

Picard, L.A. 1980. Bureaucrats, cattle, and public policy: land tenure changes in Botswana. Comparative Political Studies 13:313-356.

Robinson, S. 2005. Pastoralism in the Gorno-Badakhshan Region of Tajikistan. Nomadic Peoples 9:199-206.

Rutten, M. 1992. Selling wealth to buy poverty: The process of individualization of land ownership among the Maasai pastoralists of Kajiado District, Kenya, 1890-1990. Verlag breitenbach Publishers, Saarbrucken, Germany.

Sandford, S. 2006. Foreword. *In* J. McPeak and P. Little, editors. Pastoral livestock marketing in eastern Africa: Research and policy challenges. Intermediate Technology Publications, Rugby, U.K.

Schapera, I. 1970. Tribal innovators: Tswana chiefs and social change, 1975, 1940. Athlone, London.

Schlee, G. 1991. Traditional pastoralists: land use strategies. Pages 130-164 *In* H.J. Schwartz, S. Shaabani, and D. Walther, editors. Range management handbook of Kenya Vol. II, 1, Marsabit District. Republic of Kenya, Ministry of Livestock Development, Nairobi.

Smith, H.C. 1938. The Sukuma system of grazing rights known locally as *kupela iseso* or *ngitiri*. The East African Journal 4:129, 130.

Sneath, D. 1993. Social relations, networks and social organisation in post-socialist rural Mongolia. Nomadic Peoples 33:193-207.

Sneath, D. 2004. Proprietary regimes and sociotechnical systems: rights over land in Mongolia's "age of the market". In K. Verdery and C. Humphrey, editors. Property in question: Value transformation in the global economy. Berg, Oxford.

Sperling, L. 1987. The labor organization of Samburu pastoralism. Unpublished doctoral dissertation, McGill University, Department of Anthropology.

Stafford Smith, D.M. 2002. SCALE land-use intensification/fragmentation hypothesis. SCALE internal discussion paper.

Stahl, U. 2000. 'At the end of the day we will fight': communal land rights and 'illegal fencing' in the Otjozondjupa Region. In M. Bollig and J-B Gewald, editors. People, cattle and land: Transformation of a pastoral society in southwestern Africa. Rudiger Koppe Verlag, Koln.

Stevens, C. and J. Kennan. 2005. Botswana beef exports and trade policy. Institute of Development Studies, University of Sussex, Brighton, U.K.

Tawney, R.H. 1912. The agrarian problem in the sixteenth century. Longmans Green and Co, London.

Thompson, M. and K. Homewood. 2002. Entrepreneurs, elites and exclusion in Maasailand: trends in wildlife conservation and pastoralists development. Human Ecology 30:107-138.

Thornton, P., R. Kruska, N. Henninger, P. Kristjanson, R. Reid, F. Atieno, A. Odero and T. Ndegwa. 2002. Mapping poverty and livestock in the developing world. ILRI (International Livestock Research Institute), Nairobi, Kenya.

Thornton, P.K., S.B. BurnSilver, and K.A. Galvin. 2006. Modelling the impacts of group ranch subdivision on agro-pastoral households in Kajiado, Kenya. Agricultural Systems 87:331-356.

Tiffen, M., M. Mortimore and F. Gichuki. 1994. More people, less erosion: Environmental recovery in Kenya. John Wiley & Sons, New York.

Tsimako, B. 1991. The Tribal Grazing Land Policy (TGLP) ranches performance to date. Ministry of Agriculture, Gaborone, Botswana.

Waller, R. 1976. The Maasai and the British 1895-1905: the origins of an alliance. Journal of African History 17:529-553.

Werner, W. 1998. The evolution of land tenure in Oshikoto. *In* J. Cox, C. Kerven, W. Werner and R. Behnke, editors. The privatisation of rangeland resources in Namibia: Enclosure in eastern Oshikoto. Overseas Development Institute, London.

Werner, W. 2000. From communal pastures to enclosures: the development of land in Herero Reserves. *In* M. Bollig and J-B Gewald, editors. People, cattle and land: Transformation of a pastoral society in southwestern Africa. Rudiger Koppe Verlag, Koln.

White, J.M. and S.J. Meadows. 1981. Evaluation of the contribution of group and individual ranches in Kajiado District, Kenya, to economic development and pastoral production strategies. Ministry of Livestock Development, Nairobi.

Williams, D. 1996a. The barbed walls of China: a contemporary grassland drama. The Journal of Asian Studies 55:665-691.

Williams, D. 1966b. Grassland enclosures: catalyst of land degradation in Inner Mongolia. Human Organization 55:307-313.

Williams, D. 1997. The desert discourse of modern China. Modern China 23:328-355.

Wu, N. and C. Richard. 1999. The privatization process of rangeland and its impacts on the pastoral dynamics in the Hindu-Kush Himalaya: the case of western Sichuan, China. Proceedings of the VIth International Rangeland Congress, Vol. 1:14-21.

Yeh, E. 2005. Green governmentality and pastoralism in western China: 'converting pastures to grasslands'. Nomadic Peoples 9:9-30.

Zhaoli, Y., W. Ning, Y. Dorji and R. Jia. 2005. A review of rangeland privatisation and its implications in the Tibetan Plateau, China. Nomadic Peoples 9:31-52.

Chapter 14

COMPARING LANDSCAPE AND INFRASTRUCTURAL HETEROGENEITY WITHIN AND BETWEEN ECOSYSTEMS

Randall B. Boone[1], Shauna B. BurnSilver[1,2], and Russell L. Kruska[3]
[1]Natural Resource Ecology Laboratory, Colorado State University, Fort Collins, CO 80523, USA; [2]Department of Anthropology, Colorado State University, Fort Collins, CO 80523, USA; [3]International Livestock Research Institute, Nairobi, Kenya

Ecological research throughout much of the last century focused upon manipulative experiments on areas of a few square meters or less (Kareiva and Andersen 1988). The last quarter of the century saw the development of landscape ecology and the emergence of macroecology as a bonafide method of research and discovery (e.g., Brown 1995, Blackburn and Gaston 2002). Geographical and human-ecological research increasingly has successfully integrated human populations and their behaviors into analyses of land use change (Rindfuss and Stern 1998). Today, readily available broad-scale data, such as satellite images and global spatial databases, make comparing attributes of landscapes and the people who inhabit them uniform, thorough, repeatable, and relatively inexpensive (Roughgarden et al. 1991).

Disturbing trends in rangelands throughout the world make the need for comparisons across regions particularly pressing. Two-hundred million or more people derive a significant portion of their income from raising livestock on rangelands (De Haan et al. 1997) that comprise about 25% of the landscapes of the world (Groombridge 1992). Land use is diversifying and intensifying, including the conversion of marginally productive lands into areas of cultivation (FAO 2001). On many of these lands pastoral people are being sedentarized and grazing areas used by livestock and wildlife are being fragmented or subdivided (defined in Hobbs et al., Chapter 2). Further, a feedback can exist where, for example, conversion and

K. A. Galvin et al. (eds.), Fragmentation in Semi-Arid and Arid Landscapes: Consequences for Human and Natural Systems, 341–367.

fragmentation can cause declines in human food security, and these stressors lead residents to further fragment their land and intensify use. In other already fragmented systems, producers are now trying to re-extensify their access to lands, having identified economic and ecological costs associated with intensive livestock raising on small areas of land.

Dominant policy narratives surrounding pastoralism assume that subsistence-based animal production systems should modernize toward intensive production methods. A subsequent assumption is that natural capital resources are perfectly substitutable by economic inputs (Prugh 1999). In the context of this volume, natural capital is represented by the resource connectivity and spatio-temporal heterogeneity that characterizes unfragmented rangeland systems. Consequently, the assumption is that with the addition of economic inputs, a low-input, unfragmented and extensive livestock production system will smoothly transition into a capital intensive, input-dependent livestock production system functioning at a small scale, with associated benefits for humans, livestock, ecosystems, and national economies. However, the transition from extensive to intensive livestock systems is not free, nor has it been smooth in most cases. The inputs required to compensate for the natural capital lost through fragmentation are expensive and not readily available to many rural producers. Also, the ability of governments, especially in the developing world, to provide access to compensatory factors under market liberalism and structural adjustment programs is questionable (ADF 2003, Njenga and Davis 2003). This leaves pastoral producers in a difficult position, as they are pushed by policies and by the need to subdivide and intensify production on one hand, but are left without access to the means to bridge the productivity gap on the other. The difficulties inherent in making fragmented rangelands economically viable are illustrated by the Australian and Great Plains case studies (Stokes et al., Chapter 4; Lackett and Galvin, Chapter 5), systems in which producers are moving to re-aggregate their productive parcels, *in spite of* having high relative access to productive inputs, infrastructure services, and governmental support mechanisms that should compensate producers for losses in access to spatial scale.

Theory suggests and research has shown that the quantity of external inputs needed to support ecological and productive systems in fragmented landscapes is inversely related to the landscape and infrastructural heterogeneity (defined below) of the ecosystem (Ritchie and Olff 1999, Doncaster 2001, Ash et al. 2004, Boone and Hobbs 2004, Boone et al. 2005, Thornton et al. 2006). Lands that are vegetatively diverse provide more forage choices for livestock and wildlife; ample food, water, and habitat are more likely to be within fragments of a given size if an ecosystem is diverse. Government services and infrastructure are trade-off resources that allow producers to withstand perturbations under fragmentation. To provide some indication of how susceptible areas within and between regions would be to fragmentation, metrics may be calculated that characterize both ecological and infrastructural heterogeneity.

The goals of this chapter are to review and introduce metrics that: 1) may be used to quantify ecological heterogeneity and access to critical infrastructure within and between sites, 2) apply these metrics across sites our research team has worked in to quantify differences in ecological heterogeneity, and by implication, identify the relative importance of fragmentation, and 3) for a subset of nine sites, quantify the infrastructure assets available to producers and examine the assumption outlined above. These methods and analyses should provide an initial indication of where the loss of system heterogeneity through fragmentation would require the most external inputs. The synthetic analyses by experts across many sites reported in this volume is unique. Standardized measures of heterogeneity are needed to inform these syntheses.

1. DEFINITIONS, SCOPE, AND AREAS

There are a diverse number of landscape metrics. For example, the popular software package FRAGSTATS computes more than 100 metrics, many of which are highly correlated (McGarigal and Marks 1995, McGarigal et al. 2002). Many established metrics quantify patches within categorical maps, such as land cover or habitat maps. For example, patch richness reports the number of patches across a landscape, fractal analysis represents the dimensionality of patches, and Shannon's diversity index reports the relative abundance of different patch types (O'Neill et al. 1988, Turner et al. 1989, McGarigal and Marks 1995, reviewed in Turner et al. 2001). In contrast, our emphasis is on raster surfaces with continuous data, like elevation or values within satellite images (Musick and Grover 1991, Kiguli et al. 1999). We use several existing metrics, but also sought to create a new straightforward metric that unified spatial and temporal heterogeneity in forage resources. We also develop a metric of infrastructural heterogeneity based on economic and infrastructural data widely available from global databases, and inform these spatial analyses with qualitative information from extensive literature reviews to provide a larger political-economic context. The concept of infrastructural heterogeneity assumes that households exist within a matrix of services, resources, and accessibility options that define and limit their ability to compensate for fragmentation, and therefore affect the assumed transition towards intensive production. These analyses quantify some of the critical options that are available (or not available) for a range of households, but stop short of reflecting which services may actually be accessed on the ground. The baseline data we use are broad-brush, but the approach allows comparisons to be made across sites with different economic and ecological conditions.

Our heterogeneity metrics capture what Kolasa and Rollo (1991) term measured heterogeneity, or alternatively structural heterogeneity, rather than functional heterogeneity; we are not quantifying heterogeneity in the habitat of a single species or one type of livelihood. We provide measures of heterogeneity for a variety of sites, occupied by numerous species and people with a range of lifestyles. The metrics must be sufficiently general to inform a range of comparisons, while still capturing attributes salient to large herbivores and livestock owners. Further, we are not adding to the collection of literature that links landscape or infrastructural heterogeneity and process or function. We focus on two main components of heterogeneity, composition and configuration.

Our broader research team has worked in 22 sites around the world (Figure 14-1), some of which are cited in case-study chapters in this volume. Here we calculated landscape metrics on all 22 sites to provide a broad view of possible responses. A subset of nine sites were used in infrastructure analyses. They represent a range of current economic trajectories, land-tenure scenarios, and levels of livestock intensification. The two Northern Great Plains sites (Adams County, North Dakota and Perkins County, South Dakota) and the Northern Queensland paddocks and Victoria River District sites of Australia are private property-based, commercial livestock producers embedded in market-oriented livestock systems that enjoy relatively high levels of government support. The Vryburg, South Africa sites are split, and commercial private ranches and communal areas are analyzed separately. Both groups are enmeshed in an economy undergoing market liberalization and cutbacks in public services. Southern Kajiado District, Kenya provides an example of a largely subsistence-based but diversifying pastoral group within a communal land-use system that is in the midst of transition to private property. The Kenyan economy is engaged in a political-economic transition towards market liberalization and ongoing structural adjustment. The Moinkum Desert and Balkhash Basin sites in Kazakhstan represent a previously Socialist system that has changed to a market-based economy, and where previous state supports for pastoralism have declined.

2. MEASURES OF HETEROGENEITY

2.1 Within site heterogeneity

Metrics commonly summarize landscape heterogeneity through space *or* through time. Large herbivores of semi-arid and arid landscapes often cope with highly variable resources through *both* space and time. We sought a metric that would be: 1) able to summarize the spatio-temporal heterogeneity of an area in a single value, 2) applicable to resources selected by wild and domestic herbivores, or the people that manage them, 3) useful across spatial and temporal scales, 4) sufficiently general to be useful across all sites, and 5) straightforward. Central to the metric we developed is the movement of animals across landscapes to maximize resource access, and how that may vary as access changes. The metric may be applied to any resource that varies spatially, but forage is a dominant resource and is used here.

To calculate our herbivore-centric metric we:

1. Created raster surfaces representing forage through space and time;
2. Accrued the forage available to a sedentary animal by summing the forage at the location the animal was placed (or point in the image, see below) across the period of study;
3. Accrued the forage available to the animal if it was allowed to search within a small radius from the point it was placed for areas with the most forage available;
4. Repeated step three for increasingly larger radii, until an area larger than most herbivores will move had been searched.

The result is an increasing curve (or flat, in a perfectly homogenous landscape), a type of variogram that depicts how access to forage through time increases as animal movements are increased through space. The last step is to regress the square-root of the movement radii with the forage available for each distance (the square-root transformation tends to linearize access to greenness). The slope of the line summarizes the spatio-temporal heterogeneity of forage on the landscape surrounding the location where the animal was first placed. A highly dimensional measure of resource availability is thus integrated into a single metric.

Landscape heterogeneity metrics within sites are more easily standardized than for sites across continents. For example, sites often have available detailed land cover or vegetation maps. These maps, if well constructed, use consistent methods across the entire area so that land cover, for example, is defined using the same grain and classification throughout. As an example metric, a circular window of a given radius may be moved across a raster

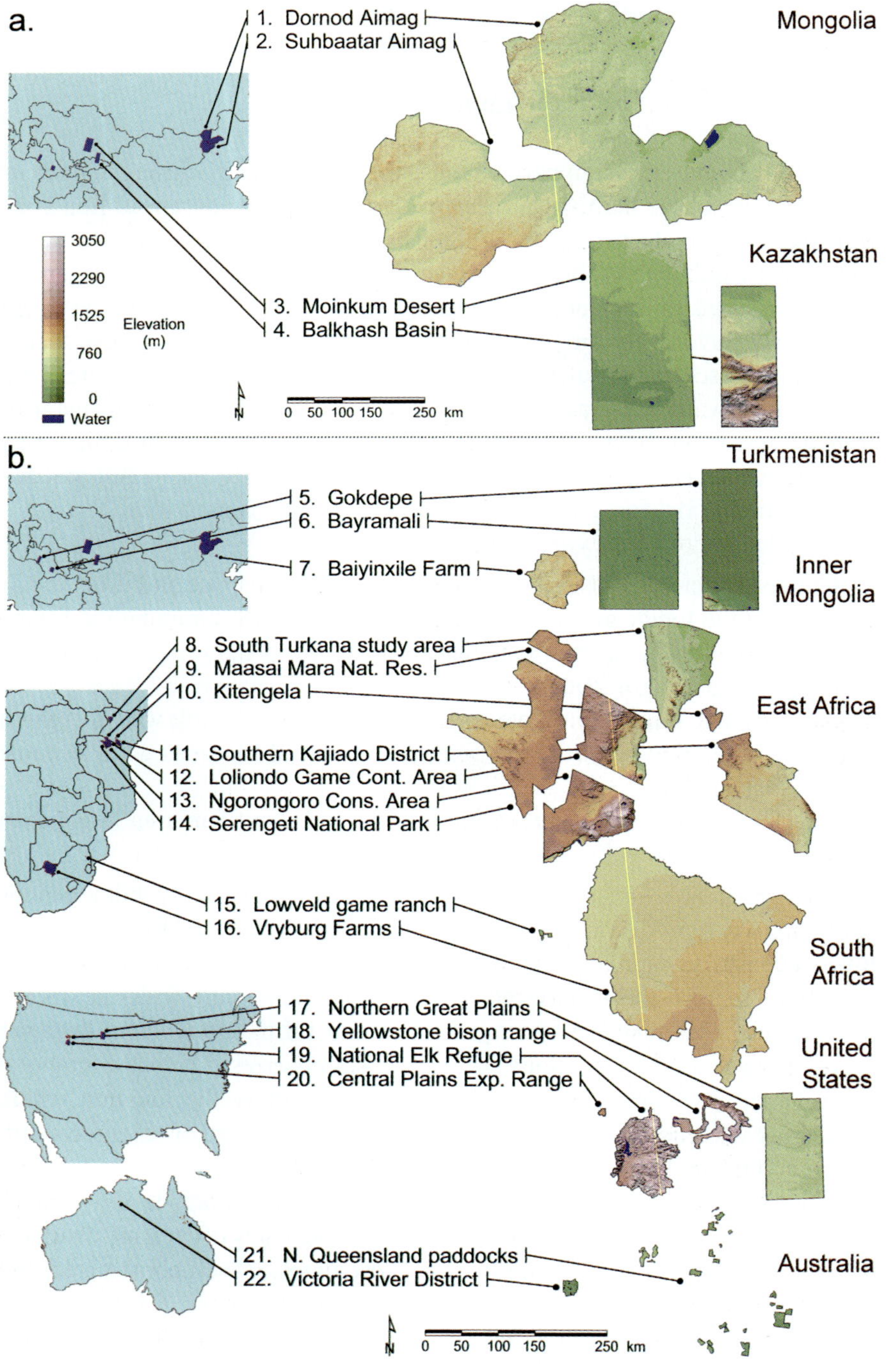

Figure 14-1. Study areas in which our team members work. The areas in the a and b panels are mapped at different scales.

image of land cover, counting the number of types within the window. Heterogeneity across many data surfaces may be summarized similarly (e.g., Kiguli et al. 1999, Fabricius et al. 2002). For example, a metric measuring the deviation of neighboring pixel values within Landsat TM satellite images was related to degradation from livestock grazing in South Africa (Tanser and Palmer 1999). Elevation has been represented in digital elevation models for decades, and recently measurements using space-borne radar (SRTM 2002) provide relatively high resolution data (each pixel represents about 90 x 90 m on the ground) for 80% of the Earth's land mass. Analyses that calculate the standard deviation of elevation for pixels within a moving window will create a surface depicting landscape heterogeneity within a study area (e.g., BurnSilver et al. 2003). A difficulty is determining the dimensions of the moving window (Fabricius et al. 2002). Ideally, the window radius is related to the area available to the organism of interest, such as the dimension of a species' home range or the grazing orbit of livestock. Alternatively, multiple resolutions may be used and each tested to see which resolution best correlates with or discriminates responses. Lastly, a radius may be selected that incorporates enough pixels (ca. 30 or more) to yield deviation estimates of sufficient precision and information without over-smoothing the results.

A measure tied to temporal landscape heterogeneity is variation in greenness across time, as shown in surfaces derived from satellite images. Soon after weather satellite images were being acquired, researchers found that a combination of images could be used to estimate plant vigor and biomass (e.g., Tucker 1979). The images, called Normalized Difference Vegetation Indices (NDVI), have since been correlated with annual net primary productivity in grasslands (Paruelo et al. 1997) and with regional stocking of livestock (Oesterheld et al. 1998). NDVI images are now generated from many sources and used widely in research.

Our herbivore-centric metric of forage heterogeneity may be calculated within sites, to quantify across space access to green forage through time. In within-site and between-site analyses, forage was represented using Spot VEGETATION NDVI images from 1998 to 2003, with 36 one-km resolution images per year. To focus on rangelands that provide forage to large herbivores, we used the MODIS land cover map to identify grasslands, savannahs, and woodlands (Lotsch et al. 2003). Southern Kajiado District, Kenya, was used in this example. For each 1 km pixel in the image and each 10-day image in the set, a custom program identified the greenest pixel within a search radius of 500 m, 1000 m, 1500 m … 29,000 m, 29,500 m, 30,000 m, and a running tally of greenness accessed was made. Figure 14-2 portrays the slope (a) of the variogram created for each pixel, and the linearity (b) of the variogram for each pixel. Darker regions in Figure 14-2a depict areas where effects of landscape fragmentation would be more minor for resident ungulates, and light regions where it would be more severe.

2.2 Between site heterogeneity

2.2.1 Landscape Heterogeneity

Many of the heterogeneity metrics useful within sites are useful to compare between sites as well. Other metrics, such as patch richness density, can be misleading if not carefully applied. Land cover mapping is often a subjective process and dependent upon specific classifications. Different teams mapping land cover in different sites are unlikely to apply the same methods. For example, one may tend toward using a smaller mapping unit, so that a smaller average patch size in their region is an artifact unrelated to ecosystem traits. Moreover, classification systems in the eastern US can be very detailed in some aspects, such as deciduous forest types, and simple in classifying deserts, whereas a system used in Asia may divide deserts into multiple categories (e.g., semi-desert, desert steppe, desert).

We used the digital elevation model cited earlier to represent variability in elevation in each of the study areas (SRTM 2002). A standard deviation can vary with the area sampled, so a standardized area was used. The mean and standard deviation (SD) of elevation within 50 circles, each five km in radius, was calculated for each area. Table 14-1 shows the average of those values, plus the area of each of the sites, and other metrics. The diversity of the Balkhash Basin (site 4, SD 135.2 m) compared to the other Middle Asian sites (sites 3, 5, 6, SD 15.2, 7.2, 8.1) is evident.

A useful climatic variability index for a site is the coefficient of variation (CV) in total annual precipitation. This index is the standard deviation in precipitation divided by the mean, multiplied by 100 to convert to a percentage, yielding a unitless measure of dispersion around the average annual precipitation. We did not calculate the CV of precipitation for sites, because the collection of rainfall data is not uniform across space or time, which can complicate comparisons. Instead, NDVI images from 1981 to 2003 were used to calculate integrated indices and their CVs. These images are acquired in a uniform way across the globe, which makes comparisons more rigorous. Interpretation of the final metrics must still be done with care, given differences between sites, such as whether or not snow cover is present part of the year. We created greenness profiles for each of the sites, then calculated integrated NDVI (i.e., summed values) and interannual CV in integrated NDVI (Table 14-1). The most variable site was South Turkana (site 8, CV 12.5%), the site where Ellis and Swift (1988) first described

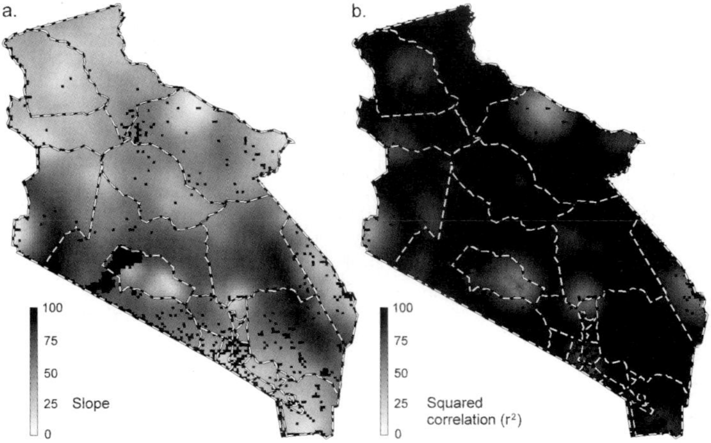

Figure 14-2. The (a) slope of the space-time variogram for southern Kajiado District, Kenya. The (b) relative fit for each km^2 is represented by r^2 values. Group ranch boundaries (dashed lines) and non-grazing lands (black 1 km^2 blocks) are shown.

live-stock population changes more closely associated with drought frequency than with primary production. Southern Kajiado District was almost as variable (site 11, CV 12.0%) as Turkana, and the Kitengela region (10, 8.4%) is quite variable as well, with high losses of livestock during drought (Reid et al., Chapter 9). Rainfall in Kajiado is highly variable, but the dominance of facultative deciduous brush-shrub habitat may make inter-annual variations in NDVI more extreme than in grass-dominated habitats.

The most recent mapping of global land cover used MODIS satellite data (Lotsch et al. 2003). The images were composed of pixels representing 926 x 926 m on earth, classified into 18 land cover types, including classes for water and unclassified pixels. Another commonly used dataset maps seasonal land cover regions (SLCR) at 1 km resolution across the globe, with between 130 and 260 types per continent (Loveland et al. 2000). We counted the diversity of land cover types in each of our study areas (Table 14-1). Standardizing the SLCR count by area (Table 14-1) quantified the low diversity in Asian steppe regions (e.g., sites 1,2,3), and high diversity of East Africa (e.g., 9,10), and the extremely diverse Yellowstone bison range (site 18). Two sites (15,20) are so small that the metrics are unreasonably high.

We applied the method described to create the herbivore-centric metric to each of the areas, generating variograms relating distance moved to resources acquired. In these analyses between sites, for each area and search radius, 1000 points were sampled, relating distance moved from that point to

Table 14-1. Heterogeneity and other metrics for areas studied by our research team.

	Study area	Area	Elevation		Int. NDVI[b]		Cover types			Variogram	Landsat
			0	σ	∑	CV	MODIS	SLCR	SLCR	Slope	Slope
		(km^2)	(m)	(m)	(index)	(%)	(n)	(n)	(n/km^2)	(ind./km)	(ind./m)
1	Dornod Aimag	123,618	818	25.1	1300	9.0	15	55	0.0004	5.42	-0.209
2	Suhbaatar Aimag	82,517	1089	28.6	1439	10.1	9	26	0.0003	4.69	-0.175
3	Moinkum Desert	71,958	456	15.2	824	6.9	13	29	0.0004	5.60	-0.250
4	Balkhash Basin	30,047	1293	135.2	1525	9.7	15	62	0.0021	9.62	-0.253
5	Gokdepe region	13,378	100	7.2	514	7.5	8	26	0.0019	6.56	-0.159
6	Bayramali region	13,298	254	8.1	737	9.6	9	19	0.0014	7.49	-0.086
7	Baiyinxile Farm	3547	1212	33.2	1482	8.6	9	10	0.0028	5.51	-0.249
8	South Turkana study area	8874	785	49.9	1235	12.5	10	30	0.0034	15.74	-0.151
9	Masai Mara Nat. Res.	1480	1594	31.4	2863	5.6	7	17	0.0115	7.89	-0.257
10	Kitengela region	453	1671	25.8	2171	8.4	6	9	0.0199	8.25	-0.246
11	Southern Kajiado District	10,741	1252	30.5	1850	12.0	14	48	0.0045	15.68	-0.211
12	Loliondo Game Cntrl Area	7269	1676	74.1	2601	4.2	14	41	0.0056	13.37	-0.245
13	Ngorongoro Cons. Area	8240	1883	136.4	2354	4.8	14	52	0.0063	16.58	-0.143
14	Serengeti National Park	13,044	1558	30.5	2820	4.6	12	37	0.0028	8.82	-0.185
15	Lowveld game ranch[a]	75	585	267.3	2601	4.2	2	3	0.0400	6.51	-0.236
16	Vryburg farms	49,222	1174	12.5	1832	7.4	12	29	0.0006	5.87	-0.296
17	Northern Great Plains	10,009	795	19.5	1985	6.5	7	13	0.0013	5.99	-0.373
18	Yellowstone bison range	2050	2438	297.7	1802	4.3	14	32	0.0156	8.85	-0.358
19	National Elk Refuge	6090	2435	158.5	1780	4.9	12	35	0.0057	12.08	-0.353
20	Central Plains Exp. Range[a]	68	1668	136.2	1444	7.8	2	2	0.0294	4.89	-0.310
21	N. Queensland paddocks[a]	1313	394	149.5	2826	6.1	7	11	0.0084	4.65	-0.284
22	Victoria River District[a]	326	167	39.4	1968	8.2	6	2	0.0061	5.65	-0.297

[a] – Areas or patches were too small to be sampled (see Methods). Elevation and standard deviation of the entire area is reported.
[b] – Int. (Integrated) NDVI indices is based upon long-term (1981-2003) coarse-resolution greenness indices. The metric ∑ is the integrated annual NDVI reflecting productivity, and the interannual coefficient of variation (CV) in integrated NDVI is shown.

the maximum greenness that could be accessed. Example variograms are shown (Figure 14-3) for four sites near each other in East Africa. The relative productivity of the areas is shown by their intercepts (β_0 in the regression), with the South Turkana study area (site 8) the least productive of the sites, and the Serengeti (14) the most productive. That said, the integrated NDVI index is a more direct measure (Table 14-1). The slopes of the lines (β_1) summarize the spatio-temporal heterogeneity of greenness at the sites. The most diverse area is Ngorongoro Conservation Area (site 13, slope 16.58). Its CV in NDVI is relatively low (Table 14-1) and droughts are not severe in the area, but it has extreme topography, spanning from areas along the shore of Lake Eyasi below 1,200 m, to mountain peaks above 3,000 m. The South Turkana study area and southern Kajiado District (site 11) have similar diversities (slopes 15.74 and 15.68, respectively), although Kajiado is more productive. The least diverse of the areas is Serengeti National Park (site 14, slope 8.82), reflecting its sweeping grasslands and mixed brush-lands. The variogram slopes for all sites are shown in Table 14-1 (all $P < 0.05$; $r^2 > 92\%$).

Our final metric quantifies land cover diversity at a fine scale, based on the correlation of pixel values across space. We used mosaiced Landsat ETM+ bands 7, 4, and 2 from the growing season in approximately the year 2000 (i.e., GeoCover 2000 from GLCF 2005). The images were generalized from their native 14.25 m pixel resolution to 57 m to speed analyses. Then a

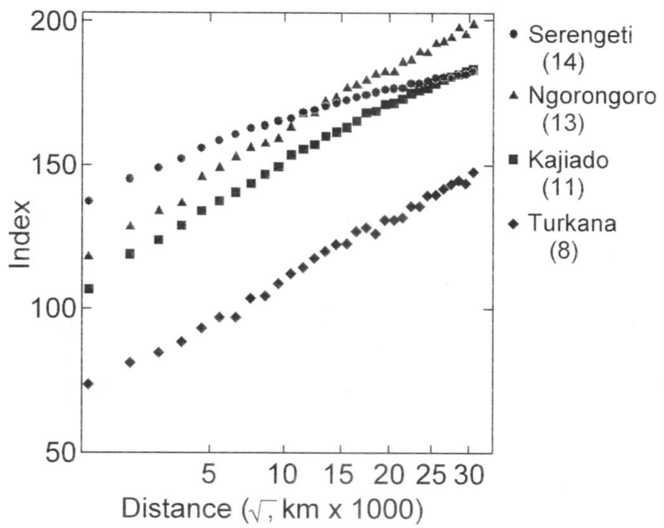

Figure 14-3. Variograms relating movement radius to greenness accessed in four East African sites. The numbers in parentheses identify the areas in Table 14-1.

principal component analysis was used to reduce the dimensionality of the images to a single band that captured the majority of variation in the original three bands (mean variation described by the first component, 72.8%, SD 9.8%, 22 sites). The metric is based on a commonly used test in spatial analyses, where the spatial autocorrelation in an image is calculated by shifting the image by progressively larger amounts (i.e., lag distances), creating a correlogram. Here, for each area, we shifted the image in eight directions (N, NE, E, etc.) by 60 m, 100 m, 150 m, 200 m … 800 m, 900 m, 1000 m, 1500 m, and 2000 m. The average correlation was calculated from the eight directions, then a correlogram was formed and the slope of the line calculated. Figure 14-4 depicts the areas with the extremes in slopes based on Landsat spatial autocorrelation: Bayramali region in Turkmenistan (site 6) and the Northern Great Plains, USA (site 17). The Bayramali region is bisected by a lush river valley, with the bulk of the area grassland and desert habitats, yielding high autocorrelation. In contrast, the agricultural areas of the Northern Great Plains are extremely diverse, with cultivated areas at different phenological stages interspersed with fallow lands, natural vegetation, riverine networks, forest pages, and developed areas. The slopes of the correlograms for all areas are in Table 14-1 (all $P < 0.05$, $r^2 > 90\%$).

2.2.2 Heterogeneity in Access to Infrastructure

Starting from an assumption that resources and productive inputs are necessary in order to compensate producers for declines in access to spatial scale, the question arises, "Which specific resources must be available to at

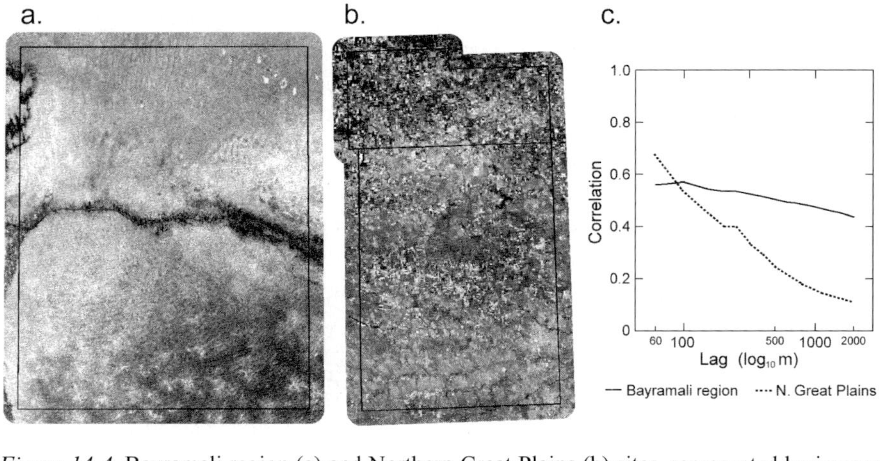

Figure 14-4. Bayramali region (a) and Northern Great Plains (b) sites, represented by images derived from Landsat ETM+ data. Bayramali has the highest autocorrelation across distances (c) of the sites we quantified, and the Northern Great Plains (c) the lowest.

least begin offsetting losses?" Access to small-scale credit for producers is identified as a significant step in raising production levels in areas as diverse as Kenya/Ethiopia (pastoralists), the Amazon (ranchers), and Albania (farmers) (Nela and Marshall 1999, Merry et al. 2004, Desta et al. 2004). However, identifying effective mechanisms to distribute credit in rural zones is problematic, and often informal credit arrangements offer more immediate options to producers (C. Kerven, pers. comm.). Njenga and Davis (2003) concur that credit facilities are important, but stress as well that strong transportation networks imply greater market linkages and opportunities to decrease poverty by increasing access to basic needs. De Wolff et al. (2000) identified that market access as well as veterinary services were strongly correlated with distance to market centers for agropastoralists in the Kenyan highlands. Fan and Zhang (2004) found that in China, infrastructure indicators such as roads, electrification density, and telephone network density played a significant role in explaining agricultural productivity levels across rural areas. In a similar vein, Dadibhavi and Vaikunthe (1990) found that in India literacy rates, density of financial institutions, road length, area irrigated, and fertilizer consumption explained trends in rural disparities. For the purposes of our analyses, examination of the pastoralism literature suggested which infrastructure types have been considered critical as a foundation for intensification of livestock production strategies. These include access to credit, livestock markets, road networks, veterinary services, disaster relief, and in some cases, input subsidies (Rutten 1992, White and O'Meagher 1995, Blench 2001, Desta and Coppock 2004).

No objective measures representing access to credit facilities and livestock markets existed for the nine study areas. We therefore calculated a metric representing infrastructure and access for each study area based on three variables: the indexed (value 0-1) mean light intensity (using satellite images described below) and road density inside each study area, summed with the inverse of the average distance to the nearest urban area with > 20,000 people. We then compared these values to each sites' human population density. Linking human density to this accessibility metric highlights where infrastructure services would be in the greatest demand under intensified livestock production. We use road density as a proxy for ease of market transport and movement, while degree of electrification and distance to urban area are proxy measures meant to reflect relative access to services that are associated with more urban market areas (Table 14-2).

All distance measures represent average distances to services within each study area. The global database Night-time Lights of the World (NOAA 1998) was used to calculate average distance to any light defined as any grid

Table 14-2. Measures used to define metric of infrastructure and access to services.

Study Sites	Mean road density (m/km^2)	Mean light intensity	Urban area distance (km)	Population density (per km^2)
North Dakota	153.00	1.52	139.83	1.02
South Dakota	101.32	2.04	140.15	0.44
Victoria River	0.00	0.00	493.94	0.15
Dalrymple Shire	3.66	0.11	130.70	2.28
Vryburg Commercial	119.00	0.23	97.49	4.20
Vryburg Communal	154.00	0.39	111.00	22.95
Kajiado	0.05	0.00	73.52	18.76
Moinkum Desert	76.33	0.02	221.93	0.70
Balkhash Basin	85.42	0.90	57.47	5.50

cell having ≥ 10% light detected within any 1 km^2 area, using ArcView GIS Software (ESRI, Redlands, CA, USA). We quantified distance to urban area based on the GRUMP (Global Rural Urban Mapping Project) urban areas database (CIESIN 2005), then calculated mean distance for the study areas and buffer zones. We calculated a road density (road length in km for each 1 km grid cell, km/km^2) based on the Digital Chart of the World edition 3 roads database (NIMA 1997). Mean population density values were initially pulled from the GRUMP database (CIESIN 2005), then cross-checked and corrected based on national level data for Australia and Kazakhstan (Akcura et al. 2002, ABS 2006, Republic of Kazakhstan 2006).

It is clear, however, that access to resources consists of more than simply presence of infrastructure on the pastoral landscape. We extended the spatial analyses described above to: 1) identify the level and type of governmental support services in place for pastoral/ranching populations, 2) identify accessibility of formal credit, 3) identify type of subsidies, and 4) identify type of veterinary services available within each of the study areas. Whenever possible, we made cross-site comparisons as to the level of services offered (high/low), but the diverse nature of the source data limited the degree to which the levels of resources were directly comparable. The type of support across sites is described, ranked and summed. To facilitate baseline comparison of how well livestock producers are doing across these diverse sites, Human Development Indicator (HDI) scores and per capita income across the nine sites are also presented. The HDI is a measure of basic needs that combines life expectancy, educational attainment, and income into a composite human development index (Watkins et al. 2005).

2.2.3 Where is the Foundation for Livestock Intensification in Place?

Our goal in this section is to compare the heterogeneity of services and productive infrastructure across the nine sites based on a literature review of available sources. Results are summarized in Tables 14-3 and 14-4.

The United States (US) has a long history of substantial government support for its agricultural sector, both in terms of policy and economic programs that contribute directly to household income and mitigate risks associated with ranching and farming in poor years. Farmers and ranchers in Perkins County, SD received $66.5 million in subsidies during the period between 1995-2004, while households in Adams County, ND received almost $72 million in support during the same period (EWG Farm Subsidy Database 2006). Beutler (2003) identified that 50% of farm net cash income in South Dakota came from government payments. Subsidies were in the form of commodity payments for specific crops, conservation programs, and disaster programs, many of which relate directly to livestock. Most households in the two counties combine cropping with livestock in some form (Lackett and Galvin, Chapter 5), so that households can draw from these sources of support simultaneously. Emergency livestock feed assistance is available in drought periods. Access to credit along formal (e.g., banking) channels for producers is high. Survey results for the Northern Great Plains sites indicated that 80% of households (N=60) currently had outstanding loans (Jennings 2000). Economically, rural households in Adams and Perkins Counties have incomes that are below the US average of $37,562 (UNDP 2003), however, incomes per capita are well above all other sites compared here. The HDI and combined infrastructure score is high.

The Australian (AU) government also has in place at national and state levels a variety of support mechanisms that are designed to help livestock producers. In 1992 the Australian government adopted a National Drought Policy that promotes a "self-reliance" approach for producers managing resources and making production decisions in dry and variable environments (White and O'Meagher 1995). If an area is declared to be under "Exceptional Circumstances," national level services available to livestock producers include six months of income support, special taxation measures, and farm management deposits that allow producers to set aside income in good years for use in bad years (DAFF 2005). State governments also offer a variety of drought services to producers (e.g., fodder and livestock transport, electricity tariffs) (Wilson et al. 2004). Formal credit services are available to producers, as are interest rate subsidies in drought periods. Veterinary services are privatized, and while there is some indication that

Table 14-3. Summary measures reflecting infrastructural access to resources on a site-wide basis.

Study countries	*Government interventions*	*Credit access formal or informal*	*Input subsidies*	
			Fodder assistance	*Veterinary inputs*
United States	Commodity subsides (4) Conservation payments Disaster payments Crop insurance	Formal: *High* (3) 80% of sampled households using credit	Emergency livestock feed assistance (1)	Privatized (2) veterinary systems
Australia "Self-reliance policy"	Drought preparation (3) Income support Government Envirofund	Formal: *High* (3) Interest rate relief (on existing loans) Farm management deposits	Freight assistance (2) scheme Fodder Livestock	Privatized (2) veterinary systems
South Africa "Greater owner responsibility"	Subsidies eliminated (0) in commercial areas Fencing Irrigation On-farm infrastructure	Formal: Commercial: *Med-High* (2) Communal: *None-Little* (1) Few rural delivery mechanisms Informal: "Owner Associations"	None (0)	*SAPs cutbacks* (1) Reduced government extensive services Privatizing services
Kazakhstan "Privatizing system"	State support lost (0.5) Breed improvement subsidies only to wealthy producers	Formal: *Low* (1) Credit only to those with collateral. Few delivery mechanisms. Informal: Kin-based loans to extended households	Fodder assistance (0.5) ends Some production loans for inputs	*Cutbacks post-* (1) *market liberalization* Privatizing services
Kenya	None (0)	Formal: *Low* (1) Only to those with title deeds or wage jobs. Few delivery mechanisms. Informal: Kin-based loans, livestock associations.	None (0)	*SAPs cutbacks* (1) Privatizing services

Summary measures: High values imply greater access to resources on a site basis: Government interventions - # program types (0-4); Credit access (Low-1, Med-2, High-3); Fodder assistance: # types (0-2); Veterinary inputs: System evolving – post cutbacks (1), privatized system in place (2).

Table 14-4. Summary measures of economic welfare and site infrastructure.

Study Sites	HDI	Per capita GDP ($PPP) [d]	Infrastructure Score Totals [e]
US-North Dakota	0.944	18,425	10
US-South Dakota		15,734	
AU-Victoria River	0.955	[a] 27,520	10
AU-Dalrymple Shire		[a] 26,161	
SA-Vryburg Commercial	0.606	[b] 5,348	4
SA-Vryburg Communal			2
KZ-Moinkum Desert [c]	0.678	3,226	3
KZ-Balkhash Basin [c]	0.709	4,436	
KE-Kajiado	0.468	629	2

[a] Average annual income per capita for Northern Territory (excluding Darwin) and Dalrymple Shire, Queensland. [b] Figure represents GDP per capita for all of North-West Province, SA. [c] Statistics are for Zhambyl District (Oblast), and Almaty District (Oblast), respectively. Sources: US: US Census Bureau 1999, US Census Bureau 2000; AU: Australia Bureau of Statistics (ABS) 2006; SA: Adelzadeh et al. 2003; KZ: UNDP Human Development Report 2001, Akcura et al. 2002, Dimitri et al. 2005, Republic of Kazakhstan: Agency of Statistics 2006; KE: UNDP 2001. [d] GDP per capita values are based on purchasing power parity (PPP), which equalizes the purchasing power of different currencies. [e] Infrastructure scores are summed from Table 14-3.

some livestock producers prefer to self-administer to their stock to lower their costs, Australia has higher per capita access to veterinary services than the US, Canada, or the United Kingdom (Frawley 2002). The HDI index and per capita Gross Domestic Product (GDP) values for Australia are high, as is the infrastructure score.

The economy of South Africa (SA) has struggled in recent years. Structural adjustment programs and efforts to liberalize market structures and sustain economic growth have led to a decline in government spending of 17.6% and 25.6% on social and economic infrastructure, respectively, since 1995 (UNDP 2003). During the Apartheid era, substantial subsidies were in place for commercial (and largely white) farmers, primarily borehole development and fencing. The end of Apartheid brought efforts to abolish the two-tiered system of service provision to white versus black areas, however, the government also eliminated all drought relief in 1994 and reduced the government role in extension and livestock marketing services (SA DOA 1998). In terms of policy, the current reorientation is towards "greater owner responsibility," reduced reliance on state subsidies, privatizing basic services, emphasis on emerging private associations of cattle owners to exchange expertise, and supporting the emergence of black commercial farmers (UNDP 2003). Access to credit is improving but remains limited, especially for black producers (UNDP 2003), and an ongoing concern is that

"Credit continues to go to the credit worthy – those who need it least" (UNDP 2003). Substantial development challenges remain. The Provincial GDP per capita figure is \$5,348, and while high for the developing sites focused on here, the figure disguises substantial inequality between groups (UNDP 2003). The HDI score for the region is the second lowest of all nine sites. Infrastructure scores for communal producers are very low, but are higher for commercial producers.

The Kazakhstan (KZ) economy underwent a dramatic shift towards the free-market in the mid-1990s, a shift that led to the break up of communal farms and the suspension of government price and infrastructure supports for the livestock and agricultural sectors. This led to the very dramatic collapse of the country's livestock economy. Breeding, veterinary and marketing services, and fodder provision were basically eliminated in a short period of time (Kerven et al. 2003). Since 1998, the Kazakhstan economy has enjoyed steady growth, but livestock producers have been faced with a depreciated asset base, limited financing options, and continued low purchase prices for their animals and products (Akcura et al. 2002). However, the government has recently begun to put substantial resources back into the livestock sector based on oil and gas revenues. One example is subsidies for pedigree livestock farms, but these subsidies go to larger producers with larger herds (WB 2004). Credit accessibility remains generally low for Kazak livestock producers and is identified as a critical bottleneck to development. However, while limited, richer producers with collateral do have greater options to access credit (Akcura et al. 2002). In 2002, the Kazakhstan government passed legislation to privatize and initially regulate veterinary services, but costs remain high for producers (WB 2004). The HDI scores are relatively high. GDP per capita and infrastructure figures are low, and differ between rural Moinkum and more centrally located Balkhash.

The Kenyan (KE) economy suffered a dramatic period of reduced growth throughout the 1980s and 1990s. In exchange for development loans and aid, the Kenyan government agreed to a series of structural adjustment mechanisms aimed at trade liberalization, reforms in the civil service, and reining in government expenditures. For the pastoral sector and rural areas in general, these measures resulted in declines in government support for education, health services, rural extension, and veterinary services (UNDP 2001). The arid and semi-arid lands of the country were particularly hard hit by these retractions in services as they already had received low development priority in the past (ADF 2003). Financial institutions are few and far between in Kajiado, loans are usually of short duration and given at high interest rates, and only wealthy pastoral producers or wage earners have the collateral needed to guarantee loans (Swift et al. 2002). Credit services in pastoral areas are perceived to have high transaction costs associated with seasonal mobility and dispersion of pastoral people. Livestock associations and lending between kin or stock associates are local-level answers to the lack of formal credit channels. Privatization of veterinary services and use of community animal health workers were strategies envisioned as a replacement for government provision of veterinary services (Mugunieri et al. 2002). However, while pastoral producers are willing to pay for veterinary services, there have been institutional delays in getting such a system up and running in rural areas and most pastoralists administer their own veterinary care. Kajiado is by far the poorest study site and has the lowest HDI and infrastructure scores across all sites.

2.2.4 Spatial Heterogeneity in Access to Infrastructure

The above review begins to address whether sufficient infrastructure and government support exists across the nine sites to support the intensification of livestock production strategies. The application of an infrastructure metric linked to human density measures additionally informs this question. Results for the Dalrymple Shire and Victoria River sites in Australia, and South Dakota in the United States indicate that although these sites are already very far along the trajectory towards intensification (Stokes et al., Chapter 4; Lackett and Galvin, Chapter 5), they have low population densities and low to medium access to services (Figure 14-5). However, we can assume that given the relatively high per capita incomes of producers in these sites, households have the means to travel the distances necessary to access services. The US North Dakota site is similar to the other US and AU sites in that population densities are very low, but physical proximity to services is high. Interestingly, we see that while there is an ongoing emphasis on achieving greater productive efficiency and intensification, producers in these areas are making concerted efforts to consolidate their parcels into larger productive units. These producers enjoy ‘ideal’ land tenure arrangements (e.g., private property), and high levels of government support, yet the trend is away from livestock production on fragmented parcels.

Although Kazakhstan sites have similar political and economic histories, producers in these sites have very different baseline infrastructure resources. The Moinkum Desert site is isolated from urban areas and electrification, and while some roads exist, they are in poor condition. On average, producers in the Balkhash Basin site are closer to Almaty City and Uzan Agach District Center, both of which have livestock markets (Laca 1998) and offer ostensibly better access to resources. Road density is relatively low, but these roads are being paved and improved (C. Kerven, pers. comm.).

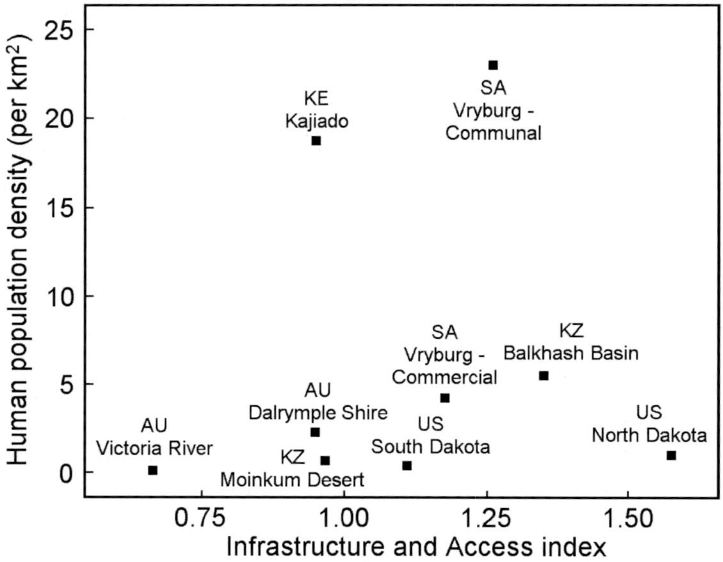

Figure 14-5. An index of infrastructure and access to economic resources versus human population density for selected sites.

Livestock producers in these sites are making efforts to re-aggregate access to rangelands that are of high quality, but also isolated from infrastructure and services (Kerven et al. 2006). However, only well-off households have the resources to regain seasonal mobility for their herds. Similarly, richer households are those more able to access livestock breed subsidies and credit options (Akcura et al. 2002, WB 2004).

The value of infrastructure and services at the South Africa Vryburg commercial site seems to be lower even than the South African communal site, but this is likely a function of lower population densities and larger ranch sizes on the commercial areas (Galvin et al., Chapter 12). South African commercial ranchers overall have medium access to resources and infrastructure, and historically commercial ranchers enjoyed much greater access to infrastructure and government support than communal producers (e.g., water point development and fencing subsidies). However, the South African government has pulled back strongly from overt inequalities in provisions, while at the same time market liberalization policies have pushed overall government support for livestock producers down. This suggests that further gains in infrastructure and support for commercial producers will be difficult.

Two study areas, Kajiado and the South African Vryburg communal areas, stand out as sites with very high population densities. Both these sites are characterized by agropastoral and pastoral activities that have livelihood strategies still focused on food security. The Vryburg communal site seems to have medium-high relative access to services and infrastructure. However, prior native homeland areas in South Africa continue to have low access to credit and other critical infrastructure. Those in Kajiado have both distance and accessibility obstacles to overcome. Population density is high, but service availability is low. Fragmentation processes are ongoing in these two sites.

3. DISCUSSION

Readers of the case studies in this volume may place the heterogeneity of the sites discussed in a global context using Tables 14-1, 14-3, and Figure 14-5. Taking the metrics in aggregate, the Balkhash Basin (site 4 in Table 14-1) is the most heterogeneous of the Asian sites, much more heterogeneous than Moinkum Desert (3), the other Kazakhstan site. Movements by pastoralists 80 years ago, as cited in Kerven et al. (2003), were short for Balkhash residents but much longer for residents of Moinkum, as predicted from the heterogeneity measures. The most homogeneous of the Asian sites is Suhbaatar Aimag, Mongolia (2). Topographically and vegetatively, Ngorongoro Conservation Area (13) is the most heterogeneous of the African sites, although the South Turkana study area (8) has the most diverse climate. Ngorongoro is used as a drought refuge by Maasai because of its diverse habitats and perennial water sources. Of the four East African sites plotted in Figure 14-3, the one we might predict most likely to support large migratory herds based on their variograms would indeed be Serengeti (14), home to world-renowned migratory ungulates. Fragmentation that truncated the migratory movements of wildebeest would have extreme effects in this homogeneous region, but the size and conservation status of the ecosystem make fragmentation less likely. In the US, neighbouring Yellowstone bison range and National Elk Refuge (18 and 19) are both heterogeneous, with Yellowstone topographically diverse, and the refuge and surrounding area diverse in forage availability, promoting migrations in wildlife that are now truncated (Lackett and Hobbs, Chapter 6). In Australia, the Northern Queensland paddocks are topographically diverse, but the Victoria River District has a more variable climate, represented by higher CV in NDVI (Table 14-1).

Our results suggest that previous assumptions of policymakers and rangeland managers that intensification on finite parcels of land should be the ultimate goal for pastoralists and ranchers may in fact be overly simplistic. Intensification combined with fragmentation may not be the evolutionary end point of livestock production strategies, particularly in arid landscapes that are heterogeneous. We see that re-aggregation (in Kazakhstan) and consolidation (in the Australia and US sites) is occurring regardless of existing infrastructure availability and support mechanisms. These sites should have had, particularly in the United States and Australia, the greatest potential to offset the costs of supporting livestock production on fragmented landscapes. Conversely, fragmentation is still either ongoing or a feature of livestock production in two sites (Kajiado and South African communal territories), which are arguably the *least* able to support intensification efforts, given the low levels of current government support, low availability of critical infrastructure, high human population densities, and the fact that fragmentation in those areas has been shown to cause declines in livestock and human well-being (Boone et al. 2005, Thornton et al. 2006). Areas where fragmentation would be of greatest concern are those that are relatively homogenous and with poor infrastructure, such as Moinkum Desert (site 3) and the Gokdepe and Bayramali regions of Turkmenistan (5,6). The collapse of the socialist government in these regions in the early 1990s led to severe recession. Privatization of land has begun in these systems and could exacerbate the need for additional inputs to offset fragmentation. Instead, pastoralists are slowly recreating their traditional seasonal movements, although distances involved are shorter (Humphrey and Sneath 1999, Kerven et al. 2003, Alimaev and Behnke, Chapter 7).

The infrastructure results clearly just scratch the surface of linkages between productive infrastructure, intensification, and fragmentation pressures but they do suggest a variety of other criteria beyond provision of basic services that define whether intensification will occur in pastoral systems. We see that other important conditions 'gatekeep' access to critical services *when they do exist* such as: household wealth levels (in KZ and SA), historical and policy factors (e.g., racial inequalities in service provision in SA), and perceptions of financial risk (e.g., lowering the provision of credit to collateral-poor pastoral populations). Similarly, within the developing economies represented (Kazakhstan, Kenya, and South Africa), improving the well-being of poor people in rural areas, where pastoral producers are, is a commonly articulated goal. However, under market liberalization and structural adjustment pressures, government spending has been skewed towards high growth sectors and high potential areas, zones that do not include rural, semi-arid pastoral lands (UNDP 2001, Akcura et al. 2002, Adelzadeh et al. 2003).

Our herbivore-centric metric of resource heterogeneity through space and time is relatively robust to differences in the extent and grain of landscapes studied (Wiens 1989), assuming a change in extent incorporates similar landscapes. The slope of curves calculated using images from different satellite sensors or data that are lower resolution are not directly comparable, but comparison among sites calculated with data defined in the same way remains appropriate. The technique also may be applied to very large or very small landscapes, if the resolution of the resource maps is high enough to depict sufficient detail for the sites. The appropriateness of the resource surfaces used will likely vary from area to area. For example, ungulate populations may be limited by forage biomass in some areas, nutritional quality in others, and disease in still other areas, but the metric provides a unified starting point for comparisons.

Landscape heterogeneity affects a myriad of ecosystem functions, such as dispersal ability (Gustafson and Gardner 1996), species packing (Nee and Colegrave 2006), fluctuations in populations (Roff 1974, Illius and O'Connor 2000, Floater 2001), and pastoral welfare (BurnSilver et al. 2003, Boone et al. 2005, Thornton et al. 2006). Infrastructural heterogeneity is equally relevant in the context of fragmentation given the need for services and infrastructure to fill the productive gap between extensive versus intensive livestock production systems. The metrics we report put the ecological and infrastructural heterogeneity of sites in the case studies in context. Our results are relevant to ongoing discussions over processes of fragmentation and its appropriateness in pastoral and ranching environments.

ACKNOWLEDGEMENTS

This work is supported by a grant from the U.S. National Science Foundation (DEB-0119618). The helpful comments of external reviewers Eric F. Lambin and María E. Fernández-Giménez are appreciated.

REFERENCES

ABS (Australian Bureau of Statistics). 2006. Latest regional profile by location name (Main Areas). http://www.abs.gov.au/AUSSTATS/.

Adelzadeh, A., C. Alvillar, G. Mhone, H. Bhorat, M. Leibbrandt, et al. 2003. South Africa: The challenge of sustainable development. Human Development Report. United Nations Development Program: 295.

ADF (African Development Fund). 2003. Kenya: ASAL-Based Livestock and Rural Livelihoods Support Project, Agriculture and Rural Development Department North, East and South Regions. Appraisal Report http://www.afdb.org/pls/portal/docs/PAGE/ADB_ADMIN_PG/DOCUMENTS/OPERATIONSINFORMATION/ADF_BD_WP_2003_162_E.PDF. 2006.

Akcura, F., S. Zhusupov, Y. Shokamanov, Z. Mukhamedkarmova, E. Gossen, et al. 2002. Rural development in Kazakhstan: Challenges and prospects. Human Development Report. United Nations Development Program Almaty, 123.

Ash, A., J. Gross, and M. Stafford-Smith. 2004. Scale, heterogeneity and secondary production in tropical rangelands. African Journal of Range & Forage Science 21:137-15.

Beutler, M.K. 2003. Impact of South Dakota agriculture 2002. South Dakota State University, Brookings.

Blackburn, T.M. and K.J. Gaston. 2002. Scale in macroecology. Global Ecology & Biogeography 11:185-189.

Blench, R. 2001. 'You can't go home again': pastoralism in the new millennium. London, Overseas Development Institute: 103.

Boone, R.B. and N.T. Hobbs. 2004. Lines around fragments: effects of fragmentation on large herbivores. African Journal of Range & Forage Science 21:79-90.

Boone, R.B., S.B. BurnSilver, P.K. Thornton, J.S. Worden, and K.A. Galvin. 2005. Quantifying declines in livestock due to land subdivision in Kajiado District, Kenya. Rangeland Ecology & Management 58:523-532.

Brown, J.H. 1995. Macroecology. University of Chicago Press, Chicago, Illinois, USA.

BurnSilver, S., R.B. Boone, and K.A. Galvin. 2003. Linking pastoralists to a heterogeneous landscape: The case of four Maasai group ranches in Kajiado District, Kenya. Pages 173-200 *In* J. Fox, V. Mishra, R. Rindfuss, and S. Walsh, editors. People and the environment: approaches for linking household and community surveys to remote sensing and GIS. Kluwer Academic Publishing, Boston.

CIESEN (Center for International Earth Science Information Network). 2005. Earth Institute at Columbia University. http://www.gateway.ciesin.org/wdc/.

Dadibhavi, R.V. and L.D. Vaikunthe. 1990. Infrastructure for rural development – A study of Regional Disparities. Journal of Rural Development 9:581-593.

DAFF (Department of Agriculture, Forestry and Fisheries). 2005. Australian Government Drought Assistance Programs. http://www.affa.gov.au/content/.

De Haan, C., H. Steinfeld, and H. Blackburn. 1997. Livestock and the environment: finding a balance. Commission of the European Communities, Food and Agriculture Organization of the United Nations, and the World Bank, Brussels, Belgium.

Desta, S. and D.L. Coppock. 2004. Pastoralism under pressure: tracking system change in southern Ethiopia. Human Ecology 32:465-486.

Desta, S., D.L. Coppock, S. Tereza, and F. Lelo. 2004. Pastoral risk management in southern Ethiopia: Observations from pilot projects based on participatory community assessments. University of California, Davis, Global Livestock Collaborative Research Support Program: 4.

de Wolff, T., S. Staal, Kruska, R., Ouma, E., Thornton, P. and W. Thorpe. 2000. Improving GIS derived measures of farm market access: an application to milk markets in the East African highlands. Fifth Seminar on GIS and Developing Countries, Los Banos, Philippines.

Dimitri, C., A. Effland, and N. Conklin. 2005. The 20th century transformation of US agriculture and farm policy. Washington D.C., United States Department of Agriculture USDA: 17.

Doncaster, C.P. 2001. Healthy wrinkles for population dynamics: unevenly spread resources can support more users. Journal of Animal Ecology 70:91-100.

Ellis, J.E. and D.M. Swift. 1988. Stability of African pastoral ecosystems: alternate paradigms and implications for development. Journal of Range Management 41:450-459.

EWG (Environmental Working Group) Farm Subsidy Database. 2006. USDA Subsidies for farms in US counties. http://www.ewg.org/farm/regiondetail.php?fips=38000&summlevel=2.

Fabricius, C., A.R. Palmer, and M. Burger. 2002. Landscape diversity in a conservation area and commercial and communal rangeland in Xeric Succulent Thicket, South Africa. Landscape Ecology 17:531-537.

Fan, S. and X. Zhang 2004. Infrastructure and regional economic development in rural China. China Economic Review 15: 203-214.

FAO (Food and Agriculture Organization of the United Nations). 2001. Pastoralism in the new millennium. FAO, Rome, Italy.

Floater, G.J. 2001. Habitat complexity, spatial interference, and "minimum risk distribution": A framework for population stability. Ecological Monographs 71:447-468.

Frawley, P.T. 2002. Review of Australia's rural veterinary services. Discussion Paper. Australian Department of Agriculture Fisheries and Forestry: 168.

GLCF (Global Land Cover Facility). 2005. Global Land Cover Facility home page. National Aeronautics and Space Administration and the University of Maryland, College Park, Maryland. http://glcf.umiacs.umd.edu/index.html; accessed May 23, 2005.

Groombridge, B. 1992. Global diversity: status of the Earth's living resources. Chapman & Hall, London, UK.

Gustafson, E.J. and R.H. Gardner. 1996. The effect of landscape heterogeneity on the probability of patch colonization. Ecology 77:94-107.

Humphrey, C. and D. Sneath. 1999. The end of nomadism? Society, state and the environment in Inner Asia. Duke University Press, Durham, North Carolina.

Illius, A.W. and T.G. O'Connor. 2000. Resource heterogeneity and ungulate population dynamics. Oikos 89:283-294.

Jennings, T. 2000. Living with uncertainty: Adaptive strategies for sustainable livelihoods in the northern Great Plains. MA Thesis, Colorado State University, Fort Collins, Colorado.

Kareiva, P.M. and M. Andersen. 1988. Spatial aspects of species interactions: the wedding of models and experiments. Pages 38-54 *In* A. Hastings, editor. Community ecology. Sinauer, Sunderland, Massachusetts.

Kerven, C., I.I. Alimaev, R. Behnke, G. Davidson, L. Franchois, N. Malmakov, E. Mathijs, A. Smailov, S. Temirbekov, and I. Wright. 2003. Retraction and expansion of flock mobility in Central Asia: costs and consequences. *In* N. Allsopp, A.R. Palmer, S.J. Milton, K.P. Kirkman, C.R. Hurt, and C.J. Brown, editors. Proceedings of the VIIth International Rangelands Congress. Durban, South Africa, August 26th-1st.

Kerven, C., I. Alimaev, R. Behnke, R.G. Davidson, N. Malmakov, A. Smailov, and I. Wright. 2006. Fragmenting pastoral mobility: changing grazing pattern in post-Soviet Kazakhstan. USDA Forest Service Proceedings RMRS-P-39: 99-110.

Kiguli, L.N., A.R. Palmer, and A.M. Avis. 1999. A description of rangeland on commercial and communal land, Peddie district, South Africa. African Journal of Range and Forage Science 16:89-95.

Kolasa, J. and C.D. Rollo. 1991. Introduction: the heterogeneity of heterogeneity: a glossary. Pages 1-23 *In* J. Kolassa and S.T.A. Pickett, editors. Ecological heterogeneity. Springer-Verlag, New York.

Laca, E. 1998. Livestock development and rangeland conservation tools for Central Asia. University of California – Davis, Davis, California.

Lotsch, A., Y. Tian, M.A. Friedl, and R.B. Myneni. 2003. Land cover mapping in support of LAI/FPAR retrievals from EOS MODIS and MISR. Classification methods and sensitivities to errors. International Journal of Remote Sensing 24:1997-2016.

Loveland T.R., B.C. Reed, J.F. Brown, D.O. Ohlen, Z. Zhu, L. Yang, and J. Merchant. 2000. Development of global land cover characteristics database and IGBP DISCover from 1 km AVHRR data. International Journal of Remote Sensing 21:1303–1330.

McGarigal, K. and B.J. Marks. 1995. FRAGSTATS: spatial analysis program for quantifying landscape structure. USDA Forest Service General Technical Report PNW-351.

McGarigal, K., S.A. Cushman, M.C. Neel, and E. Ene. 2002. FRAGSTATS: Spatial pattern analysis program for categorical maps. University of Massachusetts, Amherst, MA, USA. www.umass.edu/landeco/research/fragstats/fragstats.html.

Merry, F., A.S. Pervaze, and D.G. McGrath. 2004. The role of informal contracts in the growth of small cattle herds on the floodplains of the Lower Amazon. Agriculture and Human Values 21:377-386.

Munguneiri, L., J. Omiti, and P. Irungu. 2002. Animal health service delivery systems in Kenya's marginal areas under market liberalization: a case for community-based animal health workers. Nairobi, International Food Policy Research Institute (IFPRI) 2020 Vision Network for East Africa: 4.

Musick, H.B. and H.D. Grover. 1991. Image textural measures as indices of landscape pattern. Pages 77-103 *In* M.G. Turner and R.H. Gardner, editors. Quantitative methods in landscape ecology. Springer-Verlag, New York.

Nee, S. and N. Colegrave. 2006. Paradox of clumps. Nature 441:417-418.

Nela, N. and D. Marshall. 1999. Credit management for rural development: Albania, a special case. Public Administration and Development 19:165-177.

NIMA (National Imagery and Mapping Agency). 1997. http://webgis.wr.usgs.gov/globalgis/metadata_qr/metadata/roads.htm.

Njenga, P. and A. Davis. 2003. Drawing the road map to rural poverty reduction. Transport Reviews 23: 217-241.

NOAA (National Oceanographic and Atmospheric Administration). 1998. Stable lights and radiance calibrated lights of the world. CD-ROM. National Oceanographic and Atmospheric Administration/National Geophysical Data Center, Boulder, Colorado. http://www.julius.ngdc.noaa.gov:8080/production/html/BIOMASS/night.html.

Oesterheld, M., C.M. DiBella, and H. Kerdiles. 1998. Relation between NOAA-AVHRR satellite data and stocking rate of rangelands. Ecological Applications 8:207-212.

O'Neill, R.V., J.R. Krummel, R.H. Gardner, G. Sugihara, B. Jackson, D.L. DeAnglelis, B.T. Milne, M.G. Turner, B. Zygmunt, S.W. Christensen, V.H. Dale, and R.L. Graham. 1988. Indices of landscape pattern. Landscape Ecology 1:153-162.

Paruelo, J.M., H.E. Epstein, W.K. Lauenroth, and I.C. Burke. 1997. ANPP estimates from NDVI for the central grassland region of the United States. Ecology 78:953-958.

Prugh, T., R. Costanza, et al. 1999. Natural capital and human economic survival. Lewis Publishers, Boca Raton.

Republic of Kazakhstan: Agency of Statistics. 2006. Dynamics rows for socio-economic development of the Republic of Kazakhstan include statistical data, Agency of Statistics. http://www.stat.kz/stat/index.aspx?p=dynamic&l=en.

Rindfuss, R.R. and P.C. Stern. 1998. Linking remote sensing and social science: the need and the challenges. Pages 1-27 *In* D. Liverman, E.F. Moran, R.R. Rindfuss, and P.C. Stern, editors. People and pixels: applications of remote sensing in the social sciences. National Academy Press, Washington, D.C.

Ritchie, M. and H. Olff. 1999. Spatial scaling laws yield a synthetic theory of biodiversity. Nature 400:557-560.

Roff, D.A. 1974. Spatial heterogeneity and the persistence of populations. Oecologia 15: 245-258.

Roughgarden, J.S., S.W. Running, and P.A. Matson. 1991. What does remote sensing do for ecology? Ecology 72:1918-1922.

Rutten, M. 1992. Selling wealth to buy poverty. Verlag breitenbach Publishers, Saarbrucken.

SA DOA (South Africa Department of Agriculture). 1998. Agricultural policy in South Africa. Pretoria, Ministry for Agriculture and Land Affairs. http://www.nda.agric.za/docs/policy98.htm.

SRTM. 2002. Shuttle radar topography mission: mapping the world in 3 dimensions. US Geological Survey, Sioux Falls, South Dakota. http://srtm.usgs.gov/.

Swift, J., D. Barton, and J. Morton. 2002. Drought management for pastoral livelihoods - policy guidelines for Kenya. Natural Resources Institute, Kent, U.K.

Tanser, F.C. and A.R. Palmer. 1999. The application of a remotely-sensed diversity index to monitor degradation patterns in a semi-arid heterogeneous, South African landscape. Journal of Arid Environments 43:477-484.

Thornton, P.K., S.B. BurnSilver, R.B. Boone, and K.A. Galvin. 2006. Modelling the impacts of group ranch subdivision on agro-pastoral households in Kajiado, Kenya. Agricultural Systems 87:331-356.

Tucker, C.J. 1979. Red and photographic infrared linear combinations for monitoring vegetation. Remote Sensing of the Environment 8:127-150.

Turner, M.G., R. Costanza, and F.H. Sklar. 1989. Methods to evaluate the performance of spatial simulation models. Ecological Modelling 48:1-18.

Turner, M.G., R.H. Gardner, and R.V. O'Neill. 2001. Landscape ecology in theory and practice: pattern and process. Springer, New York.

UNDP (United Nations Development Program). 2001. Human development report. Addressing Social and Economic Disparities: Kenya: 134.

UNDP (United Nations Development Program). 2003. Millennium development goals: a compact among nations to end human poverty. Oxford University Press, Oxford, UNDP: 16.

U.S. Census Bureau. State and county quick facts, Perkins County SD and Adams County ND. 1999. http://quickfacts.census.gov/qfd/states/46/46105.html.

U.S. Census Bureau, American Factfinder. 2000. DP-3 Profile of selected economic characteristics. Census 2000 Summary File 3. South Dakota and North Dakota. http://factfinder.census.gov/servlet/QTTable?_bm=y&-geo_id=04000US46&-qr_name=DEC_2000_SF3_U_DP3&-ds_name=DEC_2000_SF3_U.

Watkins, K., H. Fu, R. Fuentes, A. Ghosh, C. Giamberardini, et al. 2005. International cooperation at a crossroads: Aid, trade and security in an unequal world. Human Development Report. United Nations Development Program: 372.

WB (The World Bank) Joint Economic Research Program. 2004. Kazakhstan's livestock sector - supporting its revival. Almaty, Kazakhstan.

White, D.H. and B. O'Meagher. 1995. Coping exceptional droughts in Australia. Drought Network News (June): 1-2.

Wiens, J.A. 1989. Spatial scaling in ecology. Functional Ecology 3:385-397.

Wilson, T., K. Olson, J. Darlington, G. Whiting, and S. Mihovilovich. 2004. Managing the drought: Queensland primary producers' strategies 2001-2004. Brisbane, Department of Primary Industries and Fisheries: 52.

Chapter 15

RESPONSES OF PASTORALISTS TO LAND FRAGMENTATION: SOCIAL CAPITAL, CONNECTIVITY, AND RESILIENCE

Kathleen A. Galvin[1,2]

[1]*Natural Resource Ecology Laboratory, Colorado State University, Fort Collins, CO 80523, USA;* [2]*Department of Anthropology, Colorado State University, Fort Collins, CO 80523, USA*

1. INTRODUCTION

Change in the world's rangelands is proceeding at an unprecedented rate. In particular, fragmentation of pastoral rangelands is occurring as population growth, "modernity" and development spurs diversification and intensification of livelihoods and as the powerful and wealthy gain access to these lands for commercial use such as industrial agriculture, conservation, or tourism (Walker and Abel 2002, Agrawal 2003, Lesorogol 2003, Woodhouse 2003). Access to resources under fragmentation may be possible in some instances and not in others. As pastoralists diversify their livelihood strategies into agriculture, business, and wage labor, their dependency on livestock often decreases. Livestock may or may not remain the main source of income, but for people who have livestock, they must still be able to access resources for their stock as long as they have them. For herders in more arid environments, livestock remain the only viable livelihood strategy. In either case, management of livestock and how to gain access to resources remains an issue. A set of rules are used, modified, and created by people during and after the process of fragmentation to gain access to grazing land resources. This chapter explores social capital, the set of rules that allow access to resources and how it can and is being used in livestock management under fragmentation.

K. A. Galvin et al. (eds.), Fragmentation in Semi-Arid and Arid Landscapes: Consequences for Human and Natural Systems, 369–389.

Much has been written about how communal land tenure is the most common and most appropriate form of pastoral property rights. There are social rules in place that direct the management practices that herders adhere to when making decisions about movement of their livestock in search of pasture and water and include limits on use of key resource areas such as dry-season grazing lands and wells (McCay and Acheson 1987, Ostrom 1990, Behnke et al. 1993, Scoones 1994, Agrawal 2003). Common property management has worked well because the people involved reduce transaction costs. Transaction costs may include searching for information, seeking partners in collective action, drawing up and enforcing contracts, and building up networks and social capital, among other things (Adger et al. 2006). Opportunity costs include the provision of social resilience through equal access, external exclusion, and economies of size with respect to labor. These are especially important under conditions of incomplete and imperfect markets, which often prevail in the arid and semi-arid regions of the world (Banks 2003).

The rules of management can quickly adapt (Berkes and Folke 1994). Behnke (1995) and others (e.g., Banks 2003) make the point that there are few clearly defined membership rules or boundaries in this situation and that this fluidity is itself strategically important in the pastoral world of spatial and temporal variability in resources.

2. PASTORALISTS AND CONNECTIVITY

How do pastoralists view forage and water? They look for resources through the needs of their livestock, that is, they look for forage and water and its quality and quantity to meet the physiological requirements of their stock. Thus, in arid and semi-arid systems where resources vary in space and time people move their livestock. This movement is analogous to a wildlife species' movements that are determined by distance between vegetation patches as well as its ability to viably use them. The ability to access resource patches is known as ecological landscape connectivity. A landscape is considered connected if it allows movement among resource patches (Taylor et al. 1993). There are two types of connectivity, structural and functional. Structural connectivity describes the extent of fragmentation in the landscape, which includes the distance between patches, sometimes called habitat continuity. Using the patches implies functional connectivity or the behavioral responses of the animals to the landscape structure. This latter notion includes factors such as vegetation types and what the landscape looks like (e.g., steep, flat, etc.). A landscape can have structural connectivity but lack functional connectivity. For example, just because a corridor structurally exists does not mean a species can use it to access another patch if that corridor is too narrow or too long (Tischendorf and Fahrig 2000). This means that the amount of movement between patches of resources is just as important for the survival of a species as the distribution of the resources. Taylor et al. (1993:571) propose that an animal's ability to utilize a resource patch will be dependent upon its ability to get there. Therefore, landscape connectivity is the degree to which the landscape facilitates or impedes movement among resource patches. Landscape connectivity generally increases chances for population viability of species (Beir and Noss 1998). Fragmentation decreases connectivity for animals.

Herders try to maintain connectivity in the ecological system through movement of their livestock. Individuals are able to move across the landscape because of social rules that permit movement. The formal and informal rules or institutions that enable people to move comprise individual and collective social capital (North 1990). The question here is what are the social institutions that enable pastoralists to move to resources for their livestock after fragmentation? Under what conditions do institutions allow movement and when is it no longer a viable endeavor? Pastoral fluid and flexible management strategies are important in uncertain environments where forage and water are patchy and ephemeral. Does fragmentation lead to diminished flexibility? When flexibility to deal with variability is diminished, pastoralists can become vulnerable to change; that is, they lose the ability to deal with that change. When institutions are no longer intact and other avenues for economic diversification and intensification are not in place, that is, when economic choices are diminished, poverty tends to increase. This may be a general trend in pastoral societies.

Flexible management strategies define social resilience in uncertain environments. Resilience is the "capacity of a system to experience disturbance and still maintain its ongoing function and controls" (Walker and Abel 2002:294). Disturbance can come from climate or something more permanent such as privatization of the land or of livestock. Disturbance causes the system components to adapt by learning (e.g., smart pastoralists) or by selection (e.g., other pastoralists go bankrupt). Individuals, their social relations, and social networks are the glue that holds together the adaptive governance (Folke et al. 2005). This chapter looks at how the "glue" has adapted under fragmentation. The chapter is organized as follows. First is a discussion of the concept of social capital as it is within this context that resources are acquired. This is followed by examples of the use of social capital among pastoralists from several areas of the world. Finally, some general patterns emerge which are discussed within the framework of resilience under change.

3. SOCIAL CAPITAL

The concept of social capital has a long and varied history in the social sciences and as a result, it has a variety of meanings. It is used to understand many things such as how people cooperate and share risk (Pretty 2002). Often it is used to understand the social organization of groups of people who see each other regularly such as kin groups or neighbors, friendship networks, or people of the same club, church or village (Falk and Kilpatrick 1999). Bourdieu (1986) suggests that social capital comprises networks of relationships where social obligation is felt. There are relations of trust, reciprocity, common rules, connectedness, and networks in social capital (Coleman 1988, Pretty 2002, Folke et al. 2005). Poverty levels are said to be less among those individuals and groups with high social capital (Narayan and Pritchett 1997).

Not all social networks are equal. *Bonding ties* between family, friends, and neighbors lead to strong local trust but can be constraining. These ties sometimes do not allow actors in the network to have access to outside information or to get help from outsiders. *Bridging ties*, on the other hand, can provide access to resources and opportunities that are outside a particular network (Newman and Dale 2005). These ties act as vertical links which help increase a network's ability to access more vertical power relationships beyond the more "horizontal" bonding ties (Adger 2003). These include linking to powerful leaders at state and national levels whose position can help locals to gain access to resources. These ties increase a household's or community's ability to adapt to change.

Governance of resources is the result of social relations and social networks of individuals and groups that result in a particular natural resource management strategy. When a group is said to be adapted, it implies that the group has the ability to change its governance over natural resources. Crisis or change triggers opportunities for learning and can create room for reorganization. Thus as common resources become fragmented, pastoralists have the opportunity to use their social capital and ability to learn to find new ways to obtain the needed resources.

Folke et al. (2005) suggest four factors that are required for dealing with rapid change and reorganization in social-ecological systems:

- Learning to live with change and uncertainty
- Combining different types of knowledge for learning
- Creating opportunity for self-organization toward social-ecological resilience
- Nurturing sources of resilience, renewal, and reorganization.

These nurturing sources include: 1) social memory of past changes in ecosystems and responses to those changes which are used to make new decisions, 2) informal social networks that need to be in place before successful adaptive governance of ecosystem management is formed, and 3) bridging organizations that are needed to connect local actors and communities with other scales or organizations. If these attributes are in place, successful social transformations toward adaptively dealing with fragmentation (or other change) can emerge. Figure 15-1 shows a simple conceptual model of the processes. Pastoralists need social capital to gain access to resources, among other things. Social capital is used to govern management of resources. If attributes such as bonding and bridging ties and the ability to learn are in place, then pastoralists have a better chance of coping with change in a manner that perpetuates human well-being and the use of resources. The next section explores social capital and its attributes among pastoral societies around the world.

4. PASTORAL FORMS OF SOCIAL CAPITAL

Most work on pastoral social capital has been focused on how people obtain and use livestock (Potkanski 1997, DeVries et al. 2006). However, these relationships are equally important in gaining access to resources. Mutual assistance networks are a good survival strategy in uncertain environments and are deeply rooted in pastoral culture through customary rules of social relationships. Just how well are they standing up today? This is explored below.

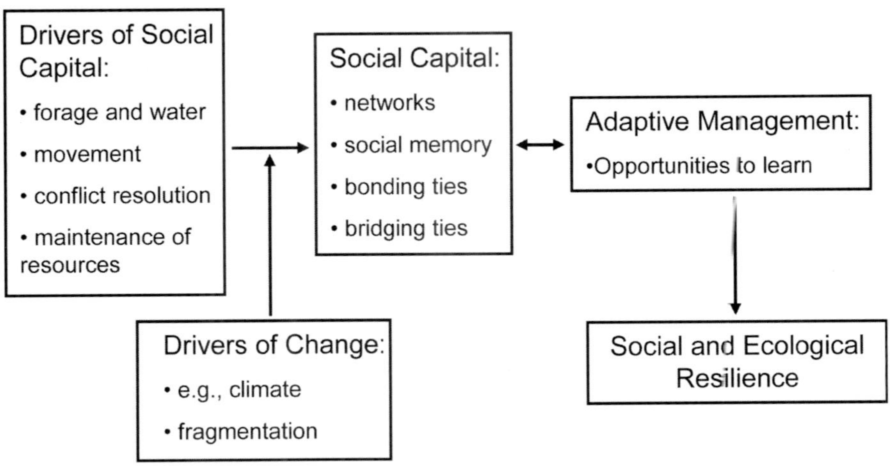

Figure 15-1. Conceptual model of resilience.

4.1 Central and East Asia

In Mongolia and the former Soviet Asia, one of the great nomadic pastoral areas of the world - what Humphrey and Sneath (1999) call Inner Asia, the cultural-economic zone of the steppes in the countries of Russia, Mongolia and China - the management of pastoral lands has long been determined by the state but mediated by customary pasture management. Mongolia, for example, has had a long history of state-established policies mixed with regional government and customary rules governing land use (Mearns 1991, 1993). Both formal and informal regulatory institutions in the past made Mongolian pastoralism sustainable, socially and ecologically. The various ruling powers (Manchu [colonial], feudal, and socialist) recognized the need for and allowed for flexibility of herders during disasters.

In pre-collective Mongolia, herders' economic status tended to determine their access to pasture resources and mobility (Fernández-Giménez 1999). Animals were owned by nobles and were herded with poorer households' livestock (Sneath 1993). The poor were less able to access means of transportation (camel carts usually) and therefore were less likely to move to new pastures while the rich were more mobile. The poor therefore were often given temporary grazing rights in exchange for labor (Fernández-Giménez 2002). During the collective period (1960-1990) in Soviet times (1912-1990), all land and most livestock belonged to the state but were allocated to collectives and governed by local leaders in conjunction with state mandates for productivity. The party decided policies, the government administration located in the province and district levels implemented these policies, and the institutions of economic production such as state farms translated policies into production actions. It was at this juncture that livestock management took place in combination with local herders (Mearns 1991). Local leaders often negotiated directly with each other (rather than the state) to gain reciprocal pasture-use privileges, especially in cases of disaster.

After the collapse of communism in 1990, collective pastoralism was dismantled. Land disputes became common and livestock were privatized (Sneath 2003). It is now common for pastoralists to access resources by out-of-season and year-round grazing of key resources, trespassing on customary winter and spring reserve pastures, and reducing the distance and frequency of seasonal nomadic moves (Fernández-Giménez 1999). Though helpful in the short-run these strategies utilize important reserve pastures and could be detrimental in the long-run.

The result has been increased pastoral poverty. Poverty today is linked to declining nomadic mobility and indirectly to out-of-season grazing and trespassing. As a result of each new political regime the migratory territory of herders is becoming smaller and smaller (see Ojima and Chuluun, Chapter

8, Fernández-Giménez 1999). The poor are turning to associations with distant kin or acquaintances to gain rights and essential pasture resources (Fernández-Giménez 2002). Fernández-Giménez (2002) and others (e.g., Templer et al. 1993) suggest that in order for Mongolian pastoralists to protect their flexibility, co-management of access to resources is necessary. This would entail shared authority between local users, regional governments, and the national government so that regulation of seasonal movements can be maintained, perhaps in a different form than in the past, but having the same set of institutions involved.

Banks et al. (2003) address community-based grassland management in western China where the state is concerned about rangeland degradation. In the 1980s, the commune system was dismantled and rangelands were allocated to whole villages or groups of households. The land was and is owned by the state, but increasingly, long-term rights to grasslands are now being granted to individual households who usually pay a fee for those rights. The thinking of government officials holds that the degradation seen today is caused by past and present overstocking, hence there are initiatives to shift from communal land ownership to individual ownership.

Whereas the government is convinced that the causes of this degradation are based on communal management practices, Banks et al. (2003) argue otherwise. Despite privatization of the rangelands, collective or group arrangements have persisted across most regions and seasonal pastures. It is not profitable to individualize the pastures due to the spatial and temporal variability of resources. Furthermore, group tenure facilitates group herding arrangements. For example, in one village individuals were granted ownership to winter pastures and herders decided to manage their private parcels through collective arrangements. The arrangements include rules that were established for when and how long community members may graze in the collective pastures. In another case in Maqu County in southwest Gansu Province, China, a pastoral development project allowed groups of ten households with individual winter pastures to re-aggregate these pastures. By pooling their pastures together and fencing the outermost boundary, pastoralists were able to lower the costs of individual fences and also share the burden of labor for herd supervision. Households in the group with more livestock have to pay households in the group with less livestock for their heavier use of the range. An adaptive and participatory approach to policy implementation seems to have made these collective/group management systems possible.

Like Fernández-Giménez (2002), Banks (2003) calls for co-management of the rangelands instead of privatization of the household ranch. In the case of China, co-management "means the sharing of responsibilities for natural resource management between national and local governments, civic organizations and local communities" (Banks 2003:2130). It is ironic that kin network groups have become more important, not less, in the post-Soviet economies and they seem to have adapted to changes in property ownership. These networks can be seen as a response to shortages and economic change. But they are likely not enough. As access to resources has become more problematic, especially for the poor, they, together with government are seen as partners in forging new ways of pasture management.

4.2 US and Australia

In the US Great Plains, a major response to fragmentation has been consolidation of farms, but often this response is not sufficient to maintain connectivity. This is because parcels under one farming operation may not be contiguous and may still be marked by fences. Nevertheless, though strictly not social, but rather economic in scope, accruing larger land holdings is allowing farmers to stay on the land through increased crop harvests. Old laws like the *Homestead Act of 1862* encouraged people to settle on parcels of land which were too small to support the farming operation during droughts and economic downturns. Over the last 70 years or so that same farm land is being controlled by a smaller number of people as individuals sell small farms to larger landowners (see Lackett and Galvin, Chapter 5). Connectivity of the land is being compensated for through technology, particularly the ability to move livestock from one pasture to another with trucks.

The Chihuauan Desert of Arizona and New Mexico has been cattle country since the late 1800s, that is, until suburban development encroached on the grazing lands in the last few decades. Traditional private ranches were once surrounded by state and/or federal lands which ranchers used for grazing their livestock. The purchase of ranches and the contemporary trend towards subdivision of ranches into 'ranchettes' has affected the ability of the remaining ranchers to access once extensive rangelands. As adjacent state and/or federal land is lost to development, private ranches are having a harder time functioning as working ranches because they no longer have access to contiguous pastures. However, ranchers have been working with scientists, environmentalists, local and federal agencies, and other ranchers to change this trend. The strategies that have been successful in retarding fragmentation include conservation easements and the purchase of development rights. Both of these mechanisms involve binding contracts that remove development potential from a ranch's private lands in exchange for money and/or other considerations (Curtin et al. 2002).

Despite fragmentation of land through formalized, private land tenure, some pastoralists in Australia are using a system called agistment in which livestock are transferred between farms to areas with better forage and water (McAllister et al. 2006). "Facilitated through a network of kin, friends, friends of friends, relatives, business partners and adversaries, agistment interactions match pastoralists who have a shortage of forage to pastoralists who have an excess" (McAllister et al. 2006:574). Agistment networks are more effective where spatial variation in resource availability is high and spatial co-variation is relatively low (McAllister et al. 2006). Basically, agistment networks provide spatial connectivity to fragmented landscapes. Stokes et al. (Chapter 4) found that Dalrymple Shire, Australia ranchers are also consolidating properties to offset rising production costs. This system is built upon trust and social networking. Cooperative management between owners seems to be restoring landscape connectivity.

4.3 East Africa

Much has been written on Maasai pastoralists and their relationships to the land, their livestock and wildlife (e.g., Western 1989, Spear and Waller 1993). Potkanski (1997) for example, describes Maasai (in the Ngrorongoro Conservation Area [NCA] and Salei Plains of Tanzania) access to resources via social capital as being carried out when the principal user of rights to grazing and browsing resources offers other Maasai use of it. This rule defines the principle that land is the property of all but that there is controlled access to collective property. Social capital is important when a herder wants to move to a new locality. If a herder moves to a new area he must have a friend or relative to visit who 'allows' the newcomer onto that collective property.

Access to water is a bit different than access to pasture because water rights can be both individually or collectively-owned. Rivers or flowing water are collectively-owned while standing water, like a well, is individually-owned. Primary users are those who inherited the water source and secondary users include patrilineal relatives first; then if any water still remains, sub-clan and finally clan members living in the area may use the resource. Non-agnates (affines and friends) can ask the owner for permission to use water, but they do not have any legal basis for demanding it. These requests are always negotiable. The social capital used by the Maasai in the NCA to move and find water and forage has to be placed however, within the context of conservation policy. Policy has effectively cut the Greater Serengeti Ecosystem into administrative parcels, most of which is off limits to pastoral land use (see Galvin et al., Chapter 11).

A very different set of changes is occurring in the Athi-Kaputei Plains in northern Kajiado Distict, Kenya in what Reid et al. (Chapter 9) call a peri-urban savanna. Here, rapid urbanization, immigration and land-tenure change among other things, have had a profound impact on pastoralism. Livestock biomass for the Maasai here has decreased by 50% over the last 25 years as land has become privatized and access to water, pasture, and salt licks have declined. People who do have livestock access small pastures on farms, in villages, towns, and cities. Some people are able to send their cattle to satellite livestock camps while others no longer share work to gain access to resources. It appears that this may be a situation where livestock networks and ties may be breaking down and this is affecting the poor to a greater extent than the rich.

In southern Kajiado District, the Maasai are in various stages of sub-division of their group ranches (BurnSilver et al., Chapter 10, Thornton et al. 2006). BurnSilver et al. (Chapter 10) stress that reciprocity is used to access the land on other group ranches that still have forage and water. So while land tenure fragments the landscape, land use through social capital provides connectivity once again. For those Maasai whose land is subdivided, means of accessing resources include households sharing their individual parcels of rangelands, rotating herds between locations, labor-sharing, grazing networks, and pasture rental (BurnSilver and Mwangi 2006). It is a peculiar unforeseen consequence of fragmentation that labor should now be a problem. As group ranches are dissolved into private ranches there is an increasing shortage of labor so grazing networks are a means by which pastoralists can again obtain that labor (BurnSilver and Mwangi 2006).

In northern Kenya, Samburu group ranches were formed, but the Samburu were not interested in changing their livestock production system until the 1980s, when a particular community privatized its previously communal land. Lesorogol (2003) describes the power struggle between young, entrepreneurial people with smaller herds who intended to gain title to parcels of land for agricultural production and the elders who opposed it. The group that wanted private ownership was exposed to more education, employment, or military service and saw the traditional society as backward and not modern. A group of charismatic elders, on the other hand, mobilized the people to overcome the difficulties to collective action and linked with powerful people outside the society. As a result, each side had equal bargaining power and all household heads were given equal size plots on the more fertile part of the group ranch for agricultural production. The less fertile, lowland area remains common property for livestock herding and new social norms are emerging that discourage the sale of the private land (not unlike the caution pastoralists use when selling livestock) (Lesorogol

2005). This is a nice example of bonding and bridging ties being used in new ways to deal with grazing land fragmentation. However, whether or not the lowland area remains viable for livestock production through time with climate, economic, and social changes remains to be seen.

In one of the most remote areas of arid land pastoralism, Turkana District, Kenya, pastoralists rely on bonding ties to secure access to resources today as in the past. However, the political and security situation has changed land use and movement patterns. This has led to a decrease in the effectiveness of traditional social networks and compromised their ability to buffer households from harsh ecological conditions and inter- and intra-tribal conflict. Traditional mechanisms that allowed people to gain access to resources relied largely upon complex social networks of friends and relatives (Gulliver 1955, McCabe 1990). Mobility was largely due to changing ecological conditions with the Turkana moving on average about 12 to 15 times per year (Galvin 1985, Dyson-Hudson and Dyson-Hudson 1999). By the late 1990s however, people were moving 100 percent of the time to avoid raids. Herders changed their pattern of small, independent moves of a few households to one of large-scale mass-migration, a defense strategy they adopted from the Karimojong (Dyson-Hudson 2000, Pike 2004). With raiding, all herds are affected equally thus the traditional mechanisms for deflecting or redistributing stress tend to break down (Gray et al. 2003). Large neighborhoods were formed and moved frequently as a unit together with armed guards and entire family units, therefore requiring new social arrangements (Pike 2004). This example shows how quickly social networks can change, but also the speed with which human resilience may shift in response to social as well as environmental conditions (Gray et al. 2003).

4.4 West Africa

For the Wodaabe of Burkina Faso, agnatic family groups help each other today as in the past (Bovin 1990). The Fulani, particularly in Mali, share or trade knowledge, labor, milk, and social contacts to gain access to resources in their transhumance movements (Turner 1999). For example, to gain access to the floodplain pastures of the inland Niger Delta of Mali, people use marriage, fostering of children, and reciprocal agreements. In this region, as is the case for many other pastoral areas, cash markets are undeveloped and therefore bartering of these assets through social contracts provides herder access to productive assets. However, it is also important for herd leaders to have political connections at the district level to maintain the rights for transhumance (Turner 1999).

Niamir-Fuller (1998) asserted that the social networks in place for Sahelian pastoralists are gradually disappearing principally by rangeland encroachment of cultivation, tourism, and wildlife conservation. The local safety nets seem to be disappearing as pastoral communities lose unity and become more settled and heterogeneous. She called for a two-tiered legal system, "an overall legal framework at the national level that officially recognizes decentralized customary rules and common property and, at the local level, flexible rules and procedures developed by a decentralized institutional system that provides a forum for negotiation and conflict resolution" (Niamir-Fuller 1998:274).

5. CONCLUSIONS

Ultimately, pastoral social capital is shaped by strong cultural, economic, and ecological imperatives. These imperatives are complex in and of themselves and differ one from another depending on the particular cultural history, ecological landscape, and current circumstances of each pastoral socio-ecological system. Furthermore, rapid change is affecting these linked components differently. Nevertheless, some generalizations emerge from the discussion above because there are features that are shared by all these socio-ecological systems. In the arid and semi-arid ecosystems where most pastoralists live, people have limited ability to manipulate the system. In the face of this limitation, they track forage and water across the landscape. Pastoralists connect the resources through management, and this is best done by movement across the landscape. Movement means that to sustain their livestock they travel within and beyond their home territory, which requires use rights and rules based in social capital (Behnke 1995, Turner 1999). The discussion below addresses common issues of social capital in a changing world under an increasingly fragmented environment.

5.1 Bonding and bridging ties

What types of social capital work to secure access to resources? The case studies suggest that what tends to work is informal local-level agreements and organizations, and formal and informal institutions at higher levels of social organization, that can help pastoralists to access resources. These are what Newman and Dale (2005) call bonding (e.g., kin groups) and bridging capital, and together they provide resilience with which to cope with change. Bridging capital brings in new and potentially novel information that can help bonding capital to acquire access to resources; this usually entails linking to the state at some level and sometimes using good scientific information, often with the help of scientists. Bonding ties do not seem to be sufficient to gain access to resources today. Another consequence of fragmentation for pastoralists is that some are better than others at 'reinventing' their social networks to get at the needed resources. But how the process of linking turns out depends on the power of groups and individuals. Lesorogol (2003) and others (e.g., Knight 1992) show that individual agency and choice play a role in changes in institutions and that understanding the relative power of actors influences institutional outcomes (Ensminger and Knight 1997). Using both old and new networks to forge new social capital seems to be the most successful. That success encapsulates what Gunderson and Folke (2005:1) call resilience.

5.2 Socio-economic stratification

Traditional bonding and network institutions tend to make wealth more evenly distributed in that the wealthy are obligated to help the poor and the poor are obliged to follow customary law. However, at the same time, traditional pastoralism tends to favor the wealthy (cf. Lesorogol 2005) in that the wealthy elders are the ones who control access to resources. Today, under rapid change many pastoralists are diversifying and intensifying their livelihoods as the case studies in this volume reveal. Diverse livelihoods ensure flexibility in changing circumstances but this process is also resulting in increased socio-economic stratification (Fratkin 1998, Broch-Due 1999, Adger 2003, Lesorogol 2003, Adger et al. 2006).

Increased wealth means that rich households do not always need friends and family as they can generally cope with the change themselves. This can put a strain on existing and past forms of social capital such as the leveling tactics of customary laws of the land. The impact is that customary rules decline so people do not feel the traditional means of social pressure. Their norms and values are transferred from the traditional society to that of the modern state or some other entity.

For example, socio-economic stratification has occurred because of privatization of pastoral resources in the Central Asian and East Asian states as they have moved from command economies of livestock production to privatized herd owners (Kerven 2003). For example, large livestock owners of Kazakstan, usually those who were senior shepherds or professionals in the state farm, are better able to get into the commercial livestock industry, occupy good pasture land, and have improved equipment and production methods. This has also occurred in Inner Mongolia, where decollectivization and privatization occurred a decade earlier with the same results (Williams 2002).

On the other end of the wealth continuum, decreased wealth or extensive poverty due to low livestock numbers also leads to pastoralists who can no longer help each other as all are equally constrained (Western and Nightingale 2002, Thompson et al. ms).

5.3 Implications for social resilience

People usually respond to constraints or change through a reservoir of knowledge that draws on past experiences of disturbances or stresses (Bradley and Grainger 2004). As people diversify or intensify livelihoods, past experiences may not always suit them well in dealing with change. New patterns need to emerge. The capacity of people both to innovate and to adapt practices to suit new conditions becomes vital (Pretty 2002). The new ideas draw on past experiences of constraints and responses to those stresses. If there are plenty of links between the various social groups then the society has high social capital (Pretty 2002). Pastoralists with the strongest social capital will be best able to withstand disturbance. "A social-ecological system with low levels of social memory and social capital is vulnerable to such changes [floods, shifts in property rights, resource failures, new government legislations, etc.]...and may as a consequence deteriorate into undesired states" (Folke et al. 2005:455). To be highly resilient, people need to be able to learn from past experiences and actively integrate the new knowledge to control their access to resources in new ways. But changes in institutions emerge from various social groups working to establish institutional arrangements that best suit their interests. Thus, social resilience is defined as the ability of groups and individuals to tolerate and respond to changes through adaptive strategies (Bradley and Grainger 2004). Increased social capital can bolster resilience of individuals and the society, and conversely, lack of social capital can lead to poverty and vulnerability (Erickson 2006).

However, it is likely that if the disturbance is very large, or what Walker and Abel (2002) call "slow variables" such as climate change, depletion of aquifers, or destruction of infrastructure, past experiences will not be helpful in dealing with the present (cf. Galvin et al. 2005). Part of the reason is that the length of a human generation and the duration of human institutional or cultural memory exist and persist at different time scales than the rate of change (Walker and Abel 2002). The faster variables such as seasonality or policy proposals benefit from social capital arrangements and rules, and self-regulatory means can adapt quickly and last a long time (Berkes and Folke 1994). We have seen from the examples above a myriad of social capital institutions, some old, some new that people are using under new circumstances. They all show to various degrees the ability to learn from past experiences and use social networks to remain resilient under change.

5.4 Implications for ecological resilience

It is clear from the examples here that the well-being of social and ecological systems are linked. Furthermore, most of the socio-ecological systems described here are non-equilibrial in nature; high spatial and temporal variability in resources is the norm (Ellis and Swift 1988, Ellis and Galvin 1994, Galvin et al. 2001). How pastoral societies adapt to these pressures under fragmentation affects ecological resilience as well as social resilience. Pastoralists keep the resource and the users negotiable because the system is non-equilibrial (Behnke 1995, Turner 1999). We have seen however, that the pastoral systems described here adapt to variability in resources in different ways. Pastoral social capital, when not interfered with by the state, has been based on bonding ties. The processes of fragmentation today requires both bonding and bridging social capital working together to maintain or acquire anew the connectivity across the landscape. The case studies in this volume also show that there is no single set of social relations that will ensure sustainability of the resource base or of the society. However individuals, their social relations, and social networks are the glue that defines the adaptive governance.

Survival of pastoral societies ultimately depends on the finite capacity of the environment to support them with forage and water, the two most essential resources for maintaining people and their livestock. Social capital and the institutions associated with it is the most important mechanism to respond to changes in these resources and is a means to adapt to the changes in an active way. The case studies demonstrate sustained collective action even following fragmentation that demonstrate creativity and resourcefulness. Successful resource management allows a certain amount of disturbance on a scale that will not destroy or disrupt the overall function of the system for the services it provides (Berkes and Folke 1998, Folke et al. 1998). Holling et al. (1998:359) state, “Flexible social systems that proceed by learning-by-doing are better adapted for long-term survival than are rigid social systems that have set prescriptions for resource use.” Flexibility has been the quintessential feature of pastoral livestock management in arid and semi-arid ecosystems (cf. Little and Leslie 1999). But social capital, social resilience and ecological resilience are all tested under change. As ever increasing rates and types of changes occur, that flexibility will remain essential.

5.5 Role of policy in supporting social capital of pastoralists

A last, but important, question is: how can policy better strengthen social capital in pastoral societies to allow pastoralists to thrive as landscapes fragment and mobility becomes more difficult? In most countries of the world, most people, including policymakers, grew up in and are most familiar with ways to manage food production in wetter, higher potential lands, where farming is the dominant form of agriculture (Leneman and Reid 2001). Accordingly, much of agricultural policy was designed to support households where people are sedentary and can easily access services and resources at central places, in towns and villages. This central place model of infrastructural development has been applied the world over in dry lands, encouraging pastoral families to settle. Pastoral families often choose to limit their mobility in order to access social services, such as health care and education, and to access markets by settling around towns (Rutten 1992, Blench 2000).

Land and natural resource polices often support establishing less flexible spatial boundaries, for example, through support of private land ownership or creation of protected areas in grazing lands that herders cannot use (Brockington et al. 2006). Strict spatial boundaries can have substantial advantages in settled land, but strongly limit the flexibility that herders need to respond to climatic variability in dry or cold grazing lands.

How then can the flexibility needed by pastoralists be supported? One way is to strengthen bridging ties of pastoral households and communities to the state, but also to other groups who have a stake in maintaining grazing lands. There are non-governmental organizations, scientists, and state and federal agencies at various levels, from the local to the global, that can forge alliances and ties that will enable pastoralists' access to resources. These types of alliances and ties might include: 1) co-management agreements on state-owned grazing lands and protected areas (as Banks et al. 2003, Fernández-Giménez 2002, and Brockington et al. 2006 suggest), 2) formal support for open dialogue and representation of pastoral groups in policy formulation (including women), 3) real engagement between scientists and communities to co-produce integrated traditional and scientific knowledge products on issues of importance to communities, and 4) state support for a decentralized institutional system that provides a forum for negotiation and conflict resolution (similar to 1 above). The greater the number of links among the various groups that can be made, the greater the social capital that is developed. All of these arrangements would go a long way to support social memory, and therefore resilience, of these coupled human-ecological systems to change.

This will be a challenge, especially as most groups and institutions function at the local scale and with privatized resources (Reid et al. 2006). Thinking about keeping resources available at a landscape scale requires new ways of thinking. While keeping resilience and vulnerability in mind, it is time to take policy cues from pastoral resource management strategies and use them as a basis for developing creative institutions and policy that really are able to cope with change.

REFERENCES

Adger, W.N. 2003. Governing natural resources: Institutional adaptation and resilience. Pages 193-208 *In* F. Berhout, M. Leach, and I. Scoones, editors. Negotiating environmental change: new perspectives from social science. Edward Elgar Publ., Northamptin, MA.

Adger, W.N., K. Brown and E.L. Tompkins. 2006. The political economy of cross-scale networks in resource co-management. Ecology and Society 10:9 (online) http://www.ecologyandsociety.org/vol10/iss2/art9/.

Agrawal, A. 2003. Sustainable governance of common-pool resources: context, methods, and politics. Annual Review of Anthropology 32:243-262.

Banks, T. 2003. Property rights reform in rangeland China: On the road to the household ranch. World Development 31:2129-2142.

Banks, T., C. Richard, L. Ping, and Y. Zhaoli. 2003. Community-based grassland management in western China. Mountain Research and Development 23:132-140.

Behnke, R. 1995. Natural resource management in pastoral Africa. Pages 145-152 *In* D. Stiles, editor. Social aspects of sustainable dryland management. John Wiley and Sons, Ltd , West Sussex, UK.

Behnke, R., I. Scoones and C. Kerven, editors. 1993. Range ecology at disequilibrium: New models of natural variability and pastoral adaptation in African Savannas. Overseas Development Institute, London.

Beir, P. and R.F. Noss. 1998. Do habitat corridors provide connectivity? Conservation Biology 12:1241-1252.

Berkes, F. and C. Folke. 1994. Investing in cultural capital for sustainable use of natural capital. Pages 128-149 *In* J. Jansson, M. Hammer, C. Folke and R. Costanze, editors. Investing in natural capital: The ecological economics approach to sustainability. Island Press, Washington, D.C.

Berkes, F. and C. Folke. 1998. Linking social and ecological systems for resilience and sustainability. Pages 1-25 *In* F. Berkes, C. Folke and J. Colding, editors. Linking social and ecological systems. Cambridge University Press, Cambridge, U.K.

Blench, R. 2000. 'You can't go home again', extensive pastoral livestock systems: issues and options for the future. ODI/FAO, London, UK.

Bourdieu, P. 1986. The forms of capital. Pages 241-258 *In* Handbook of theory and research for the sociology of education. Greenwood Press, Westport, CT.

Bovin, M. 1990. Nomads of drought: Fulbe and Woddabe nomads between power and marginalization in the Sahel of Burkina Faso and Niger Republic. Pages 29-58 *In* M. Bovin and L. Manger, editors. Adaptive strategies in African arid lands. The Scandinavian Institute of African Studies, Uppsala, Sweden.

Bradley, D. and A. Grainger. 2004. Social resilience as a controlling influence on desertification in Senegal. Land Degradation and Development 15:451-470.

Broche-Due, V. 1999. Remembered cattle, forgotten people: The morality of exchange and the exclusion of the Turkana Poor. Pages 50-88 *In* D.M. Anderson and V. Broch-Due, editors. The poor are not us: Poverty and pastoralism in Eastern Africa. James Curry Ltd, Athens.

Brockington, D., J. Igoe, and K. Schmidt-Soltau. 2006. Conservation, human rights, and poverty reduction. Conservation Biology 20:250-252.

BurnSilver, S.B. and E. Mwangi. 2006. Beyond group ranch subdivision: collective action for livestock mobility, ecological viability and livelihoods. Paper presented at: Pastoralism and Poverty Reduction in East Africa: A Policy Research Conference. International Livestock Research Institute. June 27-28. Safari Park Hotel, Nairobi, Kenya.

Coleman, J. 1988. Social capital and the creation of human capital. American Journal of Sociology 49 (supplement):S95-S120.

Curtin, C.G., N.F. Sayre, and B.D. Lane. 2002. Transformations of the Chihuahuan borderlands: grazing, fragmentation, and biodiversity conservation in desert grasslands. Environmental Science and Policy 5:55-68.

DeVries, D., P.W. Leslie, and J.T. McCabe. 2006. Livestock acquisition dynamics in nomadic pastoralist herd demography: A case study among Ngisonyoka herders of South Turkana, Kenya. Human Ecology 34:1-25.

Dyson-Hudson, N. 2000. Processes of discovery and modes of commentary through time: Karamoja and Turkana as illustrative cases. Page 190 *In* Abstracts of the 99th Annual Meeting of the American Anthropological Association. American Anthropological Association, Arlington.

Dyson-Hudson, N. and R. Dyson-Hudson. 1999. The social organization of resource exploitation. Pages 69-86 *In* M.A. Little and P.W. Leslie, editors. Turkana herders of the dry savanna: Ecology and biobehavioural response of nomads to an uncertain environment. Oxford University Press, Oxford.

Ellis, J. and D. Swift. 1988. Stability of African pastoral ecosystems: Alternate paradigms and implications for development. Journal of Range Management 41:450-459.

Ellis, J. and K.A. Galvin. 1994. Climate patterns and land use practices in the dry zones of east and west Africa. BioScience 44:340-349.

Ensminger, J. and J. Knight. 1997. Changing social norms: Common property, bridewealth and clan exogamy. Current Anthropology 18:1-24.

Erickson, P. 2006. Assessing the vulnerability of food systems to global environmental change: a conceptual and methodological review. GECAFS Working Paper 3. GECAFS International Project Office, NERC-Centre for Ecology and Hydrology, Wallingford, UK.

Falk, I. and S. Kilpatrick. 1999. What is social capital? A study of interaction in a rural community. CRLRA Discussion Paper Series. D5/1999.

Fernández-Giménez, M.E. 1999. Sustaining the steppes: A geographical history of pastoral land use in Mongolia. The Geographical Review 89:315-342.

Fernández-Giménez, M.E. 2002. Spatial and social boundaries and the paradox of pastoral land tenure: a case study from post socialist Mongolia. Human Ecology 30:49-78.

Folke, C., F. Berkes, and J. Colding. 1998. Ecological practices and social mechanisms for building resilience and sustainability. Pages 414-436 *In* F. Berkes, C. Folke and J. Colding, editors. Linking social and ecological systems. Cambridge University Press, Cambridge, UK.

Folke, C., T. Hahn, P. Olsson, and J. Norberg. 2005. Adaptive governance of social-ecological systems. Annual Review of Environment and Resources 30:441-473.

Fratkin, E. 1998. Ariaal pastoralists of Kenya: Surviving drought and development in Africa's arid lands. Allyn and Bacon, Boston.

Galvin, K. 1985. Food procurement, diet, activities and nutrition of Ngisonyoka Turkana pastoralists in an ecological and social context. Ph.D. Dissertation. State University of New York at Binghamton. 382 pp.

Galvin, K.A., R.B. Boone, N.M. Smith and S.J. Lynn. 2001. Impacts of climate variability on East African pastoralists: Linking social science and remote sensing. Climate Research 19:161-172.

Galvin, K.A., D.S. Ojima, R.B. Boone, M. Betsill and P. K. Thornton. 2005. Decision-making in Rangeland systems: an integrated Ecosystem-Agent-based Modeling Approach to Resilience and change (DREAMAR), NSF funded project, Human and Social Dynamics.

Gray, S., M. Sundal, B. Wiebusch, M.A. Little, P.W. Leslie and I.L. Pike. 2003. Cattle raiding, cultural survival, and adaptability of East African pastoralists. Current Anthropology 44(Supplement):3-30.

Gulliver, P.H. 1955. The family herds. Routledge and Kegan Paul, London.

Gunderson, L. and C. Folke. 2005. Resilience-now more than ever. Ecology and Society 10:22 (online) http://www.ecologyandsociety.org/vol10/iss2/art22/.

Holling, C.S., F. Berkes, and C. Folke. 1998. Science, sustainability and resource management. Pages 342-359 *In* F. Berkes, C. Folke, and J. Colding, editors. Linking social and ecological systems. Cambridge University Press, Cambridge, U.K.

Humphrey, C. and D. Sneath. 1999. The end of nomadism? Society, state and the environment in Inner Asia. Duke University Press, Durham, NC.

Kerven, C. 2003. Privatisation of livestock marketing and emerging socio-economic differentiation. Pages 146-166 *In* C. Kerven, editor. Prospects for pastoralism in Kazakstan and Turkmenistan. From state farms to private flocks. Routledge Curzon, NY.

Knight, J. 1992. Institutions and social conflict. Cambridge University Press, Cambridge.

Leneman, J.M. and R.S. Reid. 2001. Pastoralism beyond the past. Development 44:85-89.

Lesorogol, C.K. 2003. Transforming institutions among pastoralists: inequality and land privatization. American Anthropologist 105:531-542.

Lesorogol, C.K. 2005. Privatizing pastoral lands: economic and normative outcomes in Kenya. World Development 33:1959-1978.

Little, M.A. and P.W. Leslie, editors. 1999. Turkana herders of the dry savanna: Ecology and biobehavioral response of nomads to an uncertain environment. Oxford University Press, Oxford.

McAllister, R.R.J., I.J. Gordon, M.A. Janssen, and N. Abel. 2006. Pastoralists responses to variation of rangeland resources in time and space. Ecological Applications 16:572-583.

McCabe, J.T. 1990. Success and failure – the breakdown of traditional drought coping institutions among the pastoral Turkana of Kenya. Journal of Asian and African Studies 25:146-160.

McCay, B. and J. Acheson, editors. 1987. The question of the commons: The culture and ecology of communal resources. University of Arizona Press, Tucson.

Mearns, R. 1991. Pastoralist, patch ecology and perestroika: Understanding potentials for change in Mongolia. Institute of Development Studies Bulletin, Vol. 22, No. 4, University of Sussex, UK.

Mearns, R. 1993. Territoriality and land tenure among Mongolian pastoralists: variation, continuity and change. Nomadic Peoples 33:73-104.

Narayan, D. and L. Pritchett. 1997. Cents and sociability: Household income and social capital in rural Tanzania. Policy Research Working Paper, 41 pgs.

Newman, L. and A. Dale. 2005. Network structure, diversity, and proactive resilience building: a response to Tompkins and Adger. Ecology and Society 10:r2. http://www.ecologyandsociety.org/vol10/iss1/resp2.

Niamir-Fuller, M. 1998. The resilience of pastoral herding in Sahelian Africa. Pages 250-284 *In* F. Berkes, C. Folke and J. Colding, editors. Linking social and ecological systems. Cambridge University Press, Cambridge, UK.

North, D. 1990. Institutions, institutional change and economic performance. Cambridge University Press, Cambridge.

Ostrom, E. 1990. Governing the commons: The evolution of institutions for collective action. Cambridge University Press, Cambridge.

Pike, I.L. 2004. The biosocial consequences of life on the run: A case study from Turkana District, Kenya. Human Organization 63:221-235.

Potkanski, T. 1997. Pastoral economy, property rights and traditional mutual assistance mechanisms among the Ngorongoro and Salei Maasai of Tanzania. Pastoral Land Tenure Series Monograph 2. IIED Drylands Program, London.

Pretty, J.N. 2002. People, livelihoods, and collective action in biodiversity management. Pages 61-86 *In* T. O'Riordan and S. Stoll-Kleeman, editors. Biodiversity, sustainability and human communities: Protecting beyond the protected. Cambridge University Press, Cambridge, U.K.

Reid, R.S., T.P. Tomich, X. Jianchu, H. Geist, E. Lambin, R.S. De Fries, J. Liu, D. Alves, B. Agbola, A. Chabbra, A. Mather, T. Veldcamp, K. Kok, M. van Noordwijk, D. Thomas, and C. Palm. 2006. Linking land-use/cover science and policy: current lessons and future integration. Pages 151-171 *In* E.F. Lambin and H. Geist, editor. Land use and land cover change: Local processes, global impacts. Springer, Berlin and Heidelberg, Germany.

Rutten, M.M.E.M. 1992. Selling wealth to buy poverty: The process of individualisation of land ownership among the Maasai pastoralists of Kajiado District, Kenya, 1890-1990. Breitenbach Publishers, Saarbrucken, Germany.

Scoones, I., editor. 1994. Living with uncertainty. Intermediate Technology, London.

Sneath, D. 1993. Social relations, networks and social organization in post-socialist rural Mongolia. Nomadic Peoples 33:193-207.

Sneath, D. 2003. Land use, the environment and development in post-socialist Mongolia. Oxford Development Studies 31:441-459.

Spear, T. and R. Waller. 1993. Being Maasai. James Currey, LTD, Athens.

Taylor, P.D., L. Fahrig, K. Henein and G. Merriam. 1993. Connectivity is a vital element of landscape structure. Oikos 68:571-573.

Templer, G. J. Swift and P. Payne. 1993. The changing significance of risk in the Mongolian pastoral economy. Nomadic Peoples 33:105-122.

Thompson, M., P. Trench, E. Coast, and K. Homewood. Draft paper. Maasai pastoralists: Livelihoods at the margins. http://www.soas.ac.uk/antsocfiles/livelihoods/paper/Homewoodpaper.pdf.

Thornton, P.K., S.B. BurnSilver, R.B. Boone and K.A. Galvin. 2006. Modelling the impacts of group ranch subdivision on agro-pastoral households in Kajiado, Kenya. Agricultural Systems 87:331-356.

Tischendorf, L. and L. Fahrig. 2000. On the usage and measurement of landscape connectivity Oikos 90:7-19.

Turner, M.D. 1999. The role of social networks, indefinite boundaries and political bargaining in maintaing the ecological and economic resilience of the transhumance systems of Sudan-Sahelian West Africa. Pages 97-123 *In* M. Niamir-Fuller, editor. Managing mobility in African rangelands. FAO and Beijer International Institute of Ecological Economics, London, U.K.

Walker, B. and N. Abel. 2002. Resilient rangelands-adaptation in complex systems. Pages 293-313 *In* L.H. Gunderson and C.S. Holling, editors. Panarchy: Understanding transformations in human and natural systems. Island Press, Washington, D.C.

Western, D. 1989. Conservation without parks: wildlife in the rural landscape. Pages 158-165 *In* D. Western and M.C. Pearl, editors. Conservation for the 21st century. Oxford University Press, New York.

Western, D. and D.L.M. Nightingale. 2002. Environmental change and the vulnerability of pastoralists to drought: A case study of the Maasai in Amboseli, Kenya. UNEP.Net, Africa. United Nations Environmental Program.

Williams, D.M. 2002. Beyond great walls. Environment, identity, and development on the Chinese grasslands of Inner Mongolia. Stanford University Press, Stanford, CA.

Woodhouse, P. 2003. African enclosures: a default mode of development. World Development 31:1705-1720.

INDEX

Note: Page numbers of figures are denoted by an "f"; page numbers of tables are denoted by a "t".